Information Resources Management:
Global Challenges

Wai K. Law
University of Guam, USA

IDEA GROUP PUBLISHING
Hershey • London • Melbourne • Singapore

Acquisition Editor:	Kristin Klinger
Senior Managing Editor:	Jennifer Neidig
Managing Editor:	Sara Reed
Assistant Managing Editor:	Sharon Berger
Development Editor:	Kristin Roth
Copy Editor:	Sue Vanaerttook
Typesetter:	Diane Huskinson
Cover Design:	Lisa Tosheff
Printed at:	Integrated Book Technology

Published in the United States of America by
Idea Group Publishing (an imprint of Idea Group Inc.)
701 E. Chocolate Avenue
Hershey PA 17033
Tel: 717-533-8845
Fax: 717-533-8661
E-mail: cust@idea-group.com
Web site: http://www.idea-group.com

and in the United Kingdom by
Idea Group Publishing (an imprint of Idea Group Inc.)
3 Henrietta Street
Covent Garden
London WC2E 8LU
Tel: 44 20 7240 0856
Fax: 44 20 7379 0609
Web site: http://www.eurospanonline.com

Copyright © 2007 by Idea Group Inc. All rights reserved. No part of this book may be reproduced in any form or by any means, electronic or mechanical, including photocopying, without written permission from the publisher.

Product or company names used in this book are for identification purposes only. Inclusion of the names of the products or companies does not indicate a claim of ownership by IGI of the trademark or registered trademark.

Library of Congress Cataloging-in-Publication Data

Information resources management : global challenges / Wai K. Law, editor.
    p. cm.
  Summary: "This book addresses challenges in managing information resources in dynamic social environments across cultures, including research on key factors for social acceptance of information technology, and user adoption of information management methods. It explores new paradigms under which information resources will generate original meanings for a contemporary generation of users, with emphasis on user-centered and culture-centric information systems"--Provided by publisher.
  Includes bibliographical references and index.
  ISBN 1-59904-102-2 (hardcover) -- ISBN 1-59904-103-0 (softcover) -- ISBN 1-59904-104-9 (ebook)
 1. Knowledge management. 2. Information resources management. 3. Information networks--Management. I. Law, Wai K., 1955-
  HD30.2.I52754 2007
  658.4'038--dc22
                          2006032167

British Cataloguing in Publication Data
A Cataloguing in Publication record for this book is available from the British Library.

All work contributed to this book is new, previously-unpublished material. The views expressed in this book are those of the authors, but not necessarily of the publisher.

Information Resources Management:
Global Challenges

Table of Contents

Foreword vii

Preface ix

Section I:
IRM Challenges in Dynamic Social Environments

Chapter I
Cross-Cultural Information Resource Management:
Challenges and Strategies 1
    Xiuzhen Feng, Beijing University of Technology, P.R. China

Chapter II
Information Technology Acceptance across Cultures 25
    Amel Ben Zakour, FSJEGJ, Tunisia

Chapter III
Information-Communications Systems Convergence Paradigm:
Invisible E-Culture and E-Technologies 54
    Fjodor Ruzic, Institute for Informatics, Croatia

Section II:
IRM Challenges across Nations

Chapter IV
Critical Success Factors for IS Implementation in China:
A Multiple-Case Study from a Multiple-Stage Perspective 76
    Huixian Li, Accenture Consulting Company, P. R. China
    John Lim, National University of Singapore, Singapore
    K. S. Raman, National University of Singapore, Singapore
    Yinping Yang, National University of Singapore, Singapore

Chapter V
Developing Electronic Content for the Support of the European
Cultural Inclusion: From the Earlier "e-Europe" Initiative toward
the Future "i2010" Perspective 107
    Ioannis P. Chochliouros, Hellenic Telecommunications
        Organization S.A. (OTE), Greece and University of Peloponnese,
        Greece
    Ioannis Bougos, Hellenic Telecommunications
        Organization S.A. (OTE), Greece
    Stergios P. Chochliouros, Idependent Consultant, Greece
    Anastasia S. Spiliopoulou, Hellenic Telecommunications
        Organization S.A. (OTE), Greece

Chapter VI
Understanding the Cultural Roots of India's Technology
Development from Homi Bhabha's Post-Colonial Perspective 128
    Ramesh Subramanian, Quinnipiac University, USA

Chapter VII
Digital Culture and Sharing: Theory and Practice of a Brazilian Cultural
Public Policy 146
    Saulo Faria Almeida Barretto, Research Institute for Information
        Technology, Brazil
    Renata Piazzalunga, Research Institute for Information
        Technology, Brazil
    Dalton Martins, Research Institute for Information
        Technology, Brazil
    Claudio Prado, Research Institute for Information
        Technology, Brazil
    Célio Turino, Ministry of Culture, Brazil

Section III: Information Resources Development Challenges

Chapter VIII
Dialogue Act Modeling: An Approach to Capturing and Specifying Communicational Requirements for Web-Based Information Systems 162
Ying Liang, University of Paisley, UK

Chapter IX
Challenges in Building a Culture-Centric Web Site 192
Tom S. Chan, Southern New Hampshire University, USA

Section IV: Knowledge Management Challenges

Chapter X
Information and Knowledge Management for Innovation of Complex Technologies 211
Ning Li, University of Guam, USA
Don E. Kash, George Mason University, USA

Chapter XI
Knowledge Workers as an Integral Component in Global Information System Design 236
Michel Grundstein, MG Conseil & Paris Dauphine University, France

Chapter XII
Cultural Impact on Global Knowledge Sharing 262
Timothy Shea, University of Massachusetts Dartmouth, USA
David Lewis, University of Massachusetts Lowell, USA

Section V: Information Communication Technology Adoption Challenges

Chapter XIII
The Role of Information and Communication Technology in Managing Cultural Diversity in the Modern Workforce: Challenges and Issues 283
Indrawati Nataatmakja, University of Technology, Sydney, Australia
Laurel Evelyn Dyson, University of Technology, Sydney, Australia

Chapter XIV
From 9 to 5 to 24/7: How Technology Has Redefined the Workday 305
  Linda Duxbury, Carleton University, Canada
  Ian Towers, Carleton University, Canada
  Christopher Higgins, The University of Western Ontario, Canada
  John Ajit Thomas, Carleton University, Canada

Chapter XV
The Role of User Characteristics in the Development and Evaluation of E-Learning Systems 333
  Dianna L. Newman, University at Albany / SUNY, USA
  Aikaterini Passa University at Albany / SUNY, USA

Chapter XVI
A Model for Selecting Techniques in Distributed Requirement Elicitation Processes 351
  Gabriela N. Aranda, GIISCo Research Group, Universidad Nacional del Comahue, Argentina
  Aurora Vizcaíno, ALARCOS Research Group, Universidad de Castilla-La Mancha, Spain
  Alejandra Cechich, GIISCo Research Group, Universidad Nacional del Comahue, Argentina
  Mario Piattini, ALARCOS Research Group, Universidad de Castilla-La Mancha, Spain

Chapter XVII
Organizational Time Culture and Electronic Media 364
  Cheon-Pyo Lee, Carson-Newman College, USA

Chapter XVIII
Global Organizational Fit Pyramid for Global IT Team Selection 373
  Richard S. Colfax, University of Guam, USA
  Karri T. Perez, TeleGuam Holding, LLC, USA

About the Authors 386

Index 396

Foreword

Managing information technology (IT) on a global scale presents a number of opportunities and challenges. IT can drive the change in global business strategies and improve international coordination. At the same time, IT can be an impediment to achieving globalization. IT as an enabler of and inhibitor to globalization raises interesting questions. To what extent does IT facilitate globalization? In what ways can IT be a constraint to globalization? Is managing IT in a global context largely the same as managing IT in a domestic context? If it is not, then what aspects are different? How are differences in culture important to the implementation of information systems?

Research into these and other related topics has grown tremendously over the last 15 years. Journals focusing on global aspects of IT (i.e., global IT) were established driven by the need to understand the relationship between IT and globalization. New research tracks on global IT were created at international conferences in order to generate interest and permit research to be disseminated as soon as they are completed. Similarly, there is a growing number of books on the subject.

This book, therefore, is a very good addition to this growing collection of books. It brings together experiences from academics in countries such as Brazil, England, France, Greece, India, China, the United States, Canada, and Australia. In sharing their research and findings relating to the challenges and strategies in information resources management in the global context, I hope you will find this book a useful resource in developing further research into the subject.

Dr. Felix B. Tan
Professor of Information Systems and Head
School of Computer and Information Sciences
Associate Dean (Research)
Faculty of Design & Creative Technologies
Auckland University of Technology
New Zealand

Felix B. Tan is a professor of information systems, the associate dean (research) for the Faculty of Design and Creative Technologies at AUT University, New Zealand. Dr. Tan serves as the editor-in-chief of the Journal of Global Information Management. *He is on the Executive Council and is a fellow of the Information Resources Management Association. His current research interests are in electronic commerce, global information management, business-IT alignment, and the management of IT. Dr. Tan has published in* MIS Quarterly, Information & Management, Journal of Information Technology, IEEE Transactions on Engineering Management, *as well as other journals and refereed conference proceedings.*

Preface

Although the study of cultural differences can be traced back to the early 1980s, widespread interest on the cultural effects on technology implementation and information system development began to appear in the literature during the mid-1990s. Most of these studies investigated the influence of national cultures on technology transfer and information systems designs. A series of reports on failed information systems implementations began to surface in the late 1990s and continued to recent years. Many of the information systems failures had less to do with poor designs and more on the difficulties of fitting the information systems into the existing human systems and social environment. Many global information systems issues involved adapting information system designs to new cultural settings.

While there is still great potential value contribution from information systems, there has been insufficient attention to the significant role of the users as part of the information resource formula. An emerging trend toward user-centered information resource development requires a closer scrutiny of the relationship between users and the other information resources. An increasing number of knowledge workers will be making decisions on information systems with a superficial understanding of the complexity of information resource management. The rapid globalization process places an increasing number of users under a cross-cultural environment in which multiple cultural forces simultaneously shape the user behaviors toward information systems usage and information resources management practices. The recent tsunami of technology outsourcing intensifies the cross-cultural effects of information systems designs and constructions as well as the potential shortage of suitable information specialists at strategic locations. Millions of dollars of investments in information systems are at risk of being underutilized or rejected by users for cultural reasons.

The rapidly declining costs of technology with increasing capability provide options to users who seek information system designs driven by needs, values, usability, and economic factors, more than breakthrough technology, costs, features, and images. There is also a gradual shift toward decentralized, ubiquitous usage, with increasing emphasis on information and knowledge sharing rather than data and information preservation. This points to the need to support the generation of consumable, customized information packages vs. inconsumable, standardized information packages. This driving force has the potential to extend the usage of information resources beyond the boundary of e-commerce towards arrangements in multiplatform information exchanges and virtual networks that not only supplement but reshape the traditional architect of information systems. As mobile devices are packed with increasing capabilities to capture, store, process, and share information, there is a need to reexamine the roles and designs of information resources.

This book provides a broadened perspective to the challenges of managing information resources, especially for user-centered information development efforts. The mission is to raise awareness of the significance of the cultural issues in information resource decisions. Practical advice is included to help readers to better understand cultural barriers in the development of information resources beyond multinational and globalization applications and toward the internationalization of IRM.

Organization of the Book

This book is organized into five sections with a total collection of 18 chapters. **Section I** consists of three chapters addressing **IRM Challenges in Dynamic Social Environments**. Besides seeking ways to extend existing information resource management practices to global locations, there is also an emerging need to understand the realistic roles and acceptance of information technology under various cultures. This is especially important in the strategic deployment of information resources in order to maximize value contribution. There is also the need to understand a new paradigm under which information resources will generate new meanings for a new generator of users. Communities of users are imposing a strong influence toward the development and deployment of information technology. **Section II** consists of four chapters addressing **IRM Challenges across Nations**. The collective researches present new perspectives toward potential challenges in matching current IRM practices with social and cultural environments in several hot spots around the world, including China, Europe, India, and South America. These exploratory reports illustrate the differences of priorities and expectation for information technology under various cultural presuppositions. **Section III: Information Resources Development Challenges** addresses new challenges in designing user-centered and culture-centric information systems. Increasingly, information systems' successes hinge on effective customization rather than on adaptation of existing systems. It is challenging to create smart system design to allow an active definition of system features by users. The three chapters in **Section IV: Knowledge Management Challenges** discuss issues in knowledge sharing and knowledge management under multicultural environments. Knowledge and knowledge workers are assuming increasingly important roles in the planning, acquisition, and deployment of information resources. A better understanding of factors affecting knowledge sharing and innovation will be important in the allocation of scare resources. **Section V:**

Information Communication Technology Adoption Challenges compiles a collection of six chapters addressing various challenges in the adoption of information technology for global organizations. A recurring issue is the realization that what matters is not the capability of information technology but the manner in which information technology has been adopted and the resulting effect on the acceptance of the technology toward productive outputs. A greater challenge yet is the broadening of the definition of information technology. The widespread usage of mobile devices and electronic tools is redefining the meaning of collecting, processing, sharing, storage, analysis, and security of information. It is important to understand fully the influence of cultural presuppositions on the effective selection, planning, coordination, and support of information resources.

A brief description of each of the chapters follows.

Chapter I compares national cultural differences in order to draw attention to IRM challenges in a cross-cultural environment. The implementation of technological infrastructure and information systems merely marks the beginning of cross-cultural IRM. A key issue is to match the information perspectives between information providers and information users. The objectives and priorities of information might be framed differently by information providers who come from different cultural backgrounds. Considerations for national cultural influences on the presentation of information and interpretation of information as well as the description of information represent some of the new challenges to ensure the availability of the right information in the right form in order to meet the needs of information users.

Chapter II offers a conceptual model to explain the effect of national culture on IT adoption and usage. Information technology accepted in a specific cultural context may not be accepted in the same way in another culture. The conceptual model attempts to link pertinent cultural values to technology acceptance model (TAM) constructs and relationships. Hence, taking into account the societal environment could improve the understanding of IT acceptance behavior.

Chapter III addresses the emerging new social and cultural responsibility for information technology professionals with a new form of digital media that could reshape not only media industry but also a cultural milieu of an entire nation on a regional and global basis. As information-communications systems converge with media, including tools, services, and content, potential misuse of technologies could have a much greater influence on culture than in the past. The discussion follows on the World Library idea that is rebuilding with a new form of World Memory (World Brain), the shift from visible culture domination to the domination of invisible culture in the world of e-technologies.

Chapter IV presents, through multiple case studies on IS implementation success in China, critical success factors (CSF) that are radically different from those previously identified as relevant for the West. This exploratory work initiates a call for more in-depth investigations incorporating the dimensions of differing government policy and business environments for future research in this area. The findings also offer practitioners a set of guidelines in implementing IS applications in China.

Chapter V explores challenges following the European cultural inclusion that promoted the creation and the distribution of new forms of electronically available content. The chapter investigates diverse potential opportunities for realizing and offering innovative services, applications, and/or related facilities in the market through the proper use of modern electronic communications, especially for the promotion of cultural and social targets. Such options implicate vigorous participation both of state authorities and market industry players to launch dynamic business partnerships in parallel with efforts to improve quality of life and social cohesion in order to forward new ways of participating in society and to advance the European diversity and rich cultural heritage.

Chapter VI examines a cultural and social theory viewpoint with philosophical underpinnings to explain the roots and current happenings in the field of IT in India. The chapter introduced Homi K. Bhabha's post-colonial social theory of interstitial perspective and then discusses the application or overlay of the constructs that emanate from that to the roots of India's technology and (subsequently) IT development and its complementary effect in shaping Indian Information Systems professionals. The chapter spotlights various events and persona in India's history, including the current crop of IT professionals emerging from the subcontinent, to explain the roots and current happenings in the field of IT in India.

Chapter VII describes the Digital Culture project that has been developed in Brazil, supported by the Brazilian government. The chapter starts with the principle of sharing of information being the basis of the information society and provoking a re-analysis of the productive scenario, where users are considered as active members of the productive scenario, instead of mere consumers of the process resulting from the technical innovation carried out by business organizations. The chapter includes a discussion on concept of cultural property in the context of this new society and presents the Brazilian vision of this scenario.

Chapter VIII points to the need for a Web-based information system (WBIS) to win users by providing a user-centered and interactive Web site. A well-accepted Web site should satisfy the user requirements for effective interaction in addition to functional requirements and nonfunctional requirements. This argues for the need for new modeling approaches to be invented for WBIS analysis in order to capture and specify communication requirements for the Web site of WBIS. This chapter presents the dialogue act modeling approach created for capturing and specifying user needs for easy-to-use, interactive Web site WBIS analysis WBIS development

Chapter IX discusses the challenges in constructing a culture-centric Web site. The Internet has expanded business opportunities into global marketplaces that virtually were unreachable in the past. This is especially important to businesses that market products and services across geographical boundaries. Global Web sites must be culture-centric, taking into account the attitude, technology, language, communication, sensibility, symbolism, and interface usability of targeted communities. Web site designers must be aware of international user needs, limitations, and expectations, and must become more sensitive to cultural diversity.

Chapter X investigates the role of information and knowledge management in the innovation of complex technologies. A conceptual framework for three patterns of technological innovation (normal, transitional, and transformational) is presented, and the process of information and knowledge management in accessing and using knowledge is analyzed. Particularly, emphasis is put on the cultural impact on the information and knowledge management processes. Five case studies of evolving technologies carried out in the United States, Japan, Germany, India, and China are used to elaborate the conceptual framework and key points presented in this chapter. Lessons for managers and public policymakers concerned with facilitating the innovation of technologies are discussed.

Chapter XI, based on the MGKME, stresses the need to incorporate the employees' knowledge within the information and knowledge management systems in such a way that the knowledge of the knower should be distinguished from the flows of codified knowledge that are processed within the systems. Thus, knowledge workers assume the dual roles as a component of the system and as a user of the information systems within extended companies. However, the individual's culture may engender the incommensurability of the employees' interpretative frameworks and so have a key influence on the design, development, implementation, and utilization of the systems. A discussion is included on the challenging relationships between KM governance principles, IT governance principles, and corporate governance principles.

Chapter XII introduces how culture impacts global knowledge sharing. Effective knowledge sharing (KS), one of the four interdependent dimensions of knowledge management (KM), is particularly important in today's global environment in which national cultural differences are negotiated all the time. Knowledge sharing is described along six dimensions and national culture along four dimensions. A model is presented that provides guidelines for effectively sharing different types of knowledge within different cultural environments. Several examples are presented to illustrate the model's effectiveness.

Chapter XIII demonstrates how managers can use information and communication technology (ICT) more effectively in culturally diverse workforces. Basing an analysis on the cultural dimensions of Hofstede and Hall, the chapter compares a range of ICTs and provides a chart summarizing their strengths and weaknesses. In addition, a framework for developing ICT is proposed, and an example of its application to a global organization is presented. The study shows that none of the existing ICT tools is perfect in all situations and all cultural contexts. Therefore, managers need to provide a variety of ICTs to their employees, and developers should build flexibility into their ICT designs.

Chapter XIV explores the use of work-extension technologies such as e-mail, BlackBerry devices, portable computers, and cellular phones. The chapter presents the usage patterns of these work extension technologies by Canadian knowledge workers and describes how work is being performed in a variety of nonoffice locations outside normal working hours. Some technologies were found to lead to an increase in employee workloads and stress, while others were found to have less of an impact. Many respondents reported that information technology made them more productive and made their work more interesting.

Chapter XV presents a multi-phase cyclical model of designing, developing, and evaluating instructional technology (IT) learning systems based on inclusion of users' characteristics (experience with technology, familiarity with content, adaptability, learning style, gender, professional level). The model was developed and piloted over the course of seven years in more than 50 learning communities and has resulted in documentation of stages in which user variables interact with the process. Key elements of the model are presented in detail and supported by samples of development and related evaluation. Recommendations are included for the practice of designing and evaluating IT learning systems that meet the varied individual, cultural, and contextual needs of users.

Chapter XVI introduces a model based on techniques from cognitive psychology as a means of improving the requirement elicitation in global software development projects. The selection of appropriate technology and requirement elicitation techniques in such environments is a subject of research, because when stakeholders feel comfortable with the technology and methodologies they use, then information gathered during elicitation is expected to be more accurate. This chapter proposes reducing problems in communication by selecting a suite of appropriate elicitation techniques and groupware tools according to stakeholders' cognitive styles. It also shows how information about stakeholders' personalities can be used to make them feel comfortable and to improve their performance when working in a group.

Chapter XVII explains the electronic media diffusion process within organizations and provides a guideline to implement electronic media within organizations. The concept of time dimension culture, monochronic and polychronic, and two dimensions of media speed, production, and interaction speed are used to explain the media diffusion process within organizations. It suggests that the diffusion process and expected benefit of electronic media are significantly different, depending on national culture, organizational culture, and the characteristics of that medium. Therefore, careful examination and understanding of organizational time culture and the characteristics of media should be ahead of making a decision on electronic media adoption and implementation.

Chapter XVIII presents the Global Organizational Fit Pyramid, which provides a hierarchy of five decision levels to consider when putting together global IT teams. The Pyramid takes into account global status, social rank, experience, credential, and individual factors for the selection of global IT team members. The Pyramid also may be applicable to other global team and HR-related decision-making issues.

References

Carlson, P. A. (2000). Information technology and the emergence of a worker-centered organization. *ACM Journal of Computer Documentation, 24*(4), 204-212.

Ein-Dor, P., Segev, E., & Orgad, M. (1993). The effect of national culture on IS: Implications for international information systems. *Journal of Global Information Management, 1*(1), 33-45.

Katz, J., & Townsend, J. (2000). The role of information technology in the "fit" between culture, business strategy and organizational structure of global firms. *Journal of Global Information Management, 8*(2).

Kersten, G. E., Kersten, M. A., & Rakowski, W. M. (2002). Software and culture: Beyond the internationalization of the interface. *Journal of Global Information Management, 10*(4).

Law, W. K. (2005). Information resources development challenges in a cross-cultural environment. In M. Khosrow-Pour (Ed.), *Encyclopedia of information science and technology*. Hershey, PA: IRM Press.

Law, W. K., & Perez, K. (2005). Cross-cultural implementation of information system. *Journal of Cases on Information Technology, 7*(2), 121-130.

Lemon, W., Liebowiltz, J., Burn, J., & Hackney, R. (2002). Information systems project failure: A comparative study of two countries. *Journal of Information Management, 10*(2).

Myers, M. D., & Tan, F. B. (2003) *Beyond models of national culture in information systems research. Advanced topics in global information management*. Hershey, PA: Idea Group Publishing.

Szewczak, E., & Khosrowpour, M. (Eds.). (1996). *The human side of information technology management*. Hershey, PA: Idea Group Publishing.

Tricker, R. I. (1988). Information resource management—A cross-cultural perspective. *Information and Management, 15*(1), 37-46.

Trompenaars, F., & Hampden-Turner, C. (1998). *Riding the waves of culture: Understanding cultural diversity in global business* (2nd ed.). New York: McGraw-Hill.

Acknowledgments

The editor would like to acknowledge the help of all involved in the collation and review process of the book. In particular, I would like to acknowledge the enthusiastic responses of the contributing authors, many of whom also served as subject matter experts throughout the review process. Thanks go to all those who suggested topic ideas and provided constructive and comprehensive reviews. Special thanks also go to the publishing team at Idea Group Inc. In particular, I would like to thank our development editor Kristin Roth, who made a valuable contribution in overcoming several obstacles throughout the project, and to Mehdi-Khosrow-Pour, who motivated me to take on this project. In closing, I wish to thank my family for their love and support throughout this project.

Wai K. Law, PhD
Professor of Business Strategy and Information Systems
Guam, USA
October 2006

Section I

IRM Challenges in Dynamic Social Environments

Chapter I

Cross-Cultural Information Resource Management:
Challenges and Strategies

Xiuzhen Feng
Beijing University of Technology, P. R. China

Abstract

As modern business activities become increasingly global, so too does information resource management (IRM). To manage information resources successfully in this global environment, one great challenge facing the management team is how to deal with national cultural differences. In this chapter, national cultural differences are discussed to indicate the substance of IRM challenges in a cross-cultural environment. Many recently published cases are studied to clarify management challenges of cross-cultural ISM. Primary management issues of mismatch between information presentation and information procurement are analyzed in particular detail. Further, management solutions oriented for IRM in a cross-cultural environment are explored. Due to the lack of similar research topics, this study could supply a gap for cross-cultural IRM. The contribution of this chapter will be twofold: one is to set up a sound management mechanism for cross-cultural IRM; the other is to create sharable information resources in a cross-cultural environment.

Introduction

Organizations are increasingly aware of the potential of information resource management (IRM) for gaining a competitive advantage and sustaining their business success, because if information resources are managed properly, significant benefits can be derived from improved productivity, improved quality of decision-making, and improved performance of tasks or organizational learning curve.

IRM is defined as a comprehensive approach to the collection, storage, process, maintenance, and dissemination of electronic information as well as the exchange of information between different organizations (Brisis, 1995). IRM is also known in some literature as information management, information systems management (Bertot, 1997), and management information systems or management of information technology (Maceviciute & Wilson, 2002), and covers five types of resource management: systems support; processing data and images; conversion and transformation; distribution and communication; and, finally, retention, storage, and retrieval (Schneyman, 1985).

The rapid development and progress of information technology and Internet technology have contributed substantially to the feasibility of global business, such as manufacture, transportation, service, publication, education, and so forth. IRM accompanying various global businesses is crossing national borders and, thus, has to deal with the differences of various countries. One of these differences is culture that has the possibility of profoundly influencing IRM. According to literature, many scholars have studied IRM in relation to national cultural differences. Most of them, however, have approached it from a technology acceptance perspective, which is based on a fundamental concept that IRM supports information management by providing the technical capability and overall guidance for information management to do its job (Schneyman, 1985). For example, Robey and Rodriguez-Diaz (1989) studied how national cultural differences could affect the success in information systems implementation; Nelson, Weiss, and Yamazaki (1992) discovered that end-user computing is profoundly different in the United States compared to Japan; Shore (1994) tested the influences of culture on information systems applications; Kwon and Chidambaran (1998) studied how culture influences communication technology acceptance; Straub, Loch, and Hill (2001) examined cultural influence on transfer of information technology to developing countries; Carayannis and Sagi (2001) investigated the relationship between national cultural differences and System Design Life Cycle; Rose, Evaristo, and Straub (2002) studied culture and consumer responses to Web download time. They found that participants from polychronic societies were less troubled by download delays and perceived the delays to be shorter than did people from monochronic cultures. Recently, more research interest has moved to cross-cultural IRM, such as cross-cultural software production and use (Walsham, 2002), cross-cultural information systems adoption in multinational corporations

(Shoib & Nandhakumar, 2003), and cross-cultural implementation of information systems (Law & Perez, 2005).

For the management of cross-cultural information resources, the technology used to manipulate, manage, and transmit information resources is a significant element of IRM. It is important to understand whether the technology (or information system) can be implemented or accepted successfully in a cross-cultural environment. However, once the technology infrastructure (or information system) is in place, it doesn't mean that cross-cultural IRM is complete and accomplished, because information should be recognized as a valuable entity, independent of the technology that manipulates it (Trauth, 1989). Consequently, information management issues in the cross-cultural environment deserve more attention.

Nevertheless, the study of cross-cultural IRM related to information management is still very limited in scope. Particularly, management strategies for cross-cultural IRM are scarce. Motivated by cross-cultural IRM reality, this chapter will focus on information management challenges and strategies in the cross-cultural context. The objective of this chapter will be management strategy oriented for cross-cultural IRM. Due to the lack of similar research topics, this study could supply a gap for IRM in a cross-cultural environment. The contribution of this study is expected to provide constructive and helpful management strategies for information resource executives, information system designers, information system administers, and other similar management teams in order to manage information resources in a global context.

The chapter proceeds as follows. First, national cultural differences are analyzed to indicate fundamental differences of people's behaviors when they come from different national cultural backgrounds. Then, management issues and challenges of cross-cultural IRM concerning information management are studied in order to recognize how national cultural differences can influence people's information behavior. Subsequently, I come up with management strategies to manage cross-cultural information resources from two aspects: setting up the sound IRM mechanism and creating sharable information resources. Finally, the chapter ends with research recommendations and conclusions.

National Cultural Differences

National culture has been defined in many ways. Parsons and Shils (1951) defined culture as the shared characteristics of a high-level social system. Hofstede (1980, 1991, 2001) defined culture as the collective programming of the mind that distinguishes one group from another. Erez and Earley (1993) added that national culture is the shared values of a particular group of people, which reflects the core values and beliefs of individuals formed during childhood and reinforced throughout life (Lachman, 1983). According to Hawryszkiewycz (1994), culture has been referred to as shared values, expectations, and norms found within countries, regions, social groups, business firms, and even departments and work groups within a firm. Consequently, culture values shape people's beliefs and attitudes and guide their behaviors (Rokeach, 1973).

Hofstede's proposed dimensions of national culture are very commonly used. These dimensions allow national-level analysis and are standardized to allow multiple country comparisons (Ford, Connelly, & Meister, 2003). Many studies have confirmed the validity of these dimensions (Ronen & Shenkar, 1985; Shackleton & Ali, 1990) and employed them to account for empirical observations (Earley, 1993; Straub, 1994; Straub, Keil, & Brenner, 1997; Tan, Wei, Watson, Clapper, & McLean, 1998). Particularly, Hofstede's dimensions often are employed by researchers when international or national culture issues are discussed within information systems management and management of IT (Ford, Connelly, & Meister, 2003). In this chapter, I also employ Hofstede's model, since it has been shown as a reliable and useful tool to identify and explain the cultural differences in numerous studies across many disciplines. In addition, all the selected case study samples in this study employed this model. Accordingly, Hofstede's model could be adopted as an excellent instrument to elucidate various management challenges for cross-cultural IRM and to facilitate the development of management strategies for this research.

Hofstede created an empirical model to compare the (cultural) values of similar people (employees and managers) in different subsidiaries of the IBM Corporation in more than 64 countries. In 1968 and 1972, he organized several large-scale surveys, collectively producing more than 116,000 questionnaires in 20 different languages. Each survey included more than 100 questions related to the several values. Based on this research, Hofstede constructed four distinct national culture dimensions in 1980 and later revised them to five. These dimensions, measured by indexes, are successively power-distance, uncertainty-avoidance, individual-collective, masculinity-femininity, and long-term orientation.

Power-distance measures the acceptance of power and explains the extent to which the less powerful persons in a society accept inequality in power and consider it normal (Hofstede, 1991). In large power-distance countries, high hierarchies exist in both mental models and real-life organizations. The superiors are supposed to make decisions in the workplace without consultation of subordinates. In small power-distance countries, equality is expected and generally desired. People view themselves more as equals. In cross-cultural IRM environments, because of this difference in equality, people may have different opinions on the importance of presenting or using information between large power-distance culture societies and small distance-culture societies.

Uncertainty-avoidance explains the willingness to cope with uncertainty. It is the extent to which people within a culture are nervous by situations that they perceive as unstructured, unclear, or unpredictable and that they therefore try to avoid by maintaining strict codes of behavior and a belief in absolute truth (Hofstede, 1991). In countries with strong uncertainty-avoidance, people are more likely to shun ambiguous situations and prefer a clear structure as well as clear rules of behavior in organizations or institutions, since such structure and rules will help them to make events clearly interpretable and predictable. In countries with weaker uncertainty-avoidance, there is a more public acceptance of uncertain situations. Not only familiar but also unfamiliar risks are accepted more easily, such as those in activities for which there are no clear rules. In cross-cultural IRM environments, because of uncertainty avoidance, people may prefer different ways of presenting or acquiring information between strong uncertainty-avoidance cultures and weak uncertainty-avoidance cultures.

Individual-collective describes the priority of individualism or collectivism in a human society, which is not only a matter of ways of living together but also is intimately linked with societal norms (in the sense of value systems of major groups of the population). In collective cultures, social identity is based on group membership. Individual cultures emphasize the individual's goals and initiatives. The major differences between collective cultures and individuals can be presented as follows: The social norms of the group count rather than individual pleasure; shared group beliefs are superior to unique individual beliefs; cooperation with group members is valued rather than maximizing individual outcomes (Gudykunst & Ting-Toomey, 1988). In cross-cultural IRM environments, the individualization or personalization may affect information presentation and information acquirement.

Masculinity-femininity refers to the attitudes toward gender roles from different cultural backgrounds. In masculine societies, the traditional distinction of roles between males and females is strongly maintained; for example, males are considered to be assertive and decisive in pursuit of material success, whereas females are modest, tender, and concerned with quality of life. In feminine cultures, such distinction between males and females is less visible. Sometimes, gender roles overlap in feminine cultures because both males and females might be not only modest, tender, and concerned with the quality of life but also might be in pursuit of material success as well. In cross-cultural IRM environments, the difference between masculinity and femininity cultures might influence the way of information presentation or the way of information procurement.

Long-term orientation was defined as the fifth national culture dimension. In a low score of long-term orientation culture, people have more concern for the past and especially the present. They believe that what people have held in the past and in the present is more important than for their future. Therefore, in a low score of long-term-orientation culture, people prefer to have an immediate gratification; they desire immediate results. In the high score of long-term-orientation culture, people are more concerned about the future. They generally accept that the most important thing is to work hard and have a good foresight for the future. Therefore,

Figure 1. National cultural differences: USA, China, and The Netherlands

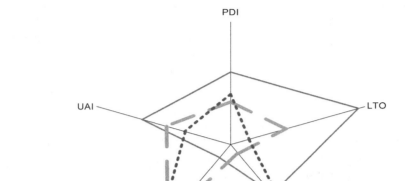

people are more patient in achieving results and goals. In cross-cultural IRM environments, the differences between low and high scores of long-term-orientation culture could influence the design of information presentation or the behavior of information acquirement.

To describe national cultural differences, the cultural scores and five cultural dimensions adopted from Hofstede's model are presented as national cultural profiles in Figure 1. I selected three countries—the USA, China, and The Netherlands—to indicate national cultural differences.

The national cultural differences among the three counties—the USA, China, and The Netherlands—are evident. For instance, The Netherlands and the USA are very close in terms of power-distance and individual-collective, which means that both The Netherlands and the USA have small power distance and high individual cultural characteristics. However, there is a huge difference between The Netherlands and the USA on the dimension of masculinity-femininity; that is, The Netherlands is a femininity culture, but the USA is a masculinity culture. Furthermore, there are huge differences between the USA and China around three culture dimensions: long-term orientation, individual-collective, and uncertainty-avoidance. China is a long-term-oriented culture, the USA is the opposite; China is a strong uncertainty-avoidance culture, the USA is a weak uncertainty-avoidance culture; China is a collective culture, the USA is individual. Nevertheless, the scores of masculinity-femininity between the USA and China are very close, which means that both the USA and China are masculine cultures. Theoretically, national cultural differences do exist and can be seen clearly in the figure of national cultural profile.

Copyright © 2007, Idea Group Inc. Copying or distributing in print or electronic forms without written permission of Idea Group Inc. is prohibited.

Management Challenges for Cross-Cultural IRM

IRM asserts that information is a resource that should be managed and used wisely to improve business processes and to gain competitive advantage in the marketplace. One particular model, Choo (1995), described information management as a continuous cycle of related activities: identifying information needs, acquiring information, organizing and storing information, developing information products and services, and distributing information and using information.

From the management point of view, individuals' thinking, attitudes, and behaviors are important in considering tasks and jobs to be done in organizations; thus, also IRM activities and stages of information management. As we know, people's thinking, attitudes, and behaviors relate to their cultural background, which is a collective phenomenon at the national level. In cross-cultural environment, the cultural distinctions at the national or societal level may be expected to exert a significant influence on IRM and, therefore, on information management. In the following subsections, management issues on cross-cultural IRM are discussed sequentially.

National Cultural Differences Influence on Information Presentation

Marcus and Gould (2000) applied Hofstede's cultural dimension and examined information presentation on various Web sites that covered industry, business, and education. To illustrate national cultural differences indicating the power-distance influence on Web sites, three university Web sites (one Malaysian and two Dutch universities) were compared. The power-distance index of Malaysia is rated 104; that is, the highest one in Hofestede's study. The Netherlands is rated 38, a correspondingly lower one. The Malaysian Web site features strong axial symmetry, a focus on the official seal of the university, and photographs of faculty and administration leaders conferring degrees, whereas the Dutch university Web site features an emphasis on students and a stronger use of an asymmetric layout.

The cultural differences between a large power-distance country and a small power-distance country are evident. The Web site information of the Malaysia university points to the power of the institution and its administration. By contrast, the Web site message of the Dutch university focuses on the important role that students play at the university. Thus, cross-cultural IRM faces a management challenge because people from different cultural backgrounds have different opinions on the importance of information, which significantly influences the type of information that should be selected and collected for information presentation.

To compare uncertainty-avoidance that influences Web site information presentation, Marcus and Gould (2000) studied two airline Web sites: (1) Sabena Airlines in Belgium with the uncertainty-avoidance index of 94 and (2) British Airways in the United Kingdom with the uncertainty-avoidance index of 35. The Sabena Web site shows a home page with very simple, clear imagery and limited choices. The British Airways Web site shows a greater complexity of content and choices with popup windows, multiple types of interface controls, and hidden content that must be displayed by scrolling.

The differences between the two airline Web sites demonstrate the national cultural differences associated with Web site design; that is, people in high uncertainty-avoidance cultures prefer the compact Web site design and people in low uncertainty-avoidance cultures show a preference for a sophisticated Web site design. Thus, cross-cultural IRM faces a management challenge, as national cultural differences could influence the way of information presentation, which is associated with developing information products and providing information service.

In the same paper of Marcus and Gould (2000), the effects of individual-collective were studied by examining national park Web sites from two countries: U.S. Glacier Bay National Park and Costa Rican National Park. The American Web site features an emphasis on the visitor and his or her goals and possible actions in coming to the park. The Costa Rican Web site focuses on nature, downplays the individual tourist, and even displays a massive political announcement that the Costa Rican government has signed an international agreement against the exploitation of children and adolescents.

The individual-collective index of the USA is 91, suggesting an individual culture. The individual-collective index of Costa Rica is 15, a collective culture. This example confirms that different orientations exist between individual cultures and collective cultures because of the interest of individuals vs. groups or organizations. Therefore, cross-cultural IRM needs to meet the management challenge, because national culture differences could influence both the way and contents of information presentation, which relates to collecting information, organizing information, and storing information.

Marcus and Gould (2000) also compared masculinity-femininity differences on the Web between the Japanese Woman.Excite Web site and the Swedish Excite Web site. They found that the Woman.Excite Web site narrowly orients the search portal toward a specific gender, which this company doesn't do in other countries. The Excite Web site makes no distinction in gender or age.

Considering that the masculinity-femininity index of Japan is 95, the highest one in Hofestede's study, this typifies Japan as a masculine culture. Sweden has the lowest value, 5, and belongs to the feminine culture. The different Web portal designs reflect national cultural differences between masculinity cultures and femininity cultures. Again, cross-cultural IRM Challenge is confronted with national culture differences that influence the way of information presentation and concern the development of information products and information services.

Another example of Marcus and Gould (2000) is long-term-orientation differences on Web sites that were examined by comparing two different versions of English and Chinese from the same German company, Siemens. The long-term-orientation score of China is 118, the highest in Hofestede's study. Germany scores 31 for long-term orientation. The English version of the Siemens Web site shows a typical Western corporate layout with a clean functional design aimed at achieving goals quickly. The Chinese version of the Siemens Web site presents many introductory messages about Siemens, such as leading technology, strong competitor, and so forth, and requires more patience to achieve navigational and functional goals.

This example clearly illustrates how national cultural differences can influence the presentation of information as well as the development of information products and services. Cross-cultural IRM faces the management challenge that national culture differences influence not only the way of information presentation but also the content of information.

National Cultural Differences Influence on Information Requirements

Choi, Lee, Kim, and Jeon (2005) conducted a qualitative cross-national study of cultural influences on user requirements of information. They adopted two cultural dimensions proposed by Hofstede and interviewed 24 people in three countries: Korea, Japan, and Finland. The study results revealed that people in the same country demonstrated similar cultural characteristics in terms of preferences on information requirements. According to Hofstede, Korean and Japanese societies have a high degree of uncertainty-avoidance, while the Finnish society has a low propensity for uncertainty-avoidance. The interview results indicated that more than 90% of Korean and Japanese participants preferred an efficient layout or space usage, a large amount of information within a screen, clear menu labeling, and secondary information about contents. In contrast, 90% of Finnish subjects did not like secondary information about contents. The different results among the Finnish, Korean and Japanese suggest that high uncertainty-avoidance users who feel threatened by ambiguous situations tend to have more and clearer messages to minimize uncertainty about the contents. The challenge of cross-cultural IRM is that national cultural differences influence information needs and information requirements in two ways—the amount of information and the clarification of information—which is highly relevant to information collection, information processing, information usage, and information disposition.

In order to investigate the national cultural differences effect on information characteristics (e.g., content, amount, and format), Choe (2004) conducted an empirical study to explore the national cultural differences influencing the information provided by management accounting information systems. He employed Hofstede's cultural dimensions and collected the sample data from Australian and Korean firms, because Australian culture was seen as very different from the Korean culture (Hofstede, 1991). The study results show that the large amount of information provided by the Korean firm relates to both the high degree of uncertainty caused by advanced manufacturing technology adoption and the strong uncertainty-avoidance of the Korean culture. By contrast, the greater amount of traditional cost control information is produced by the Australian firms, which seems to relate to the high individualism culture of Australians. This example illustrates that the cross-cultural IRM might face different types of information requirements because of national cultural differences. In other words, people and businesses do have different information requirements and needs if they come from different national cultural backgrounds.

National Cultural Differences Influence on Information Collection

Feng (2004) compared the personal home page management between a Chinese university and a Dutch university in which the Chinese culture is clearly very different from the Dutch (Hofstede, 1991). In the Chinese university, the personal home page is a standardized template designed by the personnel office. According to this template, employees provide their content to the personnel office of the university. This information is then subsequently delivered to the information management center by the personnel office when the collection of information is accomplished. This procedure is conducted from time to time as a standard routine. In the Dutch university, the personal home pages are managed by employees themselves, based on the assigned links. In this way, employees can upload the contents of their home page whenever they would like an update. Although a template is provided for guiding employees to prepare the contents of their home pages, employees are free to design their own templates and to decide what they want to include on their home pages. The key point to note here, however, is that in the survey, both Chinese and Dutch employees are willing to use their own approaches to manage their home page information.

Not surprisingly, the personal home pages of the Chinese university are very standardized, reflecting the characteristics of Chinese culture collectivity. The standardized layout of a home page symbolizes formality and seriousness of an organization. In contrast, the personal home pages of the Dutch university are more diverse in style and versatile in functions, which include working messages, personal interests and hobbies, and even family information. In this way, they reflect an important aspect of Dutch culture: individualization. The individualized style of home pages can present individual identity and personal interest. This example indicates that originations in different countries do need different approaches of collecting information because of national cultural differences. To manage cross-cultural information resources, the management challenge is to find out and adopt the proper approach of information collection.

National Cultural Differences Influence on Definition of Information

The definition of information/data is crucial for cross-cultural IRM. Although the standard data reports may be provided, users still find that the reports often do not satisfy the evolving needs of information. Especially users lacking training in Western practices refuse to accept certain information/data definitions as appropriate outcome measurements. Many users eventually attribute the unfortunate chaos to the inferior design of the data management system (Law, 2003) and information management.

Because of national cultural differences, people in one cultural background may not accept definitions of information/data from other cultural backgrounds, since they have defined and been used to the same information/data in their own but different way. Then, it is possible to misinterpret the definitions of information/data adopted from other cultural backgrounds. Therefore, cross-cultural IRM could face a management challenge: how to deal with national cultural differences if they influence definitions of information/data. This is crucial for information distribution as well as information processing, information storing, and information disposition.

Cultural Research Limitations and IRM

Over the past decade, there has been increasing interest in the information systems research literature regarding the impact of national cultural differences on the development and use of information resources. The examples presented in previous subsections are concerned with the various cultural aspects of IRM that rely on Hofstede's model of national culture. Although Hofstede's model has been accepted worldwide in order to study and explain cultural issues related to management, it is worthwhile to notice the arguments from some scholars who consider that the relationship between national cultural values and culturally-influenced, work-related values and attitudes is extremely complex and sometimes not very well explained by Hofstede's model (Mayer & Tan, 2002). Moreover, culture is something that identifies and differentiates one group or category of people from another, which is not a view that finds much support in the contemporary anthropological research literature (Mayer & Tan, 2002). This reminds IRM researchers and practitioners that current national culture research might be limited in explaining people's thinking, attitudes, and behaviors in a cross-cultural IRM environment.

Starting from this point, we should be cautious that it is even more difficult to manage cross-cultural information resources, because the traditional cultural research might not explain people's preferences on information presentation and information procurement, as well as other relevant information behaviors. Consequently, cross-cultural IRM also faces one of the most significant challenges that deal with people's information behavior relating to IRM, which may not be explained by current cultural research and study.

Management Strategies for Cross-Cultural IRM

Management challenges of cross-cultural IRM have been viewed in terms of national cultural influences. Based on the analysis in previous sections, it is believed that national cultural differences could influence every stage of information management; namely, information analysis, information collection, information process, and information disposition. Moreover, national cultural differences also could influence information-management-related activities; for example, acquiring information, organizing information, storing information, publishing information, developing information products and services, and so forth.

Although national cultural differences could influence every stage of information management and many information management activities, all aspects of information management should be based upon a consideration of information requirements and information needs in a cross-cultural environment. It is especially important to provide the acquisition of information resources just in case they are likely to be useful for cross-cultural users and to start providing access to resources just in time and easy to use. Therefore, in order to manage cross-cultural information resources properly, the provision of the right information in the right form and at the right time to meet the needs of users always should be considered as the most important objective of cross-cultural IRM.

However, people do have very different opinions on the importance of information from one culture to another, a fact that is closely associated with both the information provider and information users. In a cross-cultural environment, the key challenge of information management is the mismatch between the information provider and information users, which can be summarized as four types:

- The important information is provided but is not important for users, and therefore, the provided information is of no use to users.
- The important information required by users is not important for information providers, and therefore, it is not available for information users.

- Although the important information provided is important and exists, it is not understandable for information users because of the inconsistent or mismatched description and definition of information between the information provider and users.

- Although the important information provided is important and exists, it is not easy to access for information users because the way of information presentation from the information provider mismatches the way of information procurement from information users.

In order to ensure the availability, reliability, integration, and consistency of information resources in a cross-cultural environment, we could narrow our focus to a sharable information resource. This is to address questions such as: What is the available approach to manage cross-cultural information resources? How could we reduce the mismatch between the information provider and information users because of the cross-cultural environment? How can sharable cross-cultural information resources be created? To answer those questions, I will come up with some management strategies as solutions for cross-cultural IRM.

Management Strategy (1): Linking Requirements to Cross-Cultural Information Resource

In order to manage cross-cultural information resources, it is important to set up an efficient linkage between IRM actor and information resource in the context of a cross-cultural environment, which can be presented as the model in Figure 2. There are five entities in the model: foreign culture, native culture, IRM actor, information content, and information record. The meaning of those entities is introduced briefly as follows. In a cross-cultural environment, IRM actors who are involved in IRM might have different national cultural backgrounds. Due to the national cultural differences, these actors might have very different opinions and behave in different ways. IRM actors can be information management officers, information

Figure 2. Cross-cultural information resource management model

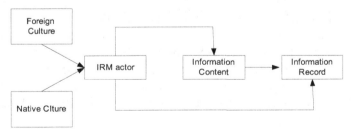

Figure 3. Closed loop model

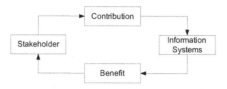

Figure 4. Closed loop-centered, cross-cultural IRM model

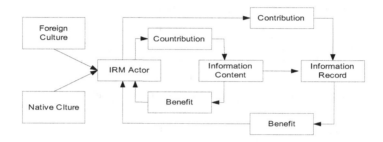

management professionals, information providers, or information customers or users. Some IRM actors might be associated with information content (or data), some of them could be related to information record (data record), and some of them might be involved in management from both information content and information record. IRM is involved with two levels of management, including information content management and information record management. Information record consists of the index message of information contents.

In order to improve IRM in a cross-cultural environment, I would like to employ the closed loop principle (Bemelmans, 2000). The principle is based on the following concept: developing, using, and maintaining information systems (IS) only will be successful in cases where the stakeholders involved have (direct) incentives to do their information systems management (ISM) tasks in an appropriate way. The best way is for stakeholders to have benefits (positive or negative incentives) for being involved in ISM. The closed loop with its entities as well as the relationships between the entities are depicted in Figure 3.

According to Bemelmans, the relationship between IS and their stakeholders can be classified into two categories: contributions and benefits. In effect, the stakeholders contribute to the relevant IS. Contributions will be delivered if and only if stakeholders experience incentives for doing so. One of the best motivators is a direct positive benefit for doing the ISM tasks in the prescribed way. The closed loop principle emphasizes the importance of creating incentives for all stakeholders in an ISM design.

In a cross-cultural IRM environment, one of the important issues is to set up effective linkage between IRM actor and information resources, which could promote the collaboration among the IRM actors and IRM activities, possibly leading to information resources integration. The solution is to employ the closed loop principle and adapt the cross-cultural IRM model as the closed loop-centered, cross-cultural IRM model, which is presented in Figure 4.

The application of the closed loop-centered, cross-cultural IRM model indicates a number of expected management improvements for cross-cultural IRM. First, this model is useful to capture the kind of information that end users or consumers need from their contribution. Moreover, it is also helpful to learn how the information user intends to use the information contained in information resources according to their contribution. In this way, the model serves as a communication channel to capture users' requirements that may relate to national cultural differences. Second, if cross-cultural IRM is armed with this model, the information provider then can try to ensure that the information delivers what is needed and when it is needed. In return, information users will be benefited and should be satisfied if they can access information resources that are as important as they thought. Third, the model can be used as a learning approach to improve cross-cultural IRM gradually based on the continuous updating contributions from both providers and users. Hopefully, the mismatch of information and information resources between information providers and users will be improved in the cross-cultural IRM environment. From time to time, the information resources, relevant management approach, and design orientation will be adapted to match the specific cross-cultural IRM environment. Accordingly, we could believe that the model of closed loop-centered, cross-cultural IRM provides a sound management mechanism by linking the relevant entities: native culture, foreign culture, IRM actors, the contribution and benefit of IRM actors, and information resources.

Management Strategy (2): Standardization

It is suggested that management should pay significant attention to sharing information content and information records as the basis for cross-cultural IRM. Due to national cultural differences, the definitions, descriptions, and presentations of information could be very different in the cross-cultural environment, as discussed in previous subsections. To share information in a cross-cultural IRM environment, management strategy 2 is to carry out standardization for cross-cultural IRM.

To eliminate different definitions and descriptions of information related to cultural differences, it is important to adopt the standard that has been accepted worldwide whenever it is available. One recommended approach is to create a standard library by collecting and storing various widely accepted key terms that possess specific meanings internationally. On the one hand, such a library is very useful for information providers because they could always prepare and modify their manuscript according to the relevant key terms in the standard library. On the other hand, the standard library will improve information presentation in a normative way. Obviously, this is a way to reduce the mismatch between information provider and information users for cross-cultural IRM, because people anywhere and everywhere could understand precisely the commonly accepted key terms.

However, it is a continuing procedure to build up a standard library. To complete such a standard library gradually, a closed loop-centered, cross-cultural IRM model in the previous subsection could be implemented to this procedure, because the feedback from information users will contribute greatly to the key terms collection. From time to time, the standard library created in this way will be complete, accurate, and reliable because of the effective involvement of information users and the proper modifications from information providers. In addition, closed loop-centered, cross-cultural IRM model can be considered as an alternative to reduce the mismatch between information providers and information users when the commonly acceptable definition or description of information does not exist.

Management Strategy (3): Integrating Cross-Cultural Information Resources

To share information in a cross-cultural IRM environment, management has to deal with the culturally influenced information resources that might be very different. The objective of cross-cultural IRM is to facilitate information sharing for cross-cultural users, especially in order for the information provider and information user to understand and accept each other mutually in a cross-cultural environment. Consequently, it is crucial to generate a sharable information resource in a cross-cultural environment by integrating cross-cultural information resources. To describe management strategies for integrating information resources, we presume that there are three culturally different information resources from partners A, B, and C, which are illustrated in Figure 5.

Figure 5. Culturally influenced information resources

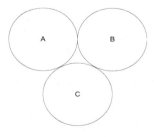

Figure 6. Common grounded information resources

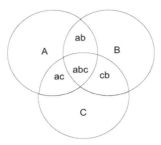

Figure 7. Integrated information resource based on minimum common ground

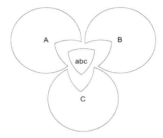

In practice, A, B, or C can be either the existing information resources from all three partners or part of the information set of requirements from any one or two of the three, and the rest will be provided by the other partner(s). To simplify the description, we will assume that A, B, and C are three culturally different existing information resources. The study results easily can be mapped to other situations.

To integrate cross-cultural information resources, we will concentrate on the cross-cultural information resources that have common ground among the information resource A, B, and C. For instance, the overlapped parts in Figure 6 are common grounds that can be named as ab, ac, cb, and abc.

- **Minimum integrating approach:** In order to generate a sharable information resource among A, B, and C in a cross-cultural environment, management can start from the common ground abc, because this part of information resource commonly exists from all three partners and is demanded by all three partners as well. Figure 7 shows a sharable integrated information resource abc for cross-cultural IRM, which can be named as the minimum integrated sharable information resource.
- **Maximum integrating approach:** In order to generate a sharable information resource among information resources A, B, and C in a cross-cultural envi-

Figure 8. Integrated information resource based on maximum common ground

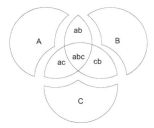

Figure 9. Integrate information resource based on takeover policy

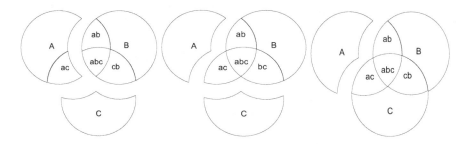

ronment, management can go further and combine all common grounds: abc, ab, ac, and cb. Additionally, the sharable information resource can include all possible combinations among those commonly grounded information resources such as abc and ab, abc and ac, abc and cb, and so forth. Figure 8 presents one of these as a sharable integrated information resource abc, ab, ac, and bc for cross-cultural IRM.

Minimum and maximum integrating approach can be named as merge policy of integrating cross-cultural information resources, which should be very useful when the common ground information is equally demanded by partners in a cross-cultural environment.

However, the common ground information resources are not always equally demanded by partners. In this case, the takeover policy can be used to generate sharable cross-culture information resources among partners A, B, and C in cross-cultural environments. Figure 9 introduces three sharable information resources for cross-cultural IRM based on the takeover policy.

When information resources A, B, and C all are required by all partners, a join policy can be used to generate a sharable information resource for cross-cultural IRM. Figure 10 indicates an integrated sharable information resource based on a join policy.

Figure 10. Integrate information resource based on join policy

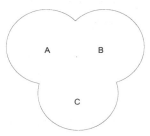

Thus, integrating information resource strategies, three policies—merge, takeover, and join—were introduced. No matter which policy is used, the standardization must be emphasized. It means that all definitions, descriptions, and presentations of information have to be commonly accepted by all partners from three different national cultural backgrounds (A, B, and C). This is the foundation to share the integrated information resource.

IRM Trends and Research Recommendations

With the rapid development of global business, the international environment of IRM is becoming more prevalent. IRM managers face more and more challenges to deal with information resources that foreigners designed; to implement information resources in foreign markets; to provide information services or information goods for foreign information users and consumers; or to procure information services from foreign providers. It is believed that in order to implement information resources and utilize information resources successfully in the global context, IRM managers have to manage national cultural differences, because the trend of cross-cultural IRM is not only popular but also inevitable and irreversible.

As mentioned at the beginning of this chapter, the objective of this chapter is to provide management solutions in order to meet the challenge of cross-cultural IRM. Based on studying national cultural differences as well as investigating management challenges of cross-cultural IRM, a number of management strategies was provided to deal with cross-cultural IRM issues. The contributions of cross-cultural IRM in this chapter can be twofold: The first is to set up a sound management mechanism based on the closed loop-centered, cross-cultural IRM model. Hopefully, this model can be used to improve, promote, and adapt information resources gradually to the specific cross-cultural IRM environment. The second is to generate sharable information resources for cross-cultural IRM based on merge, takeover, and join policy.

In order to manage cross-cultural information resources successfully, more research efforts are still needed. One recommendation is for further research to continue this study and to concentrate on practical cases or best-of-breed practices of implementing management strategies for cross-cultural IRM. It is believed that research findings of cases including both experiences and lessons will contribute greatly to IRM research as well as promote the development of international business.

Another recommendation for further research could focus on exploring key dimension(s) of Hofstede's model for cross-cultural IRM in a global environment. The purpose of such research is to find out which dimension(s) might influence IRM more than others in a cross-cultural environment. The research findings will be significantly meaningful to manage cross-cultural information resources in a proper way, since IRM managers and practitioners could pay special attention to these key dimensions.

The third recommendation is to study various effects of cultural influences on IRM in a cross-cultural environment. The relevant research questions could be addressed as follows: Which dimension(s) of Hofstede's model might influence cross-cultural IRM in a negative way? Which dimension(s) might influence cross-cultural IRM in a positive way? The goal of such research is to look for either a negative or positive effect of cultural influences on IRM in a cross-cultural environment. Both of them are equally meaningful and useful for IRM managers and practitioners. In practice, the negative effect of cultural influences on IRM could be constraint or resistance in a cross-cultural environment. Oppositely, the positive effect of cultural influences on IRM might be a driving force for cross-cultural IRM. Accordingly, the results of such research will be very useful to manage cross-cultural IRM successfully.

The final recommendation is for further research to explore other factors that may influence IRM in a global context, such as policy, legislation and law, economy, language, organizational structure, technology development, and so forth. From a global point of view, those factors can be very different from country to country. It should be interesting and meaningful to study how those factors might influence IRM in a global context. In return, the study results could promote the progress of IRM research and the development of international business.

Conclusion

IRM research covers information management, information systems management, management of information systems, and management of information technology. More attention in previous cross-cultural IRM research has been paid to technology that is used to manipulate, manage, and transmit information resources. In cross-cultural environments, information management as a significant element of IRM deserves more attention, because once technology infrastructure and information systems get implemented and accepted, it does not mean that cross-cultural IRM is complete and accomplished.

In order to manage cross-culture information resources successfully, identifying information needs and information use are crucial, since they both relate to the importance and priority of information that might be framed differently by information providers who come from different national cultural backgrounds. Providing the right information in the right form to meet the needs of information users always should take into account national cultural influences on the presentation of information and interpretation of information, as well as the description of information. The key issue of cross-cultural IRM is the mismatch between information providers and information users, because they may have very different opinions on the importance of information viewing from their own culture background. Consequently, the information provided may be useless from the user's perspective, even though the provider thought it was very important. The user may require information that is not available because the information provider thought it was not important. Meanwhile, the user requires information that does exist but is not accessible because of the different definition, description, or presentation of information between information user and provider. The mismatch would involve information collection, information process, information use, and information disposition. Especially, it would have a strong impact on developing information products and services in a cross-cultural environment.

To provide information resources just in case they are useful for cross-cultural users, the solution in this chapter is to associate users with IRM, to gain user requirements and demands from their contribution, and to benefit them by providing a useful information resource. In addition, cross-cultural IRM should pay significant attention to integrating information resources, since they form the basis for cross-cultural users to share information resources.

References

Bemelmans, T. M. A. (2000). *Bestuurlijke informatiesystemen en automatisering.* Deventer, The Netherlands: Kluwer Bedrijfsingformatie.

Bertot, J. C. (1997). The impact of federal IRM on agency missions: Findings, issues, and recommendations. *Government Information Quarterly, 14*(3), 235-253.

de Brisis, K. (1995). Government policy for information resources management and its implication for provision of information services to the public and to the experts. *Computers, Environment and Urban Systems, 19*(3), 141-149.

Carayannis, E. G.., & Sagi, J. (2001). Dissecting the professional culture: Insights from inside the IT "black box." *Technovation, 21*(2), 91-98.

Choe, J. (2004). The consideration of cultural differences in the design of information systems. *Information & Management, 41*, 669-684.

Choi, B., Lee, I., Kim, J., & Jeon, Y. (2005). A qualitative cross-national study of cultural influences on mobile data service design. *Proceedings of the Conference on Human Factors in Computing Systems* (pp. 661-670). Portland, Oregon: ACM Press.

Choo, C. W. (1995). *Information management for the intelligence organization: The art of scanning the environment*. Medford, NJ: Information Today.

Earley, P. C. (1993). East meets west meets Mideast: Further explorations of collectivistic and individualistic work groups. *Academy of Management Journal, 36*(2), 319-348.

Erez, M., & Earley, P. C. (1993). *Culture, self-identity, and work*. New York: Oxford University Press.

Feng, X. (2004). *Information systems management and culture*. Eindhoven: Universiteitsdrukkerij Technische Universiteit Eindhoven.

Ford, D. P., Connelly, C. E., & Meister, D. B. (2003). Information systems research and Hofstede's culture's consequences: An uneasy and incomplete partnership. *IEEE Transactions on Engineering Management, 50*(1), 8-25.

Gudykunst, W. B. & Ting-Toomey, S. (1988). Culture and affective communication. *American Behavioral Scientist, 31*(3), 384-400.

Hawryszkiewycz, I. T. (1994). *Introduction to system analysis and design*. Upper Saddle River, NJ: Prentice Hall.

Hofstede, G. (1980). *Culture's consequences: International differences in work-related values*. Beverly Hills, CA: Sage Publications.

Hofstede, G. (1991). *Cultures and organizations: Software of the mind*. New York: McGraw-Hill.

Hofstede, G. (2001). *Culture's consequences: Comparing values, behaviours, institutions, and organizations across nations* (2nd ed.). Thousand Oaks, CA: Sage Publications.

Kwon, H. S., & Chidambaran, L. (1998). A cross-cultural study of communication technology acceptance: Comparison of cellular phone adoption in South Korea and the United States. *Journal of Global Information Technology Management, 1*(3), 43-58.

Lachman, R. (1983). Modernity change of core and peripheral values of factory workers. *Human Relations, 36*, 563-580.

Law, W. (2003). Information resources development challenges in a cross-cultural environment. In S. Kamel (Ed.), *Managing globally with information*. Hershey, PA: Idea Group Publishing.

Law, W., & Perez, K. (2005). Cross-cultural implementation of information system. *Journal of Cases on Information Technology, 7*(2), 121-130.

Maceviciute, E., & Wilson, T. D. (2002). The development of the information management research area. *Information Research, 7*(3). Retrieved August 4, 2005, from http://InformationR.net/ir/7-3/paper133.html

Marcus, A., & Gould, E. W. (2000). Crosscurrents: Cultural dimensions and global Web user-interface design. *Interactions, 7*(4), 32-46.

Mayer, M. D., & Tan, F. B. (2002). Beyond models of national culture in information systems research, *Journal of Global Information Management, 10*(1), 24-32.

Nelson, R., Weiss, L. R., & Yamazaki, K. (1992). Information resource management within multinational corporation. *International Information Systems, 4*, 57-88.

Parsons, T., & Shils, E. A. (1951). *Toward a general theory of action*. Cambridge, MA: Harvard University Press.

Robey, D., & Rodriguez-Diaz, A. (1989). The organizational and cultural content of systems implementation: Case experience in Latin America. *Information & Management, 17*(4), 229-239.

Rokeach, (1973). *The nature of human values*. New York: The Free Press.

Ronen, S., & Shenkar, O. (1985). Clustering countries on attitudinal dimensions: A review and synthesis. *The Academy of Management Review, 10*(3), 435-454.

Rose, G. M., Evaristo, R., & Straub, D. (2002). Culture and consumer responses to Web download time: A four-continent study of mono-and polychronism. *IEEE Transactions on Engineering Management, 50*(1), 31-44.

Schneyman, A. H. (1985). Organizing information resource. *Information Management Review, 1*(1), 34-45.

Shackleton, V. J., & Ali, A. H. (1990). Work-related values of managers: A test of the Hofstede mode. *Journal Cross-Cultural Psychology, 21*, 109-118.

Shoib, G.., & Nandhakumar, J. (2003). Cross-cultural IS adoption in multinational corporations. *Information Technology for Development, 10*, 249-260.

Shore, B., & Venkatachalam, V. (1994). Prototyping: A metaphor for cross-cultural transfer and implementation of IS applications. *Information & Management, 27*(3), 175-184.

Straub, D. W. (1994). The effect of culture on IT diffusion: Email and fax in Japan and the United States. *Information System Research, 51*, 23-47.

Straub, D., Keil, M., & Brenner, W. (1997). Testing the technology acceptance model across cultures: A three country study. *Information & Management, 33*(1), 1-11.

Straub, D., Loch, K., & Hill, C. (2001). Transfer of information technology to developing countries: A test of cultural influence modeling in the Arab world. *Journal of Global Information Management, 9*(4), 6-28.

Tan, B. C. Y., Wei, K. K., Watson, R. T., Clapper, D. L., & McLean, E. R. (1998). Computer-mediated communication and majority influence: Assessing the impact in an individualistic and a collectivistic culture. *Management Science, 44*(9), 1263-1278.

Trauth, M. (1989). The evolution of information resource management. *Information & Management, 16*, 257-268.

Walsham, G. (2002). Gross-cultural software production and use: A structurational analysis. *MIS Quarterly, 26*(4), 359-380.

Chapter II

Information Technology Acceptance across Cultures

Amel Ben Zakour
FSJEGJ, Tunisia

Abstract

This chapter introduces national culture as a possible factor accounting for the differences in information technology adoption and use between countries. Based upon culture theory and the technology acceptance model (TAM), the author offers a conceptual model aiming at better understanding IT acceptance across countries of different cultures. It has been argued that six value dimensions—individualism/collectivism, power distance, uncertainty avoidance, masculinity/femininity, high/low context, and polychronism/monochronism—act as moderators of the TAM relationships. Furthermore, the author aims at helping IT designers and IT managers all over the world to understand why certain national cultural values may be congruent or not with the IT to be designed or implemented.

Copyright © 2007, Idea Group Inc. Copying or distributing in print or electronic forms without written permission of Idea Group Inc. is prohibited.

Introduction

With the globalization context, companies all over the world (developed and less developed ones) are using information technologies (IT) that have become more and more sophisticated and complex. These technologies provide huge opportunities to gain a competitive advantage since information could be obtained, processed and transmitted at very low costs (Porter & Millar, 1985). Nevertheless, in order for companies to take full advantage of IT, they have to understand the new challenge provided by these IT (Tapscott & Caston, 1994). Actually, even though organizations adopt IT that best fit their business activities, they cannot guarantee performance leveraging unless the organization members appropriately use it. According to Agarwal (1999), "Acquiring appropriate IT is necessary but not sufficient condition for utilizing it effectively" (p. 85). With the globalization incurring unlimited interconnection possibilities and an increasing number of partnerships, firms belonging to less-developed countries are investing massively in new information technologies that are expected to improve their competitiveness. Nevertheless, we notice that although these technologies are voluntarily purchased by organizations, they are not fully used or accepted in the work context at the individual level. Differences in IT usage and adoption are reported in many descriptive studies pertaining to IT implementation in different countries and contexts (Danowitz, Nassef, & Goodman, 1995; Goodman & Green, 1992). The main causes of these differences that have been identified are technical, economic, or managerial.

Our main concern in this chapter is to shed light on an issue that deserves a closer study which is the cultural factor. Indeed, the latter is expected to be an interesting explanation for the differences in IT adoption and use between countries. Furthermore, culture has been used as an explanatory factor of several managerial and organizational issues (Fisher & Smith, 2003; Thomas & Pekerti, 2003; Laurent, 1983; Hofstede, 1985; Silvester, 1990; Hernandez, 2000). The literature in the information systems (IS) field provides few studies attempting to explore the nature of the relationship between culture and IT implementation. Furthermore, most of the prior cross-cultural studies on IS hardly have focused on the individual behavior toward IT; they generally have focused on IT transfer (Hill, Loch, Straub, & El-Sheshai, 1998), on organizational characteristics related to IT implementation (Robey & Rodriguez-Diaz, 1989), or on end-user computing characteristics (Igbaria & Zviran, 1996). Moreover, these studies have focused on the concept of culture at the macro level of analysis (i.e., they attempted to compare IS-related behaviors in different countries, supposing that each country is characterized by a different set of cultural values). For example, when studying IT diffusion in Japan and the USA, Straub (1994), using Hofstede's cultural classification of 51 countries and three regions, supposed that the Japanese are high in uncertainty avoidance (ranked 7th) and the Americans are low in this dimension (ranked 43rd). Even though there is accumulated evidence pointing to national culture as a factor influencing IT adoption and acceptance, few studies have handled this issue at the individual level. Therefore, the main objective of this chapter is to provide a conceptual framework that examines the influence of national culture, though a macro-level construct, on IT adoption by integrating specific cultural value dimensions to technology acceptance model (TAM) (Davis, Bagozzi, & Warshaw, 1989). This framework is aimed at being used at the individual level since cultural dimensions characterizing national culture (i.e., individualism/collectivism, power distance, masculinity/femininity, uncertainty avoidance, high/low context of communication and polychronism/monochronism) are value dimensions considered as individual psychological dispositions and TAM is designed to capture acceptance at the individual level.

We have divided the chapter into four parts. The first part is devoted to a review of comparative and international studies in the IS field. The second part presents an overview on culture theory, which is found in the literature of various disciplines such as anthropology or cross-cultural psychology. The third part focuses on the presentation of TAM (Davis et al., 1989), judged to be the most parsimonious model of information technology usage, as well as TAM-relevant extensions. The fourth part is aimed at presenting the rationale sustaining the conceptual framework developed to explain how national culture could influence IT usage.

Studies Involving National Culture and Information Systems

Prior researches focusing on cultural issues in the IS field were conducted mainly through two main and different perspectives. According to Ronen and Shenkar (1988), the *emic* perspective "studies behavior from within a single culture-system" (p. 72), whereas the *etic* perspective focuses on universals and studies behavior "from a position outside the system and in comparison with various cultures" (p. 72). Early and Singh (1995) have provided more comprehensive classifications of international and intercultural studies. Early and Singh (1995) have proposed four research approaches to study these issues; namely, unitary form, gestalt form, reduced form, and hybrid form. The difference between these approaches stems from "examining whole systems rather than their components" (Early & Singh, 1995, p. 330).

The unitary form emphasizes the understanding of a particular cultural group using the specific constructs and concepts of this group. Actually, this kind of approach is not so much interested in cultural dimensions per se as much as it focuses on knowing how a certain behavior is manifested in a particular setting. There is no attempt to establish universals based upon such a perspective.

The gestalt form, unlike the unitary system form, focuses on the comparison of the phenomenon with a similar one in different contexts, and "constructs and relationships are derived from general principles rather than from the systems themselves" (Early & Singh, 1995, p. 330). This approach is appropriate when the researcher is concerned with the generalizability of a model across nations.

The reduced form examines the phenomenon of interest with regard to a specific feature of culture, such as the case dealing with one cultural dimension when hypothesizing about a certain behavior. Generally, comparative studies that explain the differences in the applicability of a model by the country of origin actually are using the reduced form. The cultural explanation is provided ad hoc.

The hybrid form goes beyond comparison in search of similarities and differences, and it allows researchers to explain how certain aspects of a culture can explain the phenomenon. "The reliance on nation or culture as a 'black box' is abandoned in favor of a more precise specification of theoretical relationships" (Early & Singh, 1995, p. 334).

In order to have a better visibility of cross-cultural researches in the IS field, we present in Table 1 a summary of cross-cultural studies on IS, grouped according to Early and Singh's classification.

Table 1. Overview of comparative and national studies on IS

Gestalt Form Studies			
Authors	**IS Research Area**	**Countries**	**Key Results**
Igbaria and Zviran (1996)	End-user computing (EUC)	USA Israel Taiwan	EUC characteristics (individual and organizational), perceived usefulness, effectiveness of EUC, satisfaction and acceptance vary across countries.
Straub, Keil, and Brenner (1997)	IT individual acceptance	USA Japan Switzerland	TAM explains the use of e-mail in the USA and Switzerland but not in Japan. The cultural Japanese orientation may limit e-mail use and dissociate usefulness from use.
Leidner and Carlsson (1998)	Executive Support System (ESS) adoption	Sweden USA Mexico	National culture influences the perception of the benefits of ESS on decision-making processes.
Hill et al. (1998)	IT transfer	Jordan Egypt Saudi Arabia Lebanon Sudan	Cultural factors identified in the Arab world such as family and kinship obligations, communal world view, religion, valuing the past or the concept of time, are powerful for predicting the outcomes of technology transfer.
Unitary Form Studies			
Martinsons and Westwood (1997)	Management Information System (MIS) adoption and use	China	The cultural Chinese dimensions of paternalism, personalism, and high context communications shape the use of MIS in China.

Table 1. Continued

Gamble and Gibson (1999)	Decision making and use of information	Hong-Kong	The collectivism characterizing the Chinese (paternalistic and autocratic) leads to the distortion or restriction of the information produced by the financial information system.
Rowe and Struck (1999)	Media choice	France	The use of asynchronous media is more associated with a cultural value oriented toward reactivity. The media that provide more redundancy in the feedback are more associated with relation than with task-oriented values.
Reduced Form Studies			
Straub (1994)	IT diffusion	Japan USA	Since the Japanese are characterized by a high level of uncertainty avoidance, they view e-mail as more information-rich and socially present than the Americans. The structural features of the Japanese language make e-mail difficult to use.
Rice, D'Ambra, and More (1998)	Media choice and assessment	Hong-Kong, Singapore, Australia, USA	The collectivists prefer face-to-face interaction more than the individualists do. The individualists consider telephone and business memo richer than the collectivists do. There is no difference in the perception of equivocal situations between individualistic people and collectivistic ones.
Tan, Wei, Watson, Clapper, and McLean (1998)	Computer-mediated communica-tion	Singapore USA	The impact of computer-mediated communication on majority influence depends on national culture, especially on individualism/collectivism dimension.
Hasan and Ditsa (1999)	IT adoption	West-Africa The Middle-East Australia	Differences were reported in IT adoption along eight cultural dimensions, which are Hofstede's four cultural dimensions, polychronism/monochronism, time orientation, context, and polymorphic/monomorphic.
Hofstede (2000)	IT adoption	56 countries	The higher a country scores on uncertainty avoidance, the slower it will be in adopting a new technology. IT adoption also is influenced by masculinity/femininity and individualism/collectivism dimensions.
Hybrid Form Studies			
Srite and Karahanna (2006)	IT adoption	30 countries	Hofstede's four cultural dimensions have moderating effects on TAM causal relationships.

Our current research follows more the hybrid form since we integrate to TAM a set of universal cultural dimensions deemed to characterize national culture.

Culture Theory

Even though originally rooted in anthropology, a population-level discipline, culture has been defined and researched by many other disciplines such as cross-cultural psychology. Culture has been defined according to several perspectives. Definitions go from the most complex and the most comprehensive (Kluckhohn, 1962) to the simplest (Hofstede, 1997; Triandis, 1972). According to Kluckhohn (1962), "Culture consists of patterns, explicit and implicit, of and for behavior acquired and transmitted by symbols, constituting the distinctive achievement of human groups, including their embodiments in artifacts; the essential core of culture consists of traditional (i.e., historically derived and selected) ideas and especially their attached values; culture systems may, in one hand, be considered as product of action, on the other hand, as conditioning influences upon further action" (p. 73). Hofstede (1997) defines culture as "the collective programming of the mind which distinguishes the members of one group or category of people from another" (p. 5).

Researchers in comparative and intercultural management most of the time use the concept of organizational culture or national culture. Nevertheless, they omit the fact that individual behaviors and attitudes in an organizational context could be influenced by other kinds of cultures. Indeed, culture is a multi-level phenomenon that could be approached according to different levels such as region, ethnic group, religion, language, nation, profession, firm, gender, social class (Hofstede, 1997; Karahanna, Evaristo, & Srite, 2005; Schneider & Barsoux, 2003). Culture also could be defined according to continental or political belonging (Lévi-Strauss, 1985). Furthermore, these different cultures interact with each other. For example, several ethnic groups are found in India.

In the present study, we are interested in national culture since it has been shown to influence management and organizations (Hernandez, 2000; Hofstede, 1985; Laurent, 1983; Silvester, 1990; Zghal, 1994). In a comparative study involving the USA, Indonesia, Japan, and nine countries from Western Europe, Laurent (1983) has found that managers from these countries have different behavioral patterns. For example, the Americans believe that the main goal of a hierarchical structure is to organize tasks and facilitate problem resolution, whereas for the Japanese, the Indonesians, or the Italians, the goal is to highlight the authority structure. In the same vein, Hofstede (1985) has demonstrated that the employees' perceptions of the organization and its management depend on the intensity of two national values: power distance and uncertainty avoidance. Four implicit patterns of organization conception are defined: (1) the pyramid pattern (both power distance and uncertainty avoidance) are high, which corresponds to the hierarchical bureaucracy, quite common in France and the Mediterranean countries; (2) the well-oiled machine pattern (low power distance and high uncertainty avoidance), which corresponds to an impersonal bureaucracy found, for example, in Germany; (3) the village market pattern (both power distance and uncertainty avoidance are low), which is found in Great Britain and Nordic countries; and (4) the family pattern (high power distance and low uncertainty avoidance), which is found for example in African and Asian countries. Hernandez (2000) also has shown that African management systems are influenced by a national cultural trait: paternalism. The latter corresponds to high power distance and relationships based upon mutual trust. Paternalism also has been found to be perceived positively in Tunisian culture contrarily to American culture, which is focused on contractual relationships or Dutch culture, which is focused on consensual relationships (D'Iribarne, 1989). In a comparative study between France and Germany, Silvester (1990) has confirmed the results obtained by Hofstede (1985), showing that the French organization could be assimilated to a pyramid and the German one to a well-oiled machine. For example, they found that French managers have a prescriptive kind of work, whereas German managers have cooperative kind of work through which employees could learn and build a collective identity.

Several sets of dimensions have been developed to characterize the concept of national culture. Table 2 provides an overview of the most known cultural dimensions found in several fields of studies.

The analysis of the culture frameworks in Table 2 reveals an overlap of cultural dimensions identified by the various authors. Even though the definitions of these dimensions are not fully convergent in all the cases, still they show some similarities. Table 3 gives an overview of the similarities that exist between these dimensions. The current research will use the following dimensions: individualism/collectivism, masculinity/femininity, uncertainty avoidance, power distance (Hofstede, 1997), high

Table 2. Overview of the most known cultural dimensions

Cultural Dimensions	Authors
Power Distance	Hofstede (1997)
Individualism/Collectivism	Hofstede (1997)
Masculinity/Femininity	Hofstede (1997)
Uncertainty Avoidance	Hofstede (1997)
Long-term Orientation	Hofstede (1997)
Confucian Work Dynamism	Chinese Culture Connection (1987)
Conservatism	Schwartz (1994)

Table 2. Continued

Intellectual autonomy	Schwartz (1994)
Affective autonomy	Schwartz (1994)
Hierarchy	Schwartz (1994)
Egalitarianism	Schwartz (1994)
Mastery	Schwartz (1994)
Harmony	Schwartz (1994)
Universalism/Particularism	Trompenaars and Hampden-Turner (1998)
Individualism/Communitarianism	Trompenaars and Hampden-Turner (1998)
Neutral/Emotional	Trompenaars and Hampden-Turner (1998)
Specific/Diffuse	Trompenaars and Hampden-Turner (1998)
Achievement/Ascription	Trompenaars and Hampden-Turner (1998)
Attitudes to time	Trompenaars and Hampden-Turner (1998)
Attitudes to environment	Trompenaars and Hampden-Turner (1998)
Communication context	Hall (1989); Hall and Hall (1987)
Perception of space	Hall & Hall (1987); Hall (1989)
Monochronic/polychronic time	Hall (1989)
Nature of people	Kluckhohn and Strodtbeck (1961)
Person's relationship to nature	Kluckhohn and Strodtbeck (1961)
Person's relationship to other people	Kluckhohn and Strodtbeck (1961)
Primary mode of activity	Kluckhohn and Strodtbeck (1961)
Conception of space	Kluckhohn and Strodtbeck (1961)
Person's temporal orientation	Kluckhohn and Strodtbeck (1961)

context/low context (Hall, 1989), and time perception (Hall, 1989; Trompenaars & Hampden-Turner, 1998).

Several reasons lead us to investigate national culture through these six dimensions.

First, these dimensions rely on variables that are linked more directly to social and organizational process: they focus on human values rather than on general beliefs about the way we see the world. Indeed, "culture is primarily a manifestation of core values" (Straub, Loch, Evaristo, Karahanna, & Srite, 2002); therefore, in order to better capture the relationship between culture and behavior, Triandis suggests using values (Triandis, 1972). Second, the first four dimensions (Hofstede's cultural dimensions) constitute the most used and most recognized dimensions as a whole or separately in studying cross-cultural issues in management and organizations. Because of its global coverage in terms of respondents, it seems that Hofstede's study has been unrivaled (Smith & Bond, 1999). In fact, the identification of the

Table 3. Similarities in cultural dimensions

Hofstede (1997)	Schwartz (1994)	Trompenaars & Hampden-Turner (1998)	Chinese Culture Connection (1987)	Hall (1989)	Kluckhohn and Strodtbeck (1961)
Power distance	Hierarchy / Egalitarianism	–	–	–	–
Individualism/ Collectivism	Autonomy	Individualism/ Communitarianism			Relational Orientation
Masculinity/ Femininity	Mastery/ Harmony	Achievement/ Ascription Inner-directed/ Outer-directed	–	–	Man-Nature Orientation
Uncertainty Avoidance	–	–	–	–	–
Long-term Orientation	Conservatism	Attitudes to Time	Confucian Work Dynamism	Time Perception	Time Orientation
–	–	Specific/ Diffuse	–	Space (personal space and territory)	Space Orientation (public, private)
–	–	–	–	High /Low Context	–

cultural dimensions was based upon a field study covering a sample of 40 countries in which more than 116,000 questionnaires were collected. Hofstede's work also has been validated directly or indirectly by many other researchers in different settings (The Chinese Culture Connection, 1987; Hofstede & Bond, 1984; Shackleton & Ali, 1990). The studies that have been conducted since Hofstede's work (The Chinese Culture Connection, 1987; Schwartz, 1994; Trompenaars & Hampden-Turner, 1998) exploring the national culture through values have sustained and amplified his findings rather than having contradicted them (Smith & Bond, 1999). The two dimensions related to time orientation and communication context are based upon Hall's well-established studies on intercultural communications. Indeed, cross-cultural studies of styles of communication, briefly reviewed by Smith and Bond (1999), reveal a divergence between societies in several aspects of communication and provide evidence sustaining Hall's contention about high/low context (Gudykunst & Ting-Toomey, 1988).

In the following paragraphs, we are going to give more explanations about the origins of these cultural dimensions.

Power Distance (PDI)

According to Hofstede (1997), individuals in high PDI societies accept the unequal distribution of power, which is what Schwartz (1994) calls *hierarchy*. The latter is linked to the inequality of the distribution of power, roles, and resources. Conversely, in low PDI societies, individuals do not accept this inequality. Indeed, in an organizational context, subordinates can criticize their superiors, and superiors are supposed to take into account this criticism, since everyone is free to express his or her disagreement about any issue. In a low PDI organization, the hierarchical system depicts an inequality of roles that may change over time (Hofstede, 1997). This kind of behavior is not tolerable in a society that emphasizes authority. What Schwartz identifies as egalitarianism (also egalitarian commitment) (Schwartz, 1994) can be assimilated to the concept of low PDI advocated by Hofstede.

Egalitarianism dimension is driven by values expressing the transcendence of selfish interests and leading to a voluntary commitment to promote welfare of other individuals (e.g., social justice or loyalty). Even though Schwartz explains the opposition hierarchy/egalitarianism by psychological concepts (i.e., self-enhancement/self-transcendence), Hofstede explains PDI dimension, relying more on social contexts in which individuals have learned to accept or reject the inequality of relationships between people (such as country history, ideologies, language). These two dimensions are fairly comparable and valid to characterize national culture.

Individualism/Collectivism (I/C)

Individualism concept takes its roots in the Western world. Indeed, in Britain, the ideas of Hobbes and Adam Smith about the primacy of the self-interested individual sustain this concept. In contrast, Confucianism in the Eastern world, emphasizing virtue, loyalty, reciprocity in human relations, righteousness, and filial piety, underlies a collectivistic view of the world. Even though some societies are characterized as individualistic and others as collectivistic, they have to deal with both individualistic and collectivistic orientations. These orientations co-exist, and what makes the difference among societies is the extent to which they emphasize individualistic and collectivistic values.

Individualism/collectivism has been criticized by several disciplines and especially psychology (for a review, see Kagitçibasi, 1997). Nevertheless, it is still the focal dimension in cross-cultural studies and has been used most often as an explanatory variable (Schwartz, 1994).

Hui and Triandis (1986) define collectivism as "a cluster of attitudes, beliefs, and behaviors toward a wide variety of people" (p. 240). Seven aspects of collectivism has shown to be relevant in characterizing it, which is the consideration of the implications of our decisions for other people, the sharing of material resources, the sharing of nonmaterial resources (e.g., affection or fun), the susceptibility to social influence, self-presentation and facework, the sharing of outcomes and the feeling of involvement in others' lives (Hui & Triandis, 1986).

All the dimensions pertaining to this concept of I/C (Hofstede, 1980), developed by Schwartz (autonomy/conservatism), Trompenaars and Hampden-Turner (individualism/communitarianism) and Kluckhohn and Strodtbeck (relational orientation), deal with one unique issue: the person's relationship to other people. Schwartz defines autonomy/conservatism according to the person's embeddedness in the group; meanwhile, Hofstede, Trompenaars, and Hampden-Turner define I/C, emphasizing individual vs. group interest.

Two underlying ideas support the autonomy/conservatism dimension. The first idea explains the autonomy pole: an individual is considered autonomous when he or she is holding independent rights and desires and the relationship he or she has with the others is dictated by self-interest and negotiated agreements. The second idea explains the conservatism pole: an individual is part of a group and his or her significance pertains to his or her participation in and identification with the group (Schwartz, 1994). At the individual level, Schwartz argues that there are two kinds of values: individualistic values, which are hedonism, achievement, self-direction, social power, and stimulation; and collectivistic values, which are pro-social, restrictive conformity, security, and tradition.

According to Hofstede, in individualist cultures, "the interests of the individuals prevail over the interests of the group" (Hofstede, 1997, p. 50). People are supposed to decide and achieve alone. In collectivist cultures, people prefer to decide and act in a group so that responsibility and reward will be shared. "Collectivism is associated with childrearing patterns that emphasize conformity, obedience and reliability" (Triandis, 1989, p. 16).

Individualism/Collectivism construct is considered to be the most important aspect of cultural differences because of its parsimony and its potential to explain variations in economic development (Kagitçibasi, 1997). Indeed, according to Hofstede and Bond (1988), there is a high correlation between individualism and the level of economic growth at the national level.

Masculinity/Femininity (MAS)

The MAS dimension was derived empirically from Hofstede's work on work-related values across countries. Hofstede distinguishes between cultures according to their emphasis on achievement or on interpersonal harmony. In labeling this dimension according to gender, Hofstede refers to the social and culturally determined roles associated with men vs. women and not to the biological distinction. Several other studies have identified almost the same dimension labeled as mastery/harmony (Schwartz, 1994) or achievement/ascription (Trompenaars & Hampden-Turner, 1998). All the dimensions show an obvious overlapping, since they are driven by almost the same set of opposing values. Empirically, a cross-cultural study conducted by Schwartz (1994) shows a positive correlation between MAS and mastery. In one pole, the underlying values are assertiveness, material success, control or mastery, and competition. This is the pole of masculinity (mastery, achievement). In the pole of femininity, the dominant values are modesty, caring for others, warm relationships, solidarity, and the quality of work life.

Uncertainty Avoidance

This dimension has been inferred from Hofstede's survey pertaining to the theme of work stress when he addressed his questionnaire to IBM employees. According to Hofstede, the extent to which individuals tend to avoid uncertainty can differentiate among countries. Indeed, the feeling of uncertainty is something that could be acquired and learned in the diverse institutions of a society such as family, school, or state. Each society will have its proper behavioral model toward this feeling of uncertainty. Hofstede argues that uncertainty is strongly linked to anxiety. The latter could be overcome through technology, laws, and religion (Hofstede, 1997). In strong uncertainty avoidance societies like South America, people have rule-oriented behaviors. On the contrary, in societies where uncertainty avoidance is weak, people do not need formal rules to adopt a specific behavior. In the same vein, Triandis (1989) makes the difference between loose and tight cultures. He sustains that loose cultures encourage freedom and deviation from norms, whereas in tight cultures, norms are promoted, and deviation from those norms is punished. Shuper, Sorrentino, Otsubo, Hodson, and Walker (2004) also have confirmed that countries do differ in uncertainty orientation. Indeed, they have found that Canada is an uncertainty-oriented society that copes with uncertainty by attaining clarity and finding out new information about the self and the environment, and Japan is a certainty-oriented society that copes with uncertainty by maintaining clarity and adhering to what is already known (Shuper et al., 2004)

Time Orientation

The most overlapping dimensions are attitudes to time (Trompenaars & Hampden-Turner, 1998), time perception (Hall, 1989), and time orientation (Kluckhohn & Strodtbeck, 1961), which highlight two main dimensions describing the concept of time: the structure of time (discrete vs. continuous) and the horizon of time reference (reference to the past, present, or future). According to Hall and Hall (1987), a monochronic person runs one activity at a time and associates to each activity a precise time, while a polychronic person can perform many activities simultaneously without preparing an exact schedule, or if it exists, it can be ignored. Studies conducted in this sense (Schramm-Nielsen, 2000) have shown that polychronic time is very specific to the Mediterranean countries, the Middle East, the Arab world, and Latin America. Indeed, in these countries, people assume that time is elastic, and thus, the appointment time and planning are not always respected. Actually, this behavior is considered very normal, because individuals emphasize interpersonal relationships. On the contrary, North-European and Anglo-Saxon countries usually adopt the monochronic time. In these countries, time is divided into segments and rigorously scheduled so that each activity will be assigned a specific and unique time within which it will be performed. Priorities also are classified according to time. Consequently, planning and schedules are sacred and cannot be changed. Trompenaars and Hampden-Turner (1998) called these two types of time systems the sequential and the synchronic. "Time can be legitimately conceived of as a line of sequential events passing us at regular intervals. It can also be conceived as cyclical and repetitive" (Trompenaars & Hampden-Turner, 1998, p. 126).

The time assumption underlying the remaining dimensions (long-term orientation, Confucian work dynamism, and conservatism) pertains to the temporal horizons to which persons refer to in their activities: past, present, and future. Confucian work dynamism is based upon values derived from Confucian teachings. In one pole, called by Hofstede *long-term orientation*, are found values such as perseverance and thriftiness, and in the *short-term orientation* pole are found values such as personal steadiness and stability and respect for tradition.

Schwartz identifies a value called conservatism that is similar in many aspects to the Confucian work dynamism, since it emphasizes the cohesiveness of the group and the respect for the traditional order of the group. This emphasis is depicted in at least three values characterizing Confucian work dynamism: ordering relationships, personal steadiness, and respect for tradition.

High Context/Low Context of Communication

The nature of the communication context has been highlighted only by Hall (1989) as a relevant dimension capable of discriminating between cultures. In order to better understand the concept of high context/low context of communication, it is crucial to define context and to shed light on the relationship between meaning, context, and communication. According to Hall and Hall (1987), "context is the information that surrounds an event and inextricably bound up with the meaning of that event" (p. 7). A distinction can be made between internal and external contexting. The former depends on past experience internalized into a person or the structure of the nervous system or both (Hall, 1989). The latter is based on the situation and/or setting pertaining to a particular event. Meaning is the interpretation attributed to a particular message or communication. "Words and sentences have different meanings depending on the context in which they are embedded" (Hall, 1983, p. 56). According to Hall (1989), meaning is made up of the message and the two kinds of context, which are the background and preprogrammed responses of the recipient on the one hand and the situation on the other hand. Since information, context, and meaning are the key components in defining high and low context cultures, Hall contends that "the more information that is shared, ... the higher the context" (Hall, 1983, p. 56). Therefore, high-context communications are those that rely more on context (shared or stored information) than on transmitted information. In this case, little of the meaning is conveyed in the transmitted part of the message. In high-context cultures, "simple messages with deep meaning flow freely" (Hall, 1989, p. 39). People do not need to elaborate their speech codes because of the largely shared assumptions about the rules of communication. Indeed, the emphasis is more on the manner of transmitting a message than on the message per se. For example, gestures and voice intonation can make the difference between messages of the same content. On the contrary, low-context communications rely more on the transmitted part of the message and less on the context. Americans, Germans, Swiss, and Scandinavians are found to be low-context. In such cultures, characterized by Hall (1989) as fragmented cultures, people have more elaborate codes in communication because of their lack of shared assumptions about communication rules. All the meaning they try to transmit is reflected in the information contained in the literal message. In this case, it seems obvious that there is no need to use different kinds of signals as a guide to interpretation.

Technology Acceptance Model

The technology acceptance model (TAM) is considered a tool designed to understand and measure the IT individual determinants of use. The most comprehensive TAM is designed by Davis et al. (1989). It is based on two key concepts: perceived usefulness (PU), defined as "the prospective user's subjective probability that using a specific application system will increase his or her job performance within an organizational context" (p. 985); and perceived ease of use (PEOU), defined as "the degree to which the prospective user expects the target system to be free of effort" (p. 985). According to this model, an individual who intends to use a system will have two types of beliefs (PU and PEOU), which influence behavioral intention through attitudes. In addition, PU will have a direct effect on behavioral intention. Finally, external variables will have a direct effect on the two beliefs.

In order to study individual adoption of IT, the present research will be using the TAM. The use of the latter is motivated by theoretical, empirical, and practical considerations. From a theoretical point of view, TAM takes its roots from the theory of reasoned action (TRA) (Ajzen & Fishbein, 1980). The latter provides a simple model to understand and predict a person's behavior. It goes from the principle that an individual behaves most of the time in a rational manner, takes into account available information, and assesses consequences of his or her actions implicitly or explicitly (Ajzen, 1988). According to TRA, individual behavior is influenced by behavioral intentions. These intentions at the same time are determined by attitudes toward this behavior and subjective norms. In turn, attitudes and subjective norms are dependent on behavioral and normative beliefs, respectively.

TRA is one of the most prominent theories of social psychology and the most powerful to predict a specific behavior (Agarwal, 1999; Liska, 1984). Two meta-analyses of empirical studies testing the TRA conducted by Sheppard, Hartwick, and Warshaw (1988) show that the theory has a powerful predictive utility. Moreover, it is proved to be broadly applicable to a large array of behaviors (Agarwal, 1999; Sheppard et al., 1988). A study of a large variety of behaviors, such as voting for a presidential candidate, having an abortion, or smoking marijuana, confirms the causal relationships of the model, considering attitudes and subjective norms as independent variables and intention as the dependent variable (Ajzen, 1988).

From an empirical point of view, several researches applied TAM to different kinds of IT, such as text processing application (Davis et al., 1989), computers (Igbaria, 1995), e-mail (Karahanna & Straub, 1999), the World Wide Web (Moon & Kim, 2001), with different kinds of end user populations, such as students (Davis et al., 1989; Mathieson, 1991; Szajna, 1996), employees (Karahanna, Straub, & Chervany, 1999; Lucas & Spitler, 2000), or even physicians (Hu, Chau, Sheng, & Tam, 1999), and in different organizational contexts. Other studies test TAM and compare it to competing theories such as TRA (Davis et al., 1989) or Theory of Planned Behavior (Mathieson, 1991). All of these studies show that TAM has a strong predictive power. Moreover, TAM extensions (Agarwal & Karahanna, 2000; Agarwal & Prasad, 1999; Dishaw & Strong, 1999; Lucas & Spitler, 1999; Venkatesh & Davis, 2000; Venkatesh, Speir, & Morris, 2002), the main objective of which is to extend TAM by integrating other variables hypothesized to influence directly TAM constructs or indirectly by moderating the relationships among them, support its causal relationships and confirm its predictive utility. A nonexhaustive summary of TAM extensions is presented in Table 4.

From a practical point of view, TAM gives the possibility to practitioners, especially IS designers, to identify the areas in which we can act to improve IT use. Indeed, the two key concepts of TAM (i.e., perceived usefulness and perceived ease of use) could be controlled by IT designers. Consequently, developers could make it easy to use IT and to identify more concretely managerial benefits driven by IT that they are developing.

Table 4. Summary of TAM extensions

Authors	IT Tested	Variables Added to TAM	Research Methodology
Taylor and Todd (1995)	Hardware and software in a computing resource center	- Subjective norms - Perceived behavioral control	Field study Questionnaire addressed to 1,000 students from a business school
Igbaria (1995)	Computers	- Subjective Norms - Normative beliefs and motivation to comply with the referent group - Computer anxiety - Computer knowledge - Direction support - Info center support - Organizational politics - Organizational use of the system (colleagues, subordinates, and CEOs)	Field study Questionnaire addressed to 471 managers and professionals working in 54 North-American firms
Jackson, Chow, and Leitch (1997)	Different kinds of IS	- Situational involvement - Intrinsic involvement - Prior use	Filed study Questionnaire addressed to 585 employees from organizations developing or revising their IS
Karahanna and Straub (1999)	Electronic Mail	- Social Influence - Social Presence - Perceived Accessibility - Availability of User Training and Support	Field study Questionnaire addressed to 100 users working in a worldwide American firm in the transportation sector
Venkatesh (1999)	Virtual Workplace System	- Game-based training	Experiment with business professionals: 69 in the first study and 146 in the second
Dishaw and Strong (1999)	Maintenance support software tools	- Technology-Task Fit (TTF) - Task Characteristics - Tool Experience - Tool Functionality	Field study Questionnaire addressed to programmer analysts working in three American firms, leaders in their respective fields (financial services, aerospace manufacturing, and insurance)
Lucas and Spitler (2000)	Broker workstations	- Broker strategy (work approach) - Perceived System Quality	Field study Questionnaire addressed to 41 brokers

Table 4. Continued

Venkatesh and Morris (2000)	System for data and information retrieval	- Gender - Subjective Norms - Experience	Longitudinal field study Questionnaire addressed to 445 individuals from five organizations
Agarwal and Karahanna (2000)	World Wide Web	- Personal Innovativeness - Playfulness - Self-efficacy - Cognitive Absorption	Field study Questionnaire addressed to 250 students
Venkatesh and Davis (2000)	- Scheduling information system - Windows-based environment - Windows-based customer account management system	- Subjective norms - Image - Job relevance - Output quality - Result demonstrability - Experience - Voluntariness	Three longitudinal field studies Questionnaires addressed to 48 floor supervisors in a medium-sized manufacturing firm in the first study, to 50 members of the financial department of a large financial firm in the second study, and to 51 employees from a small accounting services firm
Moon and Kim (2001)	World Wide Web	- Perceived Playfulness	Field study Questionnaire addressed to 152 students

The Influence of National Culture on IT Adoption: Conceptual Model and Research Propositions

We draw upon an extension of the TAM that integrates subjective norms (Venkatesh & Morris, 2000) as a determinant of the intention to use IT. Subjective norms reflect a person's perception of the social pressure to perform the behavior in question. The importance attributed to social norms in determining and predicting behavior varies across cultures (Triandis, 1977); therefore, we expect that integrating subjective norms will strengthen our understanding of differences in behavioral intentions and will allow a better capturing of the cultural effect on IT use.

In order to explore the cultural factor, we rely upon Hofstede's four cultural dimensions:

- **Individualism/collectivism:** Defined as the degree to which people in a society are integrated into strong cohesive ingroups.

- **Masculinity/femininity:** Defined as the extent to which a society attributes qualities such as assertiveness, material success to men. and modesty and concern about the quality of life to women.
- **Uncertainty avoidance:** Defined as the extent to which the members of a culture feel threatened by uncertain and unknown situations.
- **Power distance:** Defined as the extent to which the less powerful members of institutions and organization within a country expect and accept that power is distributed unequally.

Moreover, we rely upon Hall's work on intercultural communication, which is based on two key cultural dimensions:

- **High context/low context:** Defined as the extent to which people are aware of and pay attention to all types of situational and contextual cues when interpreting messages.
- **Polychronism (vs. monochronism):** Defined as the preference to perform many activities at the same time.

These dimensions are hypothesized to have direct effects or moderator ones on TAM constructs and relationships. Integrating cultural dimensions to TAM is an attempt to better understand the genuine influence of national culture on IT adoption at the individual level. Indeed, cross-cultural studies on IS (see Table 1) focused on IS phenomena at the organizational level, such as IT transfer. Moreover, all the authors (except Srite & Karahanna, 2006) do not conceptualize national culture at the individual level. They have just studied IT-related behaviors in different countries, supposing that each country is characterized by a different set of cultural values. Their findings can be criticized in a sense that cultural values that characterize a country could not be espoused by all the members of this country. For example, Hofstede has found that Japan is a collectivist country, but this does not mean that each Japanese person will have the same level of collectivism. Actually, Hofstede has studied work-related values at the macro level by developing national scores for each cultural value. Therefore, it is worthwhile to analyze and measure culturally espoused values when dealing with national culture at the individual level. In this study, since we are attempting to deepen our understanding of IT adoption at the individual level, we have chosen to conceptualize cultural dimensions at the individual level so that they can be integrated in TAM. The conceptual model driven by this integration is presented in Figure 1.

In the next paragraphs, we present the rationale underlying model propositions.

Figure 1. Conceptual model

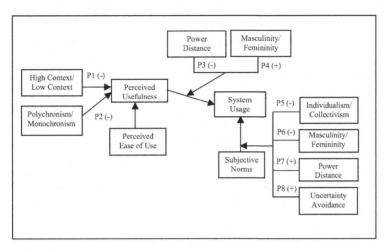

Proposition 1. Individuals espousing high-context values will have a less strong perception of IT usefulness than those holding low-context values.

Since high-context oriented individuals prefer less coded messages, the electronic communication, implying very explicit content, will not be very useful as a means of communication. It cannot inform the receiver of the context of the message. The receiver probably will seek to know more about the message by other traditional means, implying more physical cues (face expression, intonation of the voice). Martinsons and Westwood (1997) have demonstrated that the use of management information systems (MIS) in Chinese culture depends on high-context communications. "Computers do not convey the necessary richness of meaning in a high context communication environment" (Martinsons & Westwood, 1997, p. 220) Indeed, since in China, a high-context country, coding is very restricted and so messages are terse in words, using computer-based MIS will not allow a full understanding of messages. Actually, in order to better interpret the words, the Chinese rely upon other cues such as tone, dynamics, facial expression, and body language, which are not permitted when using computer-based communication.

Proposition 2. Individuals adopting monochronic time values will be using IT more than those adopting polychronic time values.

Individuals adopting the monochronic time are worried about the monitoring of time and prefer doing one thing at a time; therefore, they can view in IT a way to have control over time. Rowe and Struck (1999) argued that "people who are biased towards action see in asynchronous media the way to coordinate their tasks without interrupting them" (p. 166). In this case, we could expect that e-mail or voicemail would be appreciated highly by individuals espousing monochronic cultural values. On the contrary, individuals holding more fluid attitudes toward time are able to engage in many things at once, pay more attention to interpersonal relationships than to schedules, and accept disturbances. Therefore, they will be expected to view IT as less useful compared to individuals holding monochronic values.

> **Proposition 3.** Individuals in high power distance cultures accept less IT than individuals in low power distance cultures.

In high power distance cultures, IT could threaten the hierarchy, which reflects "the existential inequality between higher-ups and lower-downs" (Hofstede, 1997, p. 37), because it suggests decentralization. Conversely, in low power distance cultures, individuals are more interdependent, whatever their ranks in the hierarchy. Therefore, they will be more favorable to IT, which does not contradict their perception of power distribution. For example, in an interpretive study conducted by Hasan and Ditsa (1999) comparing IT adoption in Australia, West Africa, and the Middle East, they have found that in the Middle East, high in power distance, modern IT, like the Internet, was seen as a threat to the authoritarian structure of their society. In the same vein, Martinsons and Westwood (1997) have demonstrated through an explanatory theory that paternalism in Chinese culture is an influential factor regarding the use of computer-based MIS, compared to American culture.

Straub, Keil, and Brenner (1997), testing TAM in three countries (the U.S., Switzerland, and Japan), also have found that in high power distance cultures, such as Japan, the leveling effect of computer-based media is not well appreciated. In the same sense, Shore and Venkatachalam (1996) have developed a general framework, contending that the introduction of an IT application in a high power distance culture will be difficult to accommodate for superiors having authority roles and subordinates used to being controlled by superiors. Actually, a new information technology may alter lines of authority and responsibility that are clearly defined in a high power distance culture. For example, a new system may provide end users with access to previously guarded and inaccessible information, or it may establish new lines of authority through new channels of electronic communication, such as e-mail (Shore & Venkatachalam, 1996).

> **Proposition 4.** Masculinity/femininity moderates the relationship between perceived usefulness and system usage.

Perceived usefulness will be more determinant for system usage for individuals espousing masculine values, since the perceived usefulness of IT is related intimately to achievement values that predominate in masculine cultures. In feminine cultures, work environment and the quality of life are much more important than attaining goals; this is why we expect much less impact of perceived usefulness of IT on usage for individuals adopting feminine values. Leidner and Carlsson (1998) have found that the Mexicans, belonging to a masculine culture, show more interest in executive information systems (EIS) than the Swedes. More precisely, the Mexicans perceive EIS as more useful than the Swedes, because such a system allows increasing the speed of decision making, and therefore, work goals could be reached more easily. In addition, in masculine cultures in which individuals are highly assertive, media lacking social presence, such as e-mail, would not be favored (Straub et al., 1997), because it does not allow end users to be highly aggressive and assertive.

> **Proposition 5.** Individualism/collectivism moderates the relationship between subjective norms and system usage.

People in individualist cultures are more concerned with themselves than with the group. Therefore, for individuals holding individualistic cultural values, the opinions of the members of the group will not have weight in their decision to adopt IT. Conversely, people holding collectivist values will be more concerned with the maintenance of the group's cohesiveness. As a result, they will be expected to show more interest in others' opinions about IT because they want to conform to the group's behavior. In a comparative study on the applicability of the TRA across two different cultural settings, Korea and the U.S., Lee and Green (1991) have found that social norms are more important for collectivistic cultures (Korea) than for individualistic ones (U.S.) in influencing behavior.

> **Proposition 6.** Masculinity/femininity moderates the relationship between subjective norms, and system usage, such as subjective norms, will have much less influence on system usage for masculinity-oriented individuals.

In feminine cultures, individuals are expected to pay more attention to the opinions of others in behaving since they are more people-oriented than in masculine cultures in which the most important thing is goal achievement.

> **Proposition 7.** Power distance moderates the relationship between subjective norms and system usage.

In high power distance cultures, since individuals are not supposed to disagree with their superiors, their reliance upon the opinions of superiors will be more marked when assessing IT than for individuals from low power distance cultures.

Martinsons and Westwood (1997) have demonstrated through their explanatory theory that paternalism in Chinese culture is an influential factor regarding the use of management information systems compared to American culture. Paternalism is a concept very close to high power distance, since in paternalistic societies, key decisions are made by the patriarch, and management systems are very centralized.

According to Kowner and Wiseman (2003), people in tight (vs. loose) cultures in which deviation from behavior is not permitted "are more likely to obey the behavior ascribed by their social position" (p. 182). Consequently, social influence will play a greater role in shaping IT adoption for individuals belonging to tight culture than for those belonging to loose cultures.

> **Proposition 8.** Uncertainty avoidance acts also as a moderator of the relationship between subjective norms and system usage.

The social influence exerted by important persons will be much more important in determining IT use in cultures seeking to avoid uncertainty than in cultures comfortable with uncertainty. Actually, in order to deal with uncertainty and ambiguity, individuals in strong uncertainty avoidance cultures are very concerned by the establishment and respect for rules, and therefore, subjective norms will be more important as guidance to behavior than for individuals in weak uncertainty avoidance cultures that rely more on their proper competence to evaluate a situation. The process of compliance (Kelman, 1958), one of the basic processes underlying the relationship between subjective norms and behavior, will be more salient when individuals are holding high uncertainty avoidance values. Indeed, compliance supposes that the individual will conform because a social actor that is important to him or her (and having the power of reward and punishment) wants him or her to behave in a specific manner. Therefore, conforming to referent others by adopting IT will weaken the uncertainty pertaining to the nonadoption of IT.

Practical Implications

The framework we provide is an attempt to highlight the importance of culture when dealing with IT adoption in different national cultural settings. It should improve IT designers' awareness of the importance of cultural values for IT acceptance in foreign countries. Actually, the framework we offer does not suggest prescriptions to be adopted by IT designers. Rather, it encourages the latter to take into account cultural values when designing IS and/or preparing IT implementation procedures for companies around the world. Actually, IT designers should be aware that implementation failure could be caused by cultural incompatibility. As a result, they should adapt their implementation tactics so that they will be more congruent with end users' cultural systems. For example, in high power distance cultures, end users may be reluctant to use computer-mediated communication tools, because this use may lead them to contradict their superiors. Therefore, IT designers should adjust IT features to the social needs of end users.

Furthermore, the conceptual model we have proposed suggests that the key belief in TAM, which is the perceived usefulness of IT, has cultural antecedents. Indeed, the perceived usefulness of IT depends on the type of the context of communication (high or low) and on the type of time adopted (monochronic or polychronic). In addition, the well-established positive relationship between perceived usefulness and system usage is shown to be moderated by two cultural values; namely, power distance and masculinity/femininity. Finally, the intensity of the influence of subjective norms on system usage has been shown to be different, depending on the cultural values espoused by the IT users. Indeed, social influence would be most salient when end users accept and expect inequality in power, do not tolerate uncertainty, are more concerned by the group than by their selves, and are people-oriented.

Conclusion

In this chapter, we offer a theoretical perspective on the effect of national culture on IT adoption and use. Through the conceptual model we have presented, we wanted to move beyond models considering culture as a black box and offering an ad hoc cultural explanation of IT-related behaviors. First, we have identified pertinent cultural values that could offer a fruitful explanation of differences in IT- related behaviors and beliefs across countries. Then, we have provided propositions linking these cultural values to TAM constructs and relationships. We contend that high context/low context of communication and polychronism/monochronism have a direct effect on perceived usefulness of IT; power distance and masculinity/femininity have moderator effects on the relationship between perceived usefulness and system usage and individualism/collectivism; uncertainty avoidance, masculinity/femininity, and power distance have moderator effects on the relationship between subjective norms and system usage.

In the current context of internationalization, taking into account the societal environment in the understanding of IT acceptance behavior becomes more relevant in comparison with the past. Indeed, information technologies are so ubiquitous and pervasive that one should explore more their relationship with different societal contexts varying across countries. Actually, IT that is accepted in a specific cultural context may not be accepted in the same way in another culture.

The framework presented here is a first step in better understanding the relationship between national culture and IT adoption and use. It goes without saying that this conceptual model needs to be empirically tested. A field study research involving samples from different countries should be conducted.

References

Agarwal, R. (1999). *Individual acceptance of information technologies*. Retrieved October 2003, from http://www.pinnaflex.com/pdf/framing/CH06.pdf

Agarwal, R., & Karahanna, E. (2000). Time flies when you're having fun: Cognitive absorption and beliefs about information technology usage. *MIS Quarterly, 24*(4), 665-694.

Agarwal, R., & Prasad, J. (1999). Are individual differences germane to the acceptance of new information technologies? *Decision Sciences, 30*(2), 361-391.

Ajzen, I. (1988). *Attitudes, personality, and behavior*. Chicago: Dorsey Press.

Ajzen, I., & Fishbein, M. (1980). *Understanding attitudes and predicting social behavior*. Englewood-Cliffs, NJ: Prentice-Hall.

Chinese Culture Connection. (1987). Chinese values and the search for culture free dimensions of culture. *Journal of Cross-Cultural Psychology, 18*(2), 143-164.

Danowitz, A. K., Nassef, Y., & Goodman, S. E. (1995). Cyberspace across the Sahara: Computing in North Africa. *Communications of the ACM, 38*(12), 23-28.

Davis, F. D., Bagozzi, R. P., & Warshaw, P. R. (1989). User acceptance of computer technology: A comparison of two theoretical models. *Management Science, 35*(8), 982-1003.

D'Iribarne, P. (1989). *La logique de l'honneur: Gestion des entreprises et traditions nationales*. Paris: Seuil.

Dishaw, M. T., & Strong, D. M. (1999). Extending the technology acceptance model with task-technology fit construct. *Information and Management, 36*, 9-21.

Earley, P. C., & Singh, H. (1995). International and intercultural management research: What's next? *Academy of Management Journal, 38*(2), 327-340.

Fisher, R., & Smith, P. B. (2003). Reward allocation and culture: A meta-analysis. *Journal of Cross-Cultural Psychology, 34*(3), 251-268.

Gamble, P. R., & Gibson, D. A. (1999). Executive values and decision making: Relationship of culture and information flows. *Journal of Management Studies, 32*(2), 217-240.

Goodman, S. E., & Green, J. D. (1992). Computing in the Middle East. *Communications of the ACM, 35*(8), 21-25.

Gudykunst, W. B., & Ting-Toomey, S. (1988). *Culture and interpersonal communication.* Newbury Park, CA: Sage.

Hall, E. T. (1983). *The dance of life: The other dimension of time.* New York: Anchor Press Doubleday.

Hall, E. T. (1989). *Beyond culture.* New York: Anchor Books Editions.

Hall, E. T., & Hall, R. M. (1987). *Hidden differences. Doing business with the Japanese.* New York: Anchor Press Doubleday.

Hasan, H., & Ditsa, G. (1999). The impact of culture on the adoption of IT: An interpretive study. *Journal of Global Information Management, 7*(1), 5-15.

Hernandez, E. (2000, March-April-May). Afrique: Actualité du modèle paternaliste. *Revue Française de Gestion, 128,* 98-105.

Hill, C. E., Loch, K. D., Straub, D. W., & El-Sheshai, K. (1998). A qualitative assessment of Arab culture and information technology transfer. *Journal of Global Information Management, 6*(3), 29-38.

Hofstede, G. (1980). *Culture's consequences: International differences in work-related values.* Beverly Hills, CA: Sage.

Hofstede, G. (1985). The interaction between national and organizational value systems. *Journal of Management Studies, 22*(4), 347-357.

Hofstede, G. (1997). *Cultures and organizations: Software of the mind.* London: McGraw-Hill.

Hofstede, G. J. (2000). The information age across countries. *Proceedings of the 5ème Colloque de l'AIM: Systèmes d'Information et Changement Organisationnel,* Montpellier, France.

Hofstede, G., & Bond, M. H. (1984). Hofstede's culture dimensions. An independent validation using Rockeach's value survey. *Journal of Cross-Cultural Psychology, 15*(4), 417-433.

Hofstede, G., & Bond, M. H. (1988). The Confucius connection: From cultural roots to economic growth. *Organizational Dynamics, 16*(4), 4-21.

Hu, P. J., Chau, P. Y. K., Sheng, O. R. L., & Tam, K. Y. (1999). Examining the technology acceptance model using physician acceptance of telemedicine technology. *Journal of Management Information Systems, 16*(2), 91-112.

Hui, C. H., & Triandis, H. C. (1986). Individualism-collectivism—A study of cross-cultural researchers. *Journal of Cross-Cultural Psychology, 17*(2), 225-248.

Igbaria, M. (1995). An examination of the factors contributing to microcomputer technology acceptance. *Accounting, Management and Information Technologies, 4*(4), 205-224.

Igbaria, M., & Zviran, M. (1996). Comparison of end-user computing characteristics in the U.S., Israel and Taiwan. *Information and Management, 30*(1), 1-13.

Jackson, C. M., Chow, S., & Leitch, R. A. (1997). Toward an understanding of the behavioral intention to use an information system. *Decision Sciences, 28*(2), 357-389.

Kagitçibasi, C. (1997). Individualism and collectivism. In J. Berry, W. Poortinga, Y. H. Pandey, J. Dasen P. R. Saraswathi, T. S. Segall, et al. (Eds.), *Handbook of cross-cultural psychology* (Vol. 3). Boston: Allyn and Bacon.

Karahanna, E., Evaristo, R. J., & Srite, M. (2005). Levels of culture and individual behavior: An integrative perspective. *Journal of Global Information Management, 13*(2), 1-20.

Karahanna, E., & Straub, D. W. (1999). The psychological origins of perceived usefulness and perceived ease of use. *Information and Management, 35*, 237-250.

Karahanna, E., Straub, D. W., & Chervany, N. L. (1999). Information technology adoption across time: A cross-sectional comparison of pre-adoption and post-adoption beliefs. *MIS Quarterly, 23*(2), 183-213.

Kelman, H. C. (1958). Compliance, identification, and internalization three processes of attitude change. *The Journal of Conflict Resolution, 2*(1), 51-60.

Kluckhohn, C. (1962). *Culture and behavior*. New York: The Free Press of Glencoe.

Kluckhohn, F. R., & Strodtbeck, F. L. (1961). *Variations in value orientations*: Evanston, IL: Row, Peterson.

Kowner, R., & Wiseman, R. (2003). Culture and status-related behavior: Japanese and American perceptions of interaction in asymmetric dyads. *Cross-Cultural Research, 37*(2), 178-210.

Laurent, A. (1983). The cultural diversity of Western conceptions of management. *International Studies of Management and Organization, 13*(1-2), 75-96.

Lee, C., & Green, R. T. (1991, second quarter). Cross-cultural examination of the Fishbein behavioral intentions model. *Journal of International Business Studies, 22*(2), 289-305.

Leidner, D. E., & Carlsson, S. A. (1998). Les bénefices des systèmes d'information pour dirigeants dans trois pays. *Systèmes d'Information et Management, 3*(3), 5-27.

Levi-Strauss, C. (1985). *Anthropologie structurale*. Paris: Librairie Plon.

Liska, A. L. (1984). A critical examination of the causal structure of the Fishbein/Ajzen attitude-behavior model. *Social Psychology Quarterly, 47*(1), 61-74.

Lucas, H. C., & Spitler, V. (2000). Implementation in a world of workstations and networks. *Information and Management, 38*, 119-128.

Martinsons, M. G., & Westwood, R. I. (1997). Management information systems in the Chinese business culture: An explanatory theory. *Information & Management, 32*(5), 215-228.

Mathieson, K. (1991). Predicting user intentions: Comparing the technology acceptance model with the theory of planned behavior. *Information Systems Research, 2*(3), 173-191.

Moon, J-W., & Kim, Y-G. (2001). Extending the TAM for a World-Wide-Web context. *Information and Management, 38*, 217-230.

Porter, M. E., & Millar, V. E. (1985, July-August). How information gives you competitive advantage. *Harvard Business Review, 63*(4), 149-160.

Rice, R. E., D'Ambra, J., & More, E. (1998). Cross-culture comparison of organizational media evaluation and choice. *Journal of Communication, 48*(3), 3-26.

Robey, D., & Rodriguez-Diaz, A. (1989). The organizational and cultural context of systems implementation: Case experience from Latin America. *Information and Management, 17*, 229-239.

Ronen, S., & Shenkar, O. (1988). Clustering variables: The application of nonmetric multivariate analysis techniques in comparative management research. *International Studies of Management and Organization, 18*(3), 72-87.

Rowe, F., & Struck, D. (1999). Cultural values, media richness and telecommunication use in an organization. *Accounting, Management and Information Technologies, 9*, 161-192.

Schneider, S. C., & Barsoux, J. L. (2003). *Management interculture* (trans. V. Lavoyer). Paris: Pearson Education France.

Schramm-Nielsen, J. (2000, March-April-May). Dimensions culturelles des prises de décision—Une comparaison France-Danemark. *Revue Française de Gestion*, 76-86.

Schwartz, S. H. (1994). Beyond individualism-collectivism. In U. Kim, H. C. Triandis, C. Kagitçibasi, S. -C. Choi, & G. Yoon (Eds.), *Individualism and collectivism: Theory, method and applications*. Newbury Park, CA: Sage.

Shackleton, V. J., & Ali, A. H. (1990). Work- related values of managers—A test of the Hofstede model. *Journal of Cross-Cultural Psychology, 21*(1), 109-118.

Sheppard, B. H., Hartwick, J., & Warshaw, P.R. (1988). The theory of reasoned action: A meta-analysis of past research with recommendations for modifications and future research. *Journal of Consumer Research, 15*, 325-343.

Shore, B., & Venkatachalam, A. R. (1996). Role of national culture in the transfer of information technology. *Strategic Information Systems, 5*(1), 19-35.

Shuper, P. A., Sorrentino, R. M., Otsubo,Y., Hodson,G., & Walker, A. M. (2004). Theory of uncertainty orientation: Implications for the study of individual differences within and across cultures. *Journal of Cross-Cultural Psychology, 35*(4), 460-480.

Silvester, J. (1990). Système hiérarchique et analyse sociétale: Comparaison France-Allemagne-Japon. *Revue Française de Gestion* (77), 107-114.

Smith, P. B., & Bond, M. H. (1999). *Social psychology across cultures.* Boston: Allyn and Bacon.

Srite, M., & Karahanna, E. (n.d.). The influence of national culture on the acceptance of information technologies: An empirical study. *MIS Quarterly, 30*(3), 679-704.

Straub, D., Keil, M., & Brenner, W. (1997). Testing the technology acceptance model across cultures: A three country study. *Information and Management, 33*, 1-11.

Straub, D., Loch, K., Evaristo, R., Karahanna, E., & Srite, M. (2002). Towards a theory based definition of culture. *Journal of Global Information Management, 10*(1), 13-23.

Straub, D. W. (1994). The effect of culture on IT diffusion: E-mail and fax in Japan and the U.S. *Information Systems Research, 5*(1).

Szajna, B. (1996). Empirical evaluation of the revised technology acceptance model. *Management Science, 42*(1), 85-92.

Tan, B. C. Y., Wei, K., Watson, R. T., Clapper, D. L., & McLean, E. R. (1998). Computer-mediated communication and majority influence: Assessing the impact in an individualistic and collectivistic culture. *Management Science, 44*(9), 1263-1278.

Tapscott, D., & Caston, A. (1994). *L'entreprise de la deuxième ère. La révolution des technologies de l'information.* Paris: Dunod.

Taylor, S., & Todd, P. A. (1995). Understanding information technology usage: A test of competing models. *Information Systems Research, 6*(2), 144-176.

Thomas, D. C., & Pekerti, A. A. (2003). Effect of culture on situational determinants of exchange behavior in organizations: A comparison of New Zealand and Indonesia. *Journal of Cross-Cultural Psychology, 34*(3), 269-281.

Triandis, H. C. (1972). *The analysis of subjective culture.* New York: John Wiley and Sons.

Triandis, H. C. (1977). *Interpersonal behavior.* Monterey, CA: Brooks/Cole.

Triandis, H. C. (1989). The self and social behavior in differing cultural contexts. *Psychological Review, 96*(3), 506-520.

Trompenaars, F., & Hampden-Turner, C. (1998). *Riding the waves of culture* (2nd ed.). New York: MacGraw-Hill.

Venkatesh, V. (1999). Creation of favorable user perceptions: Exploring the role of intrinsic motivation. *MIS Quarterly, 23*(2), 239-260.

Venkatesh, V., & Davis, F. (2000). A theoretical extension of technology acceptance model: Four longitudinal field studies. *Management Science, 46*(2), 186-204.

Venkatesh, V., & Morris, M. G. (2000). Why don't men ever stop to ask for directions? Gender, social influence, and their role in technology acceptance and usage behavior. *MIS Quarterly, 24*(1), 115-139.

Venkatesh, V., Speir, C., & Morris, M. G. (2002). User acceptance enablers in individual decision making about technology: Toward an integrated model. *Decision Sciences, 33*(2), 297-316.

Zghal, R. (1994). *La culture de la dignité et le flou de l'organisation: Culture et comportement organisationnel—Schéma théorique et application au cas tunisien*. Tunis: Centre d'Etudes, de Recherches et de Publications.

Chapter III

Information-Communications Systems Convergence Paradigm:
Invisible E-Culture and E-Technologies

Fjodor Ruzic
Institute for Informatics, Croatia

Abstract

This chapter is on cultural aspects of information-communications systems embedded into new media environment and invisible e-technologies, and on a new age of social responsibility for information technology professionals. Besides the key issues in information technology development that create smart environment and ambient intelligence, the chapter also discusses digital e-culture and the new media role in cultural heritage. From the viewpoint of information technology, the current information-communications systems converge with media. This convergence is about tools-services-content triangle. Thus, we are confronted with a new form of media mostly presented with the term digital, reshaping not only media industry but also a cultural milieu of an entire nation on a regional and global basis. The discussion follows on the World Library idea that is rebuilding with new form of World Memory (World Brain), the shift from visible culture domination to the domination of invisible culture in the world of e-technologies predominance. From this scenario, information technology professionals coping with information systems projects, e-services development, and e-content design have more cultural responsibility than in the past when they worked within closer and inner cultural horizons and when their misuse of technologies had no influence on culture as a whole.

Introductory Remarks

The information society is, above all, an economic concept but with important social and cultural implications. The new forms of direct access to information and knowledge create new forms of e-culture. E-culture is a part of a culture. It not only concerns users but the community of information professionals as well. When one speaks of information technology, it is always from the Western point of view, whereas e-technology (especially its applications) takes place throughout the world, and every culture has a different understanding of it. The shift to an e-culture at the level of society in general is translated to the individual level, enabling cultural change to be described empirically. The term *e-culture* refers to the diffusion of new technology, its application for various purposes (especially information and communication), and shifts in related attitudes, values, and norms. E-technology may not be gnawing at the roots of our culture, but those roots are gradually absorbing it. As with all innovation, cultural or otherwise, this technology will reinvigorate, transform, and inspire older cultural forms. We are living in the era of globalization, the information economy, with borderless communities and multiple citizenships. E-culture literacy and attainment will require serious attention to new infrastructure, to the building blocks and platforms for e-culture. These are critical issues for the pursuit of information professionals' excellence, for creativity in an information society, as well as for fundamental imperatives for commerce and trade in a new media environment.

The main notion of the following text is on cultural aspects of information-communications systems embedded into a new media environment, on invisible technologies, and on a new age of cultural responsibility for information technology professionals. The key issues in information technology development that create invisible e-technologies and smart environments are under e-culture influence. From the viewpoint of information technology, the current information-communications systems converge with media. This convergence is about tools-services-content triangle. Thus, we are accepting a new form of media mostly presented with the term *digital*, reshaping not only media industry but also a cultural milieu of an entire nation on a regional and global basis. The discussion follows on the new e-technology and information-communications systems convergence as the basis for defining pervasive computing and positive e-technologies. The findings at the end of this chapter explain the process of a fundamental cultural shift from the computer-based information technology to the computerless (invisible) e-technologies in which the e-culture is the essential factor of the success. The discussion section is about the role of information technology professionals coping with information systems projects, e-services development, and e-content design. They have more social responsibility than in the past when they worked within closer and inner cultural horizons and when their misuse of technologies had no influence on culture as a whole.

Background on Information Communications Systems and New Media

One of the most valuable and essential processes that humanity can engage in and which is, therefore, essential to look at in terms of information technologies, is the process of self-determination. The principal of self-determination of people was embodied as a central purpose of the United Nations in its 1945 charter. The purposes of the United Nations are to develop friendly relations among nations based on respect for the principle of equal rights and self-determination of nations, and to take other appropriate measures to strengthen universal peace. Resolution 1514 (XV) of December 14, 1960, containing the Declaration on the Granting of Independence to Colonial Countries and Peoples, stated that all nations have the right to self-determination; by virtue of that right, they freely determine their political status and freely pursue their economic, social, and cultural development (United Nations, 1960). In the 1990s, these issues continued to be highly relevant as numerous people around the world strove for the fulfillment of this basic right of self-determination. The UN General Assembly in 1995 again adopted a resolution regarding the universal realization of the right of nations to self-determination. Thus, the General Assembly reaffirmed the importance for the effective guarantee and observance of human rights and of the universal realization of the right of nations to self-determination (United Nations, 1995). By this, we see that self-determination is tied to all aspects of life: political, economic, social, and cultural. It is ultimately about how we choose to live and allow others to live together on this planet. Furthermore, information technology plays a key role in current economic and social affairs, so the information technology specialists/professionals have much more social responsibility than other professions. Information and communication technologies and networking infrastructures are playing an expanding role in supporting the self-determination of people and emergent nations. Access to information and the facilitation of communication provides new and enhanced opportunities for participation in the process of self-determination. It gives the potential to enhance political, economic, social, educational, and cultural advancement beyond the scope of traditional institutions and forms of governance.

The next step in recognizing cultural and social dimensions of information technology on the international scene is regarding the Council of Europe document, Declaration of the Committee of Ministers on human rights and the rule of law in the information society (Council of Europe, 2005). The Declaration recognizes that information and communication technologies are a driving force in building the information society with the convergence of different communication media. It also stressed that building societies should be based on the values of human rights, democracy, rule of law, social cohesion, respect for cultural diversity, and trust between individuals and between nations, and their determination to continue honoring this commitment as their countries enter the Information Age.

Vannevar Bush (1945) predicted that the advanced arithmetical machines of the future would be (a) electrical in nature, (b) far more versatile than accounting machines, (c) readily adapted for a wide variety of operations, (d) controlled by instructions, (e) exceedingly fast in complex computation, and (f) capable of recording results in reusable form. The new computer devices as smart devices, linked through communications systems, are creating new forms of information-communications systems. Thus, the new form of information appliances and ubiquitous information technology creates the basis for the concept of an information-processing utility. Based on interactive and ubiquitous carriers of information, the first generation of new information systems evolved to provide easy communication over time and space barriers. Thus, the new information systems are media. They are virtual communication spaces for communities of agents interested in the exchange of goods and knowledge in a global environment. Further promising technologies are pervasive computing and augmented reality. The vision of pervasive computing is, to some extent, a projection of the future fusion of two phenomena of today: the Internet and mobile telephony. The emergence of large networks of communicating smart devices means that computing no longer will be performed by just traditional computers but rather by all manners of smart devices. From these notions, it is evident that information-communications systems open the way to information society development. The information society is based on the new (digital) media that provides vast opportunities for information/content networking. New organizational networks are built, cutting across national borders and interests. The networks themselves increasingly may take precedence over nation-states as the driving factor in domestic and foreign affairs. At the same time, native communities have been actively engaged in creating and utilizing such networks with increasing participation and sophistication.

We are entering the era of new media. New media are tools that transform our perception of the world and, in turn, render it invisible or visible. Information technology (IT) professionals must understand new problems, considering the role of e-technologies in the integration and interaction between cultures. It is apparently true with tera architecture of the sensor networks that will transform business, healthcare, media, and e-culture itself. A new form of information-communications systems boosts intelligent networks with the majority of computers that are invisible and disposable. The IT professionals have the challenge in turning all that data into useful and meaningful information and in resolving cultural and privacy issues that accompany pervasive networked computing and ambient intelligence. IT professionals are confronted with the stage when e-technologies extract analytic values from social networks turning information issued by sensors and other data sources into knowledge management systems.

Defining new media is hard work. If we begin to use voice or books in an innovative fashion, we have just made old media into new media. Whatever we define as new media now would be old media as soon as we add innovations. We cannot define new media strictly based on the use of new technology for distance communication, since technology is always changing. What is new media today will be passé tomorrow. If we try to define new media by process rather than by structure, we are still in trouble. Whatever we define new media as today no longer will be valid tomorrow as technology changes the structures and processes. Trying to define the limits of change is a futile effort due to the very nature of change. This means that the regulation of new media is also an exercise in utility. The experiences with first-generation media platforms showed that in order to take advantage of the potentials and chances offered by new media, we need to explore their features and learn how to use them effectively and to build them efficiently. In short, we need to develop innovative concepts, frameworks, and methodologies for the design, realization, and management of the new media. The new media offer unprecedented opportunities and potentials for positively changing almost any aspect of our lives. The growing importance of new media and the demand for appropriate platforms have given rise to the development of innovative technologies and components for such media. Consequently, we can now observe the first generation of media platforms and the first management approaches for such platforms.

The evolution in convenient, high-capacity storage of digital information is one of the enabling technologies for new media. Disk drives that allow local storage, retrieval, and manipulation of digital content are increasing in capacity and falling in price. The current TV experience will evolve into a highly personalized process. Consumers have access to content from a wide variety of sources tailored to their needs and personal preferences. New business models and opportunities for the various providers in the value chain will evolve in an organic market focused on addressing individuals directly with new services. This will allow content providers to respond more effectively to audience needs. Digital media and the emerging communication technologies have created an overabundance of programs and information available from which each consumer can choose. The consumer will need new solutions enabling smart and active decision making over viewing preferences, such as a personal filter for the multitude of choices, dynamically adapting to changing needs and preferences.

Communications technology is available for the support of highly complex interenterprise service networks that support new services (Negroponte, 1996). Altogether, this creates a new view on the product, emphasizing the utility of the package (product and services) instead of the product itself. Analyzing the lifecycle of a product is crucial for synthesizing and specifying new types of benefits for a customer. Therefore, modern manufacturers have to provide benefits to customers. Questions are what could be the benefits and what kind of utility may be beneficial for the customer. Based on that exercise, they have to come up with appropriate business concepts based on new media e-culture.

The Culture and Invisible Culture

To cope with the new culture space in the context of information-communications systems embedded into new media environment, there is a need for basic definitions on culture, new media, and digital e-technologies. The great advances in culture come not when people tried to impose the values of one culture to the exclusion of all others, but rather when modern individuals try to create structures that are more exciting by combining elements from different cultures. The current information technology is capable of recording universal standards and particulars around the world, and it opens new ways and sources for creativity and global cultural heritage. Hence, information technology must reflect the full range of human existence, the values, the culture, and the entire knowledge. At the same time, the new e-culture is born interacting with e-technologies, and it exists in new cultural ecology. This new cultural ecology stimulates the development of a new trio (triple convergence) consisting of e-technologies, e-culture, and e-society.

The culture is a shared set of manifest and latent beliefs and values (Sackmann, 1991). It helps people to categorize and predict their world by teaching them about habits, rules, and expectations from the behaviors of others. Culture also molds the way people think—what their motivations are, how they categorize things, what inference and decision procedures they use, and the basis on which they evaluate themselves. Most of other definitions are too narrow. Sociologists have focused on behaviorist definitions of culture as the ultimate system of social control. In this system, people act appropriately and monitor their own standards and behaviors. Thus, the culture consists of the learned ways of group living and group responses to various stimuli; sociologists describe the content of the culture as the values, attitudes, beliefs, and customs of a society. Media theorists have explored the interplay of culture and technology, which has led to an emphasis on some aspects of culture. The new approaches are considering cultural ecology as consisting of new media in which various types of media are translated into a common digital form that is accessible within a single framework.

Figure 1. Triple context of culture in information society

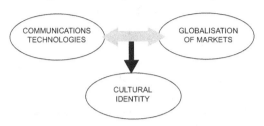

Today's networked media allow each user to participate actively in the creation of cultural expressions, which we perceive simultaneously and with immediate proximity. The new culture is emerging due to the use of digital technology (e-technology). At the same time, there is a strong relation between values promoted by the new digital e-culture and the traditional moral values created by the major world cultures. These notions open the new contextual approaches on culture in the information society. The culture related to the information society is about three contextual elements (Figure 1.)

The forces of globalization and technology development are paradoxical by nature, offering both threats and opportunities for cultural diversity. Yet, the information society is currently perceived only as an economic imperative in a new environment shaped by rapid information technology developments, based on visions shaped primarily by technologist and business concerns and priorities. The prevailing options embedded in these visions, such as globalization based on cultural homogenization, are questionable not only from a political and social standpoint but also in economic terms. Citizens around the world are becoming increasingly concerned about the way accelerating processes of globalization and technological innovation are leading to cultural homogenization and immense concentrations of financial power. Globalization generally is seen to be a phenomenon driven primarily by economic interests. As such, it has neither moral content nor values. Therefore, it could be independent from culture.

Cultural Diversity and the Information Society

Cultural diversity potentially can become a key asset in the information society, despite the fact that the culture could be defined as an obstacle. It is clear from the previous discussion on globalization that the economic forces of globalization pose a serious threat to cultural identity. Information technologies are not only the tools that accelerate the pace of globalization, but they are also becoming the key means of access to any good or service. One could thus argue that cultural diversity is an obstacle. Cultural diversity is essentially a question of communication, both

Figure 2. Tools-services-content triangle as the basis for information society

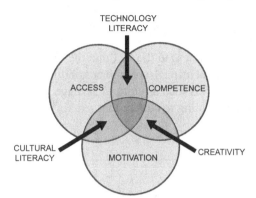

internally to one's own culture and externally with distributed cultures. Thus, one of the central issues is not only access to new e-technologies but also the capability of manipulating new media in order to participate actively in communicational exchanges.

Technology literacy is needed to operate IT effectively. Cultural literacy refers to the ability of an individual (or community) to relate to the services made by one's own cultural heritage and with access to another culture in a positive way. This means learning from both similarities and differences, being able to reject some aspects, and accepting others. Cultural literacy thus lies at the heart of the possibility of communication in a context of cultural diversity. Content is about Creativity. Creativity is the factor lying at the intersection of motivation and competence, and it covers both the individual and the collective levels. Collective creativity is of greater importance if we are aiming for a shift toward an information society in which a given community will depend on the collective creativity of its social and economic individuals and organizations (with accepting collective memory and brain).

Visible Culture

Western culture has had a strong bias toward the so-called fine arts, such as painting and sculpture. These expressions of culture, which are continuously visible, were more significant than the performance arts (theatre, film, music), which are only visible when they are being played. One of the key elements of great visible culture is uniqueness. Thus, the challenge of universal standards has brought the question of uniqueness back to the fore. To communicate internationally, we need global standards that bring the risk of reducing everything to one mode of expression.

The world of telephony offers an interesting case in point. We clearly need standards and uniform rules for telephones, yet every conversation on those telephones still can be different (Veltman, 1997). In this imperative to record the particular as well as the universal, creativity is only one essential element. The major cultures of the world owe much of their greatness to the fact that they have a recorded tradition, which stabilizes the corpus but also ensures the possibility of a cumulative dimension, which is reflected in terms such as cultural heritage. Even so, there are many skills in the craft tradition relating to culture that remain oral and invisible.

Invisible Culture

Many people favor material culture because it is visible and easily recognized. Culture is about more than objects in visible and tangible places. For example, many computer users are accustomed to thinking of computers as tools for answering questions. We need to think of them as tools for helping us to understand which questions can be asked, to learn about contexts when and where questions are not asked, of knowing that there are very different ways of asking the same thing. If software continues to be dominated by one country and if the so-called wizards of those programs all rely on the questions of that single country, then many potential users of computers inevitably will be offended, and it is likely that they will not use the programs.

The base assumption is that culture for us is invisible (invisible culture). As workers do not know that they are participants accepting entire organization values, the culture of an organization is invisible (Cooke & Lafferty, 1989). Yet, it is all-powerful. Therefore, this assumption is important for invisible e-culture. We are working in organizations that actually drive our behavior and performance in a way that most of the time is not visible to us. A definition of culture includes the way we do things.

A way to look at culture is actually to have some outside expert or outside person come in and question the way things are going, which can begin to give some insight to the people inside the organization about their own culture (Cleary & Packard, 1992). However, the culture is also considering exchange of ideas, thoughts, and beliefs, and it helps people realize that things can be done differently somewhere else. The success in one culture does not mean success in another culture, and there are actually many ways to succeed, change, and live. People realize that they can act on some other level of culture and act on their own culture.

E-Culture

We could define culture as the beliefs, behaviors, languages, and entire way of life of a particular time or group of people. Culture includes customs, ceremonies, art, inventions, technology, and traditions. The term also may have a more specific aesthetic definition and can describe the intellectual and artistic achievements of a society. The new world economy develops in e-culture and characterizes with the fast, open access to information and the ability to communicate directly with nearly anyone anywhere (Kanter, 2001). This sets e-culture apart from traditional environments.

In a first approximation, one could say that an e-culture is emerging from the convergence of communication and computing along with globalization and the penetration of e-technology in the smallest corners of our lives. The advent of information and communication technology goes hand in hand with changes in attitudes, skills, and behaviors that play a central role in daily life. The advent of an e-culture is correlated with terms of a broad definition of culture. This concerns the culture of a society with both invisible and material characteristics. E-technology as a part of the cultural information may be classified as e-invisible culture, but the outputs of that technology (information appliances) may range among the material (visible) cultural products.

The shift to an e-culture at the level of society is translated to the individual level enabling cultural change to be empirically described. The term *e-culture* refers to the diffusion of new technology; its application for various purposes (especially information and communication); and shifts in related attitudes, values, and norms. The human thinking and behavior are changing gradually by information and communication technology. E-technology may not be gnawing at the roots of our culture, but those roots gradually are absorbing it. As with all innovation, cultural or otherwise, this technology will reinvigorate, transform, and inspire older cultural forms. We are living in the era of globalization, the information economy, with borderless communities and multiple residencies. E-culture literacy and attainment will require serious attention to new infrastructure and to the building blocks and platforms for e-culture. These are critical issues for the pursuit of information professionals' excellence, for creativity in an information society, as well as for fundamental imperatives for commerce and trade in new media environments.

E-Technologies and Ubiquitous Information (Digital) Appliances

The Internet is without precedent because of two key features: its interactive and communicative natures. It is not a commodity in the sense that you can go out and buy a TV. You cannot go out and buy a net. The key word here is interactivity. Interactivity implies a dialogue of some kind, a changing response based on changing stimuli. There is much talk of interactive Web sites, but even the best of

Figure 3. Scenario of universal personal information appliance

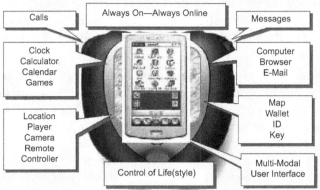

these choose from a preprogrammed set of possibilities in order to give the illusion of being interactive. You have interactivity over the telephone, and you have interactivity in a face-to-face dialogue. However, you do not have interactivity in traditional analogue television. The prosperity of the net is that it permits simultaneous interactivity with thousands and, perhaps, millions of people worldwide. This is a first in the history of humankind.

With the new and upcoming information-communications systems with pervasive and personal appliances, there will be a huge number of networked intelligent devices and information appliances functioning as self-organizing and managed networks (Figure 3).

In the near future, information communications systems with invisible networked devices, sensors, and appliances will transform businesses, public administration, public services, and the way we communicate within digital networks. Digital networks through the new form of information transport by ultra wideband and WiMax technologies will boost the intelligent networks development around the entire globe. The computer is becoming invisible and everywhere simultaneously. This is the beginning of the invisible computer era.

Parallel to this development is that of networking, which conceivably could result in all the invisible computers in the world being networked into a single virtual computer. This would lead to the evolution of a computer that would be everywhere and nowhere at the same time. Technology itself and on its own is not a cultural determinant. Technology is the invention of a particular culture—a cultural expression. The relationship between culture and technology is not linear and monodirectional but rather multidimensional and hyperspatial. Ten years from now, the computer as we know it today will be an anachronism, a device consigned to museums. Instead, the digital information and services once delivered via conventional computers will be available through almost everything we touch. At the heart of this next generation of computing is the network. It will be pervasive and personal. Looking out a decade or two, every person and thing could be instrumented with sensors that feed data into the content base and take actions on behalf of the client.

The new term, *speckled computing*, goes in that way. It offers a radically new concept in information technology that has the potential to revolutionize the way we communicate and exchange information (Arvind, 2005). Computing with Specknets will enable linkages between the material and digital worlds, and it is the beginning of truly ubiquitous computing. As the once-separate worlds of computing and wireless communications collide, a new class of information appliances will emerge. Where once they are used regularly, the post-modern equivalent might not be explicit after all. Rather, data sensing and information processing capabilities will fragment and disappear into everyday objects and the living environment. At present, there are sharp dislocations in information processing capability—the computer on a desk, the PDA/laptop, the mobile phone, smart cards, and smart appliances. However, Speckled Computing, the sensing and processing of information, will be highly diffused. The person, the artifacts, and the surrounding space become computational resources and interfaces to those resources. Surfaces, walls, floors, ceilings, articles, and clothes will be invested with a computational activity for rich interactions with the computational resources.

The information appliance is the natural outcome in the evolution of information processing. That is why they were foreseen a long time ago. Digital computers started out as expensive mainframes accessible to a few. The next step was the personal computer that individuals could own, and it was incorporated into everyday human activities. Thus, it was essential to have as much functionality in the personal computer as possible. Information technology is making feasible small and inexpensive devices that are smart. This helps to push intelligence closer to the people, the ultimate customers of information technology that accept some of the hidden cultural attributes of the entire community.

In a smart environment, the interaction end points simply could not be cognitively or physically visible. In essence, the user may have no idea that they are engaging in a computer-mediated communication. A smart environment is a composite space made from many individual objects. These objects are either fixed or mobile. The term *invisible* means that a technology has become so natural (common) and so comfortable that we use it all the time without ever thinking of it as a technology or a number of linked technologies. These invisible technologies are taken for granted. Since they are no longer technological, we can afford to think of them as customary, as the day-to-day workings of our world. As it continues to become more acceptable and as people come into the any-information system more proficient with these new tools, computer technologies will become increasingly invisible.

The exact example of the ubiquitous e-technologies environment is under way through the ubiquitous city in South Korea. New Songdo City (U-city), being built on a manmade island, will feature pervasive computer technology throughout, driven by RFID tags and CDMA wireless communication. Although many Western observers would find the lack of privacy disquieting, Asian countries are more interested in the technological potential of such environments. U-life will become its own brand, its own lifestyle. Residents will enjoy full videoconferencing calls between neighbors, video on demand, and wireless access to their digital content and property from anywhere in Songdo. At the same time, privacy is also encountered: all information services will be anonymous, and they will not be linked to user identity.

Invisible E-Technologies' Interaction with Invisible E-Culture

Our relationship with the manmade world is dominated by the paradigm of the device. This paradigm creates an illusory separation between the technological means (the machinery, the medium) and the technological ends (the commodity, the message). Technological progress generally is devoted to increasing the availability of technological goods, to make them everywhere available instantly without risk or hassle. At the same time, we want the machinery to become invisible. For various reasons, we repress ourselves from seeing the machinery and our dependence upon it. Our conscious awareness focuses on the message, and we refuse to acknowledge the medium (Pribram, 1971). Our perception of the world and our place in it are inextricably mediated through technology and the device paradigm. Our discovery of ourselves (identity through self-determination) is technical and complicated.

We are entering a new era of computing, often referred to as ubiquitous or pervasive computing. Ubiquitous computing consists of information appliances, specialized and easy-to-use devices that avoid the complexity of the computer. The future belongs to information appliances. When technology changes rapidly, greater ease of use serves to attract more users and developers, creating new frustrations. The most we can do is ameliorate the spread of the information appliance products and services. To do this, it appears necessary to recognize that flexibility and ease of use are in an unavoidable conflict and that the optimal balance between those two factors differs among users. Therefore, systems should be designed to have degrees of flexibility that can be customized for different people. Information appliances will be popular, since they will provide many new services for which the desktop computer is ill-suited and will do so in user-friendly ways. However, they will introduce their own complexity, and the level of frustration with information technology will not decrease. This is a result of the conflict between usability and flexibility. The information appliance market will be anything but mature for a long time to come. The emphasis in information processing has been and is likely to continue to be on development of novel applications. When the available information is stored on computers, it is important for information management applications to be able to model users' interpretations of their data and to capture the possibly different meanings, semantics links, and relationships with which users associate the information units available. This is in correlation to one's personal culture. For this purpose, various Personal Information Management tools are being developed to assist the user with navigation/browsing over various forms of personal digital data. As an example of capturing, organization, and archiving new media content, the MyLifeBits project is very explanatory. MyLifeBits has the aim to store in digital form everything related to the activities of an individual, providing full-text search, text and media annotations, and hyperlinks to personal data (Gemmell, Aris, & Lueder, 2005).

Information technology should mature to the humane technology of appliances in which the technology of the computer disappears behind the scenes into task-specific devices that maintain all the power without the difficulties. This could be explained by the technology of radio. Thus, computers should evolve the way radio receivers did (Norman, 1998). However, there is a problem of motivation, beliefs, knowledge, and frustration dealing with the information technology. This is considering culture (dominantly invisible culture). The problem is that with information appliances and by invisible technologies, we are confronted with the services that must be wel- understood and stable. We will not see this scenario with information appliances, not for a long time. In a world with a huge potential in services, content, and navigation, we cannot know how people will want to use information appliances.

Careful design that is focused on human factors and incorporates powerful processors and software can provide information appliances that are a delight to use. However, once the number of devices to be connected increases and wireless communication with WiFi networks expands, the difficulties will increase. Building complicated systems that work is hard. Building ones that work and are user-friendly is much harder. Further, it is necessary to balance the demand for user friendliness with the demand for more features. A tradeoff between flexibility and ease of use is unavoidable. The problem is that we should not be thinking just of individual information appliances. We have to be concerned with the whole system, which is likely to be complex. The problem is also how to balance flexibility and ease of use in a way that can be customized for people with different needs. This problem is especially focused on multimedia home systems. The home information appliance environment is likely to be more complicated than the office environment today. In addition, many users will be less knowledgeable about information technology than the typical office worker will (Ronfeldt, 1992). Therefore, it will be essential to outsource the setup and maintenance of home computing and electronics to experts. This notion opens new ways for information technology professionals that are accessing information appliances environment (this environment asks for new interface design, navigation methods, and computational power over networked appliances). Hence, there is convergence of culture and technology in use, be they visible or invisible.

Positive E-Technologies

Rheingold (2000) argues that the technology that makes virtual communities possible has the potential to empower ordinary citizens at a relatively small cost. E-technology potentially can provide citizens and professionals advantage and power, which is intellectual, social, commercial, and political. At the same time, civil and informed people must understand the advantages that e-technology provides. They must learn to use it wisely and constructively, as it cannot fulfill its positive potential by itself. Thus, the positive e-technologies should be developed as reduction technologies that make target behaviors easier by reducing a complex activity to a few simple steps. One of the most explicit theories that attempts to describe our natural inclination to do a cost/benefit assessment is expectancy theory. This theory posits that behavior results from expectations about what alternatives will maximize pleasure and minimize pain. E-technologies also should be self-monitoring technologies to perform tedious calculations or measurements, helping people achieve goals or outcomes. Ideally, these technologies work in real time, giving users immediate feedback on a performance or status. When people can take immediate action on a persuasive message, psychologists have found the message more persuasive than when presented at other times. The recent effort on real-time speech translation is an obvious example of these notions. Researchers from the International Center for Advanced Communication Technologies (interACT), a joint venture of Carnegie Mellon and the University of Karlsruhe, have developed a wearable system that allows real-time speech translation. The system consists of sensors that detect mouth muscle movements, translates that to a spoken language, and then retranslates that into other languages. It will make communication and cultural learning more likely, since people using this technology will be empowered to come together when they otherwise would not interact.

Culture is omnipresent in all technological advancements over the course of history, whether it is the result of intrinsic societal dynamics or the extrinsic factors of the environment. As history clearly documents, whenever technology changes, some pressing force of culture has had an effect on it. Moreover, there is a sort of invisible complimentary system between culture and technology; that is, whenever technology changes, the culture will adapt its way of life to fit the technology. For example, with the invention of the technology necessary for agriculture, cultures worldwide changed their hunting and gathering way of life in order to use the new technology and expand its horizons. This would be expected with the information technology, too. In essence, culture indeed influences human technology, but technology also simultaneously molds the way in which cultures function.

Persuasive Agenda

Like human persuaders, persuasive interactive technologies can bring about positive changes in many domains, including health, business, safety, and education (Dillard & Pfau, 2002). With such ends in mind, the new area of information technology development is created under the term *captology*. Captology focuses on the design, research, and analysis of interactive computing products created for the purpose of changing people's attitudes and behaviors. The fact that people respond socially to computer products has significant implications for persuasion. It opens the door for computers to apply a host of persuasion dynamics that are described collectively as social influence. These dynamics include normative influence and social comparison as well as less familiar dynamics such as group polarization and social facilitation.

Just as the term *software* shifts the emphasis from media/text to the user, the term *information behavior* also can help us to think about the dimensions of cultural communication, which previously went unnoticed. These dimensions always have been there, but in an information society, they have rapidly become prominent in our lives and, thus, intellectually visible. Today, our daily life consists of information activities in the most literal way: checking e-mail and responding to e-mail, checking phone messages, organizing computer files, using search engines, and so forth. In the simplest way, the particular way people organize their computer files, use search engines, or interact on the phone can be thought of as information behavior. Of course, according to a cognitive science paradigm, human perception and cognition, in general, can be thought of as information processing. While every act of visual perception or memory recall can be understood in information processing terms, today there is much more to see, filter, recall, sort through, prioritize, and plan. In other words, in our society, daily life and work largely revolve around new types of behavior activities that involve seeking, extracting, processing, and communicating large amounts of information. Information behaviors of an individual form an essential part of individual identity. They are particular tactics adopted by an individual or a group to survive in information society. Just as our nervous system has evolved to filter information existing in the environment in a particular way that is suitable for information capacity of a human brain, so we evolve particular information behaviors in order to survive and prosper in an information society. In today's world of information, people suddenly are shifting their attention to the Web for their computing needs.

Levy (1998) contends that communication in the virtual world can cultivate collective intelligence, which can encourage the development of intelligent communities. He states that sharing of information, knowledge, and expertise in e-communities can promote a kind of dynamic, collective intelligence, which can affect all spheres of our lives. He contends that the virtual world can foster positive connections, cooperation, bonds, and civil interactions. In e-groups or e-communities, which are flexible, democratic, reciprocal, respectful, and civil, this collective intelligence can be enhanced continually. Researchers in science, education, business, and industry are pooling their collective intelligence, knowledge, and data in collaborative memories. These are virtual centers in which people in different locations work together in real time, as if they were all in the same place. Science, education, commerce, and industry have become increasingly global. Collaboration, which is efficient, maximizing, and timesaving among distance researchers in these fields, has become more critical.

This new e-culture paradigm within the Web users' community opens the ways for Web 2.0 e-technologies platform comprising the set of principles and practices based dominantly on the user behavior and cultural values of collaboration. Most users find that Web 2.0 sites are extremely useful, because they are always available (whenever they need it and anywhere they go) with their information. Web 2.0 is the network as platform, spanning all connected devices. Web 2.0 applications are those that make the most of the intrinsic advantages of that platform:

- Delivering software as a continually updated service that gets better, the more people use it.
- Consuming and remixing data from multiple sources, including individual users, while providing their own data and services in a form that allows remixing by others.
- Creating network effects through architecture of participation.
- Going beyond the page metaphor of Web 1.0 to deliver rich user experiences.

One of the key aspects of Web 2.0 is that it connects people so that they can participate effortlessly in fluid conversations and dynamic information sharing. At the same time, information appliances and computing devices are giving people permapresence on the Web. Before now, the user consciously had to go to cyberspace by sitting at a desktop computer and looking at it through a display. Web 2.0 applications will become invisible as they become more popular, and there also would not be such a phrase as "going on the Web." Moreover, if the network is omnipresent and invisible, we do not need the term *cyberspace* anymore. Web 2.0 is also more human and a social one labeled with social interactions like conversation, sharing, collaboration, publishing, which could be supported by the corresponding processes (blogging, tagging, sharing, publishing, networking) and content formats (blogs, wikis, podcasts, folksonomies, social software). In addition, Johnson and Kaye (2004) stated that the Web would become a trustworthy place and the users would take it with much more reliance and credibility.

Computers Influence our Thoughts and Actions

Although culture is mostly learned, it is bound by necessity to a particular setting or context of its behavioral and material articulation. Culture is both conservative and adaptable. Culture is articulated symbolically and has the function of symbolically integrating the diverse moments and spaces of culture into a coherent sense of order. This format is emerging throughout the social field as a format of technology (the point-to-point Internet, file sharing, grid computing, blogs), and as a third mode of production producing hardware, software (often called open sources software) and intellectual and cultural resources (wetware) that are of great value to humanity (GNU/Linux, Wikipedia).

Cognitive scientist Clark (2003) believes that we are liberating our minds, thanks to our penchant for inventing tools that extend our abilities to think and communicate, starting with the basics of pen and paper and moving on to ever more sophisticated forms of computers and e-technologies. He declares that we are, in fact, human-technology symbionts, or natural-born cyborgs, always seeking ways to enhance our biological mental capacities through technology. The persuasive e-technologies are in front of us to solve the problem of difficulties in utilization of the computer, which complexity is fundamental to its nature. We have to start over again to develop information appliances that fit people's needs and lives (Norman, 1998). In order to do this, companies must change the way they develop information system products. They need to start with an understanding of people: user needs first, technology last. Companies need a human-centered development process, even if it means reorganizing the entire company.

People are more readily persuaded by computing technology products that are similar to them in some way (Fogg, 2003). Although people respond socially to computer products that convey social cues, in order to be effective in persuasion, hardware and software designers must understand the appropriate use of those cues. If they succeed, they make a more powerful positive impact. If they fail, they make users irritated or angry. With that in mind, when is it appropriate to make the social quality of the product more explicit? In general, it is appropriate to enhance social cues in leisure, entertainment, and educational products, especially with smart mobile devices (Rheingold, 2002). Users of such applications are more likely to indulge, accept, and perhaps even embrace an explicit cyber social actor. When perceived as social actors, computer products can leverage these principles of social influence to motivate and persuade.

Discussion on New Forms of Cultural Responsibility of IT Professionals

Culture and ethics are a very important part of our everyday life in information society. The invention of new e-technologies tends to bring many different dilemmas into the lives of the creators and the people who use them. Some technologies have been created without choice, and we must make sure we fully understand how to use them properly. The introduction and use of new technologies require a check against the moral structure of the society and the ethical beliefs of the individuals that will feel the effects of such an addition to their lives (Postman, 1992). This belief should be the foundation of innovation so that members of the society can have a strong, viable, and ethical solution to satisfy their wants and needs and to extend their capabilities. Ubiquitous computing and smart environments will be characterized by massive numbers of almost invisible miniature sensing devices that potentially can observe and store information about our most personal and intimate experiences. The new forms of direct access to information and knowledge create new forms of e-culture. It concerns not only users but the community of information professionals as well.

Technology can be a powerful tool for change, especially when used responsibly. Responsible IT management should be an important part of any socially responsible enterprise's strategies, policies, and practices. Users should have information technology choices that can and should reflect organizational, community, and national values and social responsibility. These notions are considering IT professionals' activities; they should create applications that guarantee accessibility. Accessibility to information via the information-communications systems should not be inhibited by disability or resource limitations, and the design solutions should be for the user experience. Usability of information technology solutions requires attention to the needs of the user (information consumer). The information-communications systems create new psychological demands from human. They ask us to bring a greater capacity for innovation, self-management, and personal responsibility. They also demand social responsibility of information technology professionals. Information technology is the wave of today and the future. Society must adapt to it by creating responsible rules, norms, ethics, and knowledge workers that will enhance its rapid growth.

Many firms acting on the global scene via information-communications systems are committed to incorporating socially responsible projects into their policies and activities. Corporate social responsibility is a development that is here to stay for the long term as a part of corporate policy influencing the company's involvement in the well being or development of local as well as global communities (Furnham & Gunter, 1993). Information technology firms are in a unique position to distribute their high-tech expertise and cultural values in development projects. The new information society environment poses a new relation between values promoted by the new digital civilization and the traditional moral and cultural values created by the major world civilizations. Computer technology and ethical egoism are the products of secular research within a free market capitalist society. The majority of non-Western societies and some Western, as well, follow ethical rules created within traditional culture. These rules are centered on guiding the individual in properly fulfilling his or her role within the society, which means the superiority of the society over the individual. The changes that information technology is bringing to people's lives are revolutionary, and one of the features of every revolution is that it is at the same time both a process of creation and of destruction. The revolutionary process itself is a very rapid one, which means that there is little or no time for a methodical and deep reflection on it while the process is actually in progress. These points ask for more attention from information technology professionals to cope with the culture exposed through the visible objects. They also should implement invisible culture elements when designing new information services. One possible way of minimizing the harm could be through incorporating the experiences of the process of intercultural dialogue into the process of creating a global e-culture of the information society.

References

Arvind, D. K. (2005). Speckled computing. In *Proceedings of Nanotech 2005, Anaheim, CA* (Vol. 3, pp. 351-354). Cambridge, MA: Nano Science and Technology Institute.

Bush, V. (1945). As we may think. *The Atlantic Monthly, 176*(1), 101-108.

Clark, A. (2003). *Natural-born cyborgs: Minds, technologies, and the future of human intelligence.* Oxford, UK: Oxford University Press.

Cleary, C., & Packard, T. (1992). The use of metaphors in organizational assessment and change. *Group and Organizational Management, 17*, 229-241.

Cooke, R., & Lafferty, J. (1989). *Organizational culture.* Plymouth, MI: Human Synergistics.

Council of Europe. (2005). *Declaration of the committee of ministers on human rights and the rule of law in the information society.* Strasbourg, France: Office of Publications.

Dillard, J. P., & Pfau, M. (2002). *The persuasion handbook: Developments in theory and practice.* London: Sage Publications.

Fogg, B. J. (2003). *Persuasive technology: Using computers to change what we think and do*. San Francisco: Morgan Kaufmann Publishers.

Furnham, A., & Gunter, B. (1993). Corporate culture: Definition, diagnosis, and change. *International Review of Industrial and Organizational Psychology, 8*, 233-261.

Gemmell, J., Aris, A., & Lueder, R. (2005). Telling stories with MyLifeBits. In *Proceedings of the IEEE International Conference on Multimedia 2005,* Amsterdam, The Netherlands (pp. 1536-1539). Piscataway, NJ: IEEE Publications.

Johnson, T. J., & Kaye, B. K. (2004). For whom the Web toils: How Internet experience predicts Web reliance and credibility. *Atlantic Journal of Communication, 12*(1), 19-45.

Kanter, R. M. (2001). *Evolve: Succeeding in the digital culture of tomorrow*. Boston: Harvard Business School Press.

Levy, P. (1998). *Becoming virtual: Reality in the digital age*. New York: Plenum Publishing.

Negroponte, N. (1996). *Being digital*. London: Coronet.

Norman, D. (1998). *The invisible computer: Why good products can fail, the personal computer is so complex, and information appliances are the solution*. Cambridge, MA: MIT Press.

Postman, N. (1992). *Technopoly: The surrender of culture to technology*. New York: Alfred A. Knopf.

Pribram, K. (1971). *Languages of the brain*. Englewood Cliffs, NJ: Prentice-Hall.

Rheingold, H. (2000). *The virtual community: Homesteading on the electronic frontier*. Cambridge, MA: MIT Press.

Rheingold, H. (2002). *Smart mobs: The next social revolution*. Cambridge, MA: Perseus Publishing.

Ronfeldt, D. (1992). *Cyberocracy is coming*. London: Taylor & Francis.

Sackmann, S. A. (1991). Uncovering culture in organizations. *Journal of Applied Behavioral Science, 27*, 295-317.

United Nations. (1960). *General assembly resolution 1514 (XV), declaration on the granting of independence to colonial countries and peoples.* 947th Plenary Meeting. New York: UN Publishing Office.

United Nations. (1995). *General Assembly resolution 49/148, universal realization of the right of peoples to self-determination.* Forty-Ninth Session. New York: UN Publishing Office.

Veltman, K. H. (1997). Why culture is important in a world of new technologies. In *Proceedings of the Panel on Cultural Ecology, 28th Annual International Institute of Communications Conference,* Sydney, Australia. Sydney: International Institute of Communications.

Section II

IRM Challenges across Nations

Chapter IV

Critical Success Factors for IS Implementation in China:
A Multiple-Case Study from a Multiple-Stage Perspective

Huixian Li
Accenture Consulting Company, P. R. China

John Lim
National University of Singapore, Singapore

K. S. Raman
National University of Singapore, Singapore

Yin Ping Yang
National University of Singapore, Singapore

Abstract

Whereas the literature indicates critical success factors (CSF) for IS implementation success, it is the thesis of this chapter that in situations involving government policy and business environments that are radically different from the West, there exist different and context-specific CSF; the focus of this chapter is on Chinese businesses. This chapter presents a multiple-case study, guided by a multi-stage perspective, consisting of eight computer applications in four representative companies; for each company, two applications—finance and manufacturing requirements planning (MRP)—were studied. Findings suggest 10 factors shown critical for IS implementation success in the Chinese organizations. We highlight and discuss five novel factors especially relevant to the Chinese context. This exploratory work initiates a call for more in-depth investigations incorporating the dimensions of differing government policy and business environments for future research in this area. The findings also offer practitioners a set of guidelines in implementing IS applications in China.

Introduction

China's government has identified utilization of information technology (IT) as a key mechanism to achieving the country's modernization goals. A widespread acquisition and assimilation of information systems (IS) started in late 1970s among Chinese organizations. This continued into the 1990s and was further manifested after the country's entry into WTO in December, 2001. It has been reported that more than two-thirds of Chinese companies had implemented IS by 2001 (Chinese Economy and Trade Information Center, 2001). An even greater role of IS in Chinese companies is anticipated as they need to become more competitive in order to face challenges in the global marketplace. Nonetheless, the success rate of information systems implementation in China has been relatively low (Lu, Qiu, & Guimaraes, 1988). In general, IS implementations in Chinese firms are not as successful as in firms of Europe or North America (Zhu & Ma, 1999). Both academics and practitioners are keen on acquiring insights into this phenomenon.

China is a radically distinct context in both ethnic and cultural terms; it is highly conceivable that Chinese social behavioral patterns and business dynamics intervene with system implementation processes and that key success factors are deemed different than the Western context. Previous work on IS implementation in China (Dologite, Fang, Chen, Mockler, & Chao, 1998) largely has drawn on factors suggested in the existing (Western) literature and has tested them using surveys (Reimers, 2002; Zhang, Lee, Zhang, & Chan, 2002). Shanks, Parr, Hu, Corbitt, Thanasankit, and Seddon (2000) have investigated critical success factors (CSF)[1] for IS implementation in China in a case study using a stage model. However, the setting of the study, which involved a joint venture between a Chinese partner and a Japanese manufacturer, may not be a typical setup for Chinese businesses. Our literature review also extends to 30 studies on success factors for IS implementation published by key Chinese journals. The factors mainly fall into the following categories: management support, business process re-engineering (BPR), project management, and social and economic environment (Li, 1999; Lin & Zhu, 2002). The recommended approach for IS implementation is to utilize IS strategically to facilitate the modernization process of Chinese companies by adjusting foreign technology and management experiences to the Chinese economic and social contexts (Lin & Zhu, 2002; Qi, Huo, Chen, & Wang, 2000; Zhang, 2001). Nevertheless, these studies are limited by being primarily normative and are mostly summaries of anecdotal accounts drawn from the industry. Empirical testing and validation are in want.

This chapter presents a case analysis to fill in the gap by identifying the critical factors for successful IS implementation in Chinese organizations. It intends to provide insights pertinent to IS implementation through efforts in synthesizing the related literature and conducting an in-depth empirical investigation in four organizations. Based on the findings, the chapter examines the interrelationships between the CSF and their sociostructural context and proposes a more wholesome model in order to understand the IS implementation process. Our chapter is organized as follows. The next section provides background of the subject and its relevant literature and deliberates on the research framework guiding this qualitative work. We present the main thrust of the chapter in a form of case study work that we carried out in several Chinese firms. The following section discusses the research findings and proposes an improved research framework for IS implementation issues in China. The last part concludes the chapter by highlighting the key contributions arising from the case study as well as future research opportunities.

Background

Overall Structure: The Stage Perspective

Several models have been developed to study the IS Implementation process, and the stages have been defined and named in different ways. These models are based on the organizational innovation prospective, which views innovation as a process comprising three stages: initiation (generation), adoption (acceptance), and implementation (Pierce & Delbecq, 1977; Thompson, 1969). The three-stage model stops at the point where the innovation is available. Kwon and Zmud (1987) adapted the three-stage model to IS and extended it to six stages: initiation, adoption, adaptation, acceptance, use, and incorporation. Cooper and Zmud (1990) fine-tuned the six stages into initiation, adoption, adaptation, acceptance, routinization, and infusion. The initiation, adoption, and adaptation stages are the same as the three stages of the organizational innovation perspective discussed previously, and the last three stages are post-implementation activities.

Recently, Markus and Tanis (2000) developed a four-stage model for enterprise information systems implementation. The four stages are chartering, project, shakedown, and onward and upward. The chartering stage comprises decisions leading to the funding of an enterprise system. The project stage covers the activities intended to get the system up and running in one or more organizational units. The shakedown stage is the organization's coming to grips with the enterprise system. Lastly, the onward and upward stage continues from normal operation until the system is replaced with an upgrade or a different system. During this stage, the organization finally is able to ascertain the benefits of implementing IS.

In sum, research on IS implementation has shifted its focus from the IS life cycle to the processes that are affected by IS. While each IS implementation is being examined carefully, researchers are more interested in investigating the benefits that IS could bring to organizations from a business point of view. Based on a comparison on the stage models, this study adopts the four-stage model developed by Markus and Tanis (2000) that is particularly pertinent to enterprise information systems implementation. The following points incorporate the justification of our choice for the four-stage model, which is deemed more appropriate than the other two for the purpose of this study.

- The perspective of the model is in line with the aim of this study. The model views IS success in terms of its contributions to company performance, which is promising to fulfill the purpose aimed at an IS success model for Chinese organizations.

- By combining goals and actions with external forces over time, this model could help to understand the IS implementation in China, which is influenced not only by key organizational actors but also by macro-economy concerns.

- This model emphasizes the role of planning, which is very important for this research. Many Chinese organizations, executives, and technical specialists differ in their approaches to decision making about whether, why, and how to undertake IS from their counterparts in Western countries (Franz & Robey, 1987; Martinsons & Westwood, 1997; Nandhakumar, 1996). The effectiveness of their decision-making approaches profoundly influence the outcomes of IS implementation. Therefore, planning activities should not be missed in this study.

- It is both representative of the literature and conceptually economic, as it was built on the history evolution of enterprise information systems (Markus & Tanis, 2000) from the 1970s until the 1990s and the findings of IS implementation research.

- It suggests a set of success criteria for each stage, which could help to measure the outcome of each stage, thus arriving at CSF for practitioners (Rockart, 1979).

- The research tries to understand the CSF and their context over time. Markus and Tanis' (2000) model has proved effective for this purpose (Nah, Lau, & Kuang, 2001).

Although this model is based on research on enterprise information systems, our view is that internal procedures of enterprise information systems and packaged information systems share considerable similarities. Further, we argue that this approach is potentially valuable, because most Chinese organizations that have implemented IS applications are striving to develop highly integrated enterprise information systems in the long term (Chinese Economy and Trade Information Center, 2001). The theoretical and methodological issues of enterprise information systems, which are highly integrated, have been studied in detail in a Western context. The findings of these studies can offer valuable guidelines for implementing integrated IS in China.

IS Success Measures

To date, most claims of success (rarely failure) of IS projects in China are made by IS vendors or company management and often are biased for their own interest without any clear success measures. In this research, consistent with Markus and Tanis' (2000) model of stage-specific outcomes, a key indicator has been identified for each stage of implementation. For the chartering stage, an important outcome is organizational fit between the information system's objectives and organizational objectives (Zwass, 1998). For the project stage, the outcome of interest is meeting requirements, the degree to which the system meets user requirements established during the feasibility study (Soh, Yap, & Raman, 1992), and completion extent, the extent to which a project is completed on time and within budget. For the shakedown stage, the outcome of interest is use in terms of the percentage of full functions being used for the intended purposes (DeLone & McLean, 1992, 2003). Finally, for the onward and upward stage, it is paramount to look at net benefits in terms of net contribution toward realizing the goals of the organization implementing the IS (DeLone & McLean, 2003), which can be viewed as the overall success outcome of IS implementation. The matching between Markus and Tanis' (2000) success recipe and IS success measures is summarized in Table 1.

Success Factors in IS Implementation

It is generally suggested that IS success factors vary, depending on whether the context has to do with customized application development or package implementation, and between small-scale, dedicated packages and large, integrated systems (Parr, Shanks, & Darke, 1999). As the current research focuses on packaged IS, CSF were synthesized from studies of success factors for implementation of packaged systems (e.g., ERP systems) published in international and Chinese journals (Akkermans &

Table 1. Stage-specific success measurements based on Markus and Tanis (2000)

Stage	Recipe for Success	Success Measure
Chartering	Success occurs when executives make sound decisions about investing in enterprise systems and bring the organization into alignment with these decisions.	**1. Organizational Fit** Alignment between the information system's objectives and organizational objectives (Zwass, 1998)
Project	Success occurs when 1. The project team faithfully executes a sound project plan and appropriately responds to technical and human challenges that arise during the project. OR 2. The project team appropriately modifies the project plan to match changing business and organizational conditions	**2.1 Meeting Requirements** The degree to which the system meets user requirements established during feasibility study (Soh et al., 1992) **2.2 Completion Extent** The extent to which the project is completed on time and within budget. For the shakedown stage (Soh et al., 1992)
Shakedown	1. The organization is well-prepared to accept and use a system and related infrastructure of sufficient quality to meet business needs. OR 2. Appropriate measures are taken to fix technical and organizational problems arising during shakedown quickly and effectively.	**3. Use** The percentage of full functions being used for the intended purposes (DeLone & McLean, 2003)
Onward and Upward	1. Benefit from use of the system combining with favorable competitive conditions. AND 2. The future evolution of the enterprise system and related infrastructure is well-managed.	**4. Net Benefits** Net contribution toward realizing the goals organization implementing the IS (DeLone & McLean, 2003)

van Helden, 2002; Zhang et al., 2002). This is a reasonable starting point, as Chinese companies are shifting toward a free economy, following the business strategy in the developed countries, and many information systems adopted in China originate from foreign, developed countries (Zhang et al., 2002). According to Rockart's (1979) advice on the need for a guiding model or framework covering some main dimensions in order to explore CSF for a certain phenomenon, the synthesizing effort was guided mainly by two preliminary frameworks (Kwon & Zmud, 1987; Zwass, 1998), which group the factors into four major areas: individual factors related to both manager and users, technology factors, task-related factors, and structural factors. Table 2 provides the summary with examples of relevant references.

The Role of Government Policy

In China, one key element of Chinese economy is the role of government policy. The Chinese government has actively guided IS application development and uti-

Table 2. Summary of CSF in existing literature on IS success

Factors	Descriptions		References
Individual	Management support	Support of senior managers in terms of adequate resources, policy, and so forth	Bingi, Sharma, Maneesh, and Godla (1999); Cameron and Meyer (1998); Holland and Light (1999); Zhang et al.(2002); Zwass (1998)
	User involvement	Participation of end users	Alavi and Joachimsthaler (1992); Ives and Olson (1984); Shanks et al.(2000); Sumner (1999); Zwass (1998)
	User training and Motivation	Equipping and motivating users with appropriate knowledge of IS and the skills to operate	Bingi et al. (1999); Cameron and Meyer (1998); Olson (2004); Sumner (1999); Zhang et al.(2002); Zwass (1998)
Technological	Organizational fit	Fitness between the information system's objectives and organizational objectives	Hong and Kim (2002); Zhang et al.(2002); Zwass (1998)
	External expertise (vendor and consultant)	Adequacy and quality of vender support; consultant effectiveness	Shanks et al.(2000); Thong, Yap, and Raman (1996)
Task-Related	Clear goals	Clearly defined and well-understood goals	Bingi et al. (1999); Cameron and Meyer (1998); Holland and Light (1999); Parr et al.(1999); Shanks et al. (2000); Sumner (1999)
	Change management	Effort to change management process to fit the IS process	Zhang et al. (2002); Zwass (1998)
	Project management	Managing the information systems development appropriately and effectively	Bingi et al. (1999); Cameron and Meyer (1998); Holland and Light (1999); Shanks et al. (2000); Sumner (1999); Zwass (1998)
Structural	Centralization	Degree of concentration of decision-making activities	Reimers (2000); Steers (1977)

lization in organizations. It has adopted a variety of policies to promote the use of IS as a key objective in achieving modernization goals (Tate & Maier, 1987). It also has allocated funds for R&D on IS development project. Anecdotal evidence and industry experience suggest that behavior in IS implementation in China could not be explained fully without considering the factors associated with the governmental policies. Therefore, we have accorded explicit attention to this factor by including it in the model as an antecedent factor in explaining how the diverse interests of IT stakeholders may affect implementation efforts.

Figure 1. Research framework

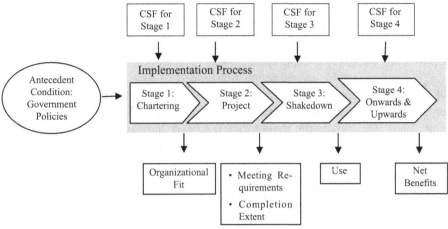

The Guiding Framework

Figure 1 depicts our research framework based on the previous discussion. The four stages are embedded within the implementation process. CSF and success measures are shown to be stage-dependent. As an important dimension, the Chinese government policy is positioned as an antecedent condition in the chain of effects leading to IS success.

Research Approach

Consistent with the research objective, a multiple-case study approach is adopted (Yin, 1994). This approach is most appropriate when the phenomenon under investigation is broad and complex, the existing body of knowledge is insufficient to permit posing of causal questions, and the phenomenon cannot be studied outside its context (Bonoma, 1985). IS implementation in China aptly fits these three criteria. First, the way Chinese organizations utilize IS is very intricate and unique (Martinsons & Westwood, 1997); it warrants understanding from empirical effort in its natural setup. Second, studies on IS implementation in Chinese corporation mainly have been conceptual and anecdotal in nature, implying that it is almost impossible to form causal research hypotheses. Third, the cultural element is deemed to be influential in the Chinese business context and environment and, hence, plays an important role without which IS implementation understanding cannot be meaningful and adequate.

A Pilot Study

To empirically validate the research framework, we started our case investigation with a pilot study involving two systems in one IT company. The main purpose of the pilot study was to test the appropriateness of the research framework as well as to refine the interview guide and data collection procedure. Two applications—finance and personnel administration—from a typical Chinese corporation (company X) were selected. Company X is a Chinese government holding company listed in Hong Kong. It has 300 staffs in the Beijing (R&D center) and has offices and development centers in Hong Kong and in major cities throughout China. The 2002 interim turnover for the period ending June 30 is HK $183 million. The company's business covers four major areas: information security products and solutions, geographic information systems, e-finance, and enterprise/government information platforms. Before the field visit, information on the background of company X and the two applications was gathered from published reports and e-mail correspondence. A list of interviewees and a timetable subsequently were communicated to the secretary of the chief information officer (CIO), who arranged a two-week field-study schedule.

The main channel to get information during the field study was interviews, complemented by other sources of information, including archival records, on-site observation, and the company's intranet. Interviewees included the senior manager responsible for initiating the two applications, the IT manager, the project manger, two key users, and two end users.

The pilot work has helped us to prepare for the in-depth case studies by revealing the following points. First, the Chinese economy and cultural context can influence the IS implementation process in multiple aspects. Second, the interview guide was useful but time consuming and ineffective in eliciting CSF. While the guide addressed the main issues that may arise during the IS implementation process, some of the topics were not relevant to specific interviewees. For example, end users usually had no idea about the criteria for IS selection. Correspondingly, the interview guide was revised and tailored to different informants, respectively. As well, we have revisited the guiding framework with a closer attention on the possible influences of Chinese-related factors.

The Case Study

Eight cases involving four companies were selected along three dimensions: financial vs. manufacturing applications, Chinese traditional management vs. foreign management, and foreign vs. local software products (see Table 3). For each company, the finance application and MRP stock/production/distribution application were selected. Chinese enterprises are encouraged as a matter of national policy in order

Table 3. Sampling matrix

Management Approach		Chinese Management: Company A	Chinese Management: Company B	Foreign Management: Company C	Foreign Management: Company D
Applications	Finance	Foreign Product (2A)	Local Product (4B)	Foreign Product (6C)	Local Product (8D)
	Manufacturing	Local Product (1A)	Foreign Product (3B)	Foreign Product (5C)	Local Product (7D)

to improve their competitiveness through strategic use of information systems. The two most popular application packages are financial management information systems and manufacturing management information systems, such as MRP. This research environment provided an adequate setting to identify both internal and external issues through interorganization comparison and intraorganizational comparison.

Firm ownership, Chinese or foreign, is associated with a number of variables describing the implementation process (Reimers, 2002). One characteristic of ownership that most likely is linked to the IS implementation process in China is the management approach, due to the cultural differences between Chinese management and Western management. In this study, we choose two companies built on the Chinese traditional management approach (companies A and B) and two companies (C and D) built on foreign management practices. Further, the IS available in the Chinese market can be divided into two categories according to their embedded business process: local product, which is Chinese business process oriented; and foreign product, which is Western business process oriented. These two types of software may have different impacts on the IS implementation process.

Table 4 provides a description of the applications and related companies. The cases are referred to in the form of case number/firm name. For example, the fifth case is labeled 5C because it was conducted in company C.

Data Collection

Interview notes, recorded tapes, individual communications, archival records, and documentation served as the case database. They were gathered in several steps. The first step involved collecting data on IS applications and their backgrounds by means of published reports, telephone, e-mail, and the Internet. Key informants included the CEO, the CIO or IT manager, project manager, functional managers whom the application may concern, at least two IT support staffs, all key users, and at least one end user for each main function. During the field visits, semi-structured and unstructured interviews (tape-recorded) were conducted.

Numerous informal conversations took place between the researcher and the actors concerned; comprehensive documentary sources also were consulted after interviews. Less formal interviews (via e-mail or telephone calls to follow up on points of interest) frequently were conducted during the data analysis process.

Data Analysis

This research utilizes causal maps to uncover the CSF for IS implementation, the interrelationships between CSF, and their context. Causal map is a set of variables and

Table 4. Description of cases

Case	Description	Company
1A	Case 1 is an MRP for production, stock, and distribution. It was developed and installed by the branch software company of company A and implemented in year 2000.	Company A is a listed private holding company. It produces fresh meat and related products. It has 40 subsidiary companies throughout the world, sales offices in Los Angeles, Moscow, Holland, and Hong Kong, and 200 direct sale outlets throughout China. More than 18,000 employees generated sales turnover of 7 billion RMB (around $800 million U.S.) during 2001.
2A	Case 2 is a finance management system. It was developed and installed by the branch software company of company A and implemented in year 2000.	
3B	Case 3 is a manufacturing and sales system implemented by the IT department based on the database of Oracle 9i. It was introduced in year 2000 to support material purchase, production, warehouse, and distribution management.	Company B was once the sales office for a foreign group corporation located in Shanghai. In October 2000, it set up the manufacturing division to assemble metering/magnetic pumps, which has become the manufacturing center in the Asia-Pacific area for the foreign group corporation. With a total workforce of just more than 40, company B yielded more than 400 million RMB of sales in 2001.
4B	Case 4 is a finance management system developed by local vendors and implemented in year 2002.	
5C	Case 5 refers to BAAN IV Distribution and Manufacture applications. It was implemented in year 2002 to support purchase, production, and warehouse and sales management.	Company C was formerly a Chinese stat-owned company. In 1995, it assimilated foreign direct investment (FDI) and became the only manufacturing base in Asia of a world group company. Started with automatic door series, it has begun to expand its production line, aiming at manufacturing all elevator products. It has three sales offices in China and a worldwide client base. A total of 300 employees generated more than 200 million RMB in year 2001.
6C	Case 6 refers to BAAN IV finance application for corporate finance management. It was implemented from February 2002 to October 2002.	
7D	Case 7 is local vendor developed MRP II system implemented by company D in year 1999 to support its production and warehouse management.	Company D is a Chinese foreign joint venture founded in 1984 as the first branch company of a worldwide famous elevator company. It is a manufacturer, installer, and service provider of elevators, escalators, moving walkways, and other horizontal transportation systems. Its 2,000 employees have contributed to the revenue of about 400 million RMB in year 2001.
8D	Case 8 is a finance management system developed by a government sponsored by the government. It was implemented in year 2002.	

connections that are inferred by the researcher after several exposures to a stream of experiences. It provides evidence supportive of causal relationships through research design and multiple studies (Miles & Huberman, 1994). It has been found to be an effective way to study CSF for IS development (Butler & Fitzgerald, 1999). This approach comprises three parts/steps. The first step was the review of individual success factors (Bullen & Rockart, 1986). CSF elicited from each interviewee were reviewed against the guiding dimensions of CSF (Kwon & Zmud, 1987). To avoid bias in studies on IS development (Lee, Goldstein, & Guinan, 1991), both the user's and developer's world views captured at the beginning of the interview process were apprehended. Data triangulation techniques were employed in order to enhance the validity of the CSF (Patton, 2002), and the interviewees were checked against each other in seeking a pattern of success factors. This step resulted in a set of application-specific CSF (Bullen & Rockart, 1986; Rockart, 1982).

The second step was the cross-case stage-specific comparison. The CSF and the stage-specific outcomes were first measured. For factors that had been examined in existing literature (e.g., clear goals), we adopted the published scales of measurement (Parr et al., 1999). New factors that emerged during the interview were investigated by asking more questions during the interview and further validated using multiple cases (e.g., accordance with government policy). Then, a value was assigned to each of the new CSF. As an illustration, end-user involvement had been investigated by asking two questions: "Were end users involved in the development at any stage?" and "What was the nature of their involvement?" A value of yes or no was recorded based on the interpretation of the recorded data. The success of all stages for the eight applications then was measured. For example, the success of the third stage was investigated by asking questions such as, "What percentage of the functions addressed in the plan is in use now?" The tape-recorded data were interpreted by the researchers. The measuring process was done through interpretive analysis of the qualitative data using the technique of content and constant comparative analysis (Patton, 2002). For example, a value of "S" was given for successful stages in which more than 50% was in use or positive comments like "mostly" were given; value "F" signifies failed stages in which less than 50% of the functions was in use or negative comments like "we rarely use it" were made. Another example is measuring net benefit, the measurement for the fourth stage. It was measured by balancing the cost of the system and its contributions to the company in terms of cost savings, expanded markets, time savings, and so forth. In this way, the outcomes of all stages were categorized into two classes (successful/unsuccessful for "project" and "shakedown") or three classes (successful/moderately/unsuccessful for "chartering" and "onward and upward"). Finally, a process of comparison was undertaken to determine which independent variables are directly connected to the stage success. Limited by the size of the sample available and the nature of the case study, we relied on simple inspection of data rather than on statistical techniques to determine how various factors influenced the dependent variable (Reich & Benbasat, 1990).

Figure 2. Process of data analysis

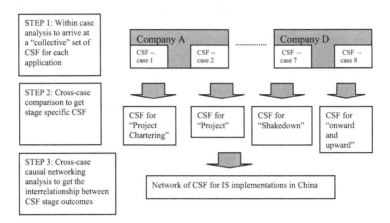

The third step was the cross-case causal map analysis (Miles & Huberman, 1994). Based mainly on the findings in case 1A, the interrelationships among CSF and their effects on stage outcomes were modeled. The model was modified by taking other cases into consideration, resulting in a generic causal map. The causal map utilizes the stage-specific CSF identified in STEP 2 and interlinks the key factors to reflect the relationships among the CSF themselves. Figure 2 is a summary of this process. We dedicate the next section to elaborate on each of the CSF as well as depict the causal network model.

Discussions and Findings

The findings on CSF for IS implementation in Chinese companies were grouped into two clusters: one cluster comprises the factors elicited from this study that are particularly relevant, although not exclusive, for Chinese companies; the other cluster includes factors that bear similarity with those synthesized from the literature and have proved relevant for IS implementation in China.

China-Specific CSF

First-Hand[2] Support in Terms of Financial Resources and Personal Advocacy

Generally speaking, the implementation process of finance applications was not as straightforward as that of the manufacturing application. In prior systems, financial information was the privilege of senior managers. Many nonstandard transactions could greatly convenience senior managers. Most senior managers objected to the foreign system (application in 6C) or product by inexperienced, local vendors (application in 4B). These systems formalize the financial process and public financial information. On the other hand, products by experienced local vendors (7D) had a special design that could help senior managers retain their interests. The evidence as previously provided indicates that political factors should be considered to have a good understanding of the IS implementation process in China.

While management support was found to be indispensable to the successful implementation of IS, it is the lack of first-hand support that differentiates unsuccessful IS implementations from successful ones. The first-hand (FH) figure exists widely in China's governmental agencies as well as organizations. Conventionally, the FH figure can be the Communist Party Committee Secretary in government-related organizations. It is worth noting that FH need not correspond with top management (although there may be situations in which the correspondence exists); in fact, FH in a company may exercise greater influence than the top person in the management hierarchy. In this way, FH support differs from top management support. In successful or moderately successful implementation, according to the cases, FH allocated sufficient funds at the chartering stage as well as during the onward-and-upward stage. They personally and enthusiastically advocated the project throughout the company. For example, FH of company A allocated 5 million RMB (about US$600, 000) and 1 million RMB (about US$130,000) for the applications in Case 1A and Case 2A, respectively. In addition, the FH invited IS experts to educate and train functional managers in IS-related knowledge. Indeed, FH is such a crucial enabler of IS success in China that the lack of his or her support invariably would result in less than desirable implementation outcomes. In this scenario, the FH of company B for application 4B displayed a lukewarm endorsement by allocating a total fund of only 200,000 RMB for the software; as well, he showed little interest in the implementation process. It turned out that the application that the IT department installed could not deliver any net benefit. In conclusion, the FH support in terms of financial resources and personal advocacy is essential for IS implementation success.

Guanxi[3] *between and among Members in Steering Committee*

Guanxi in steering committee is found to be an important factor that differentiates the successful project from the failed project. Committee members for successful projects had good *guanxi*, which makes the committee members trust each other and strive for the accomplishment of the project. *Guanxi* is found to be built on the ties and similarities in aspects pertaining to family and relatives, friends, schooling history, working history, dialects, and so forth. The *guanxi* factor is important for fostering a shared (group) identity that causes the parties involved to be more conscientious in performing group tasks; it thus was found to be an important success factor for

IS implementation. For example, the vendor and the IT manager for the application in Case 8D were close friends. The IT manager demonstrated a great deal of trust in the CEO of the vendor company; he decided to adopt the system without comparing it to those of any other vendors. The CEO of the vendor, on the other hand, was determined to devote the maximum capability to make the system successful. The main incentive was that he wanted to keep and even enhance the relationship with the IT manager through the project, which he perceived to be more valuable than the financial profit from the project itself. Likewise, most committee members for applications in 1A and 2A had a special *guanxi* connection with each other that assured their efforts. For instance, one functional manager, the key user for implementing 1A, said, "I am doing far from what I should do as a key user, mainly for friend, the IT manger. I have gone along with him very well and it is he who asked me to join this steering committee to help out." Comparably, the performance of the steering committee without good *guanxi* turned out far less than desirable. Without personal ties, the committee for applications in cases 3B and 4B did not have a feeling of *group*, and it was hard to fulfill the basic requirement of the project, which was considered one of the main reason leading to the failure of the project.

IS frequently disturbs the distribution of intraorganizational power among the key actors, where power is defined as the ability to get one's way in the face of opposition or resistance to one's desires (Pfeffer, 1981). The failure of the application in 3B was attributed, to a certain extent, to an anticipated and resented redistribution of power among sales executives and senior managers brought by the IS; the element of good *guanxi* is obviously absent. As mentioned by a member of the implementation committee, the implementation of finance application 4B was perceived as a means to curb power of the sales executives. They had dominating power on product pricing and sales channels, which gave access to underground benefits from sales. However, application in 4B introduced a standard pricing and sales channel management system, which the sales executives refused to use. This is an indication of a bad *guanxi* being a contributing reason that leads to unsuccessful IS implementation.

Internal Communication

Data from interviews suggest that implementation is more likely to fail when people from different departments do not communicate regularly. Successful applications were accompanied by regular internal communication, while reasons behind the failed implementations could be traced to the fact that the key informants did not meet in time to discuss the problems arising from the implementation process. This could be highlighted by the experience of one functional manager from company B: "I knew almost little about the function of the system until the time when IT department told me 'hi, here is the system, try it out.'" Inevitably, the function of the application in 3B hardly could satisfy the users, leading to the failure. Likewise, the failures of systems 4B and 7D also were traced back to the insufficiency of internal communication.

Accordance with Government Policy

The effect of government policy was obvious in the finance applications. The most successful finance application 8D was a product by a local vendor sponsored by the government. The in-house-developed applications of 2A and 4B and the foreign product of 6C were not as successful as 8D. One reason for the success of the local software was that the government-sponsored vendors were kept informed of the government policy changes, so they could upgrade their systems in time. "Our vendor is really amazing, they often upgrade our system according to the change in the government finance policy even before we know the changes," as put by the finance manager of company D. In contrast, the in-house developers and foreign vendors turned out to be inert to those changes, as claimed by the accountant of company C: "The government policy is changing, but the system is very hard to change; so, we have to change our report now and then, but it takes a very long time to change as required by the new policy, so we have to turn back to chapter report. As a result, we have to generate far more reports than before the system was implemented." Therefore, it should be reasonable to claim the necessity for the system to be in line with government policy.

Adjusting IS to the Changing Business Environment

The necessity of system upgrading according to the fast-changing business environment in China is obvious from the study. Successful implementations were adjusted continuously in accordance with environmental change, while failure to adjust caused problems. For example, application 7D was implemented only three years ago; however, its benefits were shrinking, as evaluated by the manufacturing manager of company D: "… now, it could hardly satisfy our production line, if it would not be upgraded or replaced, it will become an obstacle rather than facilitator for the production line."

CSF Corresponding to Literature

As previously stated, the factors that emerged from the literature were assessed first through interpretive analysis of the qualitative data using the technique of content and constant comparative analysis (Patton, 2002). Then, a process of comparison was made to determine which independent variables are directly connected to the stage success. Simple inspection of and analysis of data determined how various factors influenced the dependent variable (Reich & Benbasat, 1990).

Table 5 gives an overview of the comparison. The blank area means that the factors are not pertinent to the stage; for example, key users usually do not participate in the chartering stage or the onwards-and-upwards stage.

Key User Involvement

Key user rather than end user should be involved in the planning stage as well as the shakedown stage in the model. It was indicated that key users should have a good knowledge of their own businesses as well as be ready to learn new IS knowledge and techniques. Case 3B and Case 4B had no appointed key users; according to the data, this was one of the important reasons for their failures. The delay of develop-

Table 5. Factors influencing IS implementation outcomes

Factors	Stage Outcomes												
	Organizational Fit (Chartering Stage)				Completion Extent and Meeting Requirements (Project Stage)			System Use (Shakedown Stage)			Net Benefit (Onward and Upward Stage)			
	Factor value	S	M	U	Factor value	S	U	Factor value	S	U	Factor value	S	M	U
	No. of cases	3	3	2	No. of cases	5	3	No. of cases	5	3	No. of cases	3	2	3
Key user involvement					Yes (6)	5	1	Yes (6)	5	1				
					No (2)		2	No (2)		2				
External expertise—vendor/consultant	Yes (6)	3	1	2	Yes (6)	4	2	Yes (6)	3	3	Yes (6)	1	2	3
	No (2)		2	0	No (2)	2	0	No (2)	2	0	No (2)	2		0
Clear goal	Yes (7)	3	3	1										
	No (1)			1										
Change management					Yes (1)		1	Yes (1)		1				
					No (7)	5	2	No (7)	5	2				
Project management					Yes (6)	5	1							
					No (2)		2							

Note: S—Successful; M—Moderately Successful; U—Unsuccessful

ing the application in 7D was traced back partly to the lack of business knowledge of key users, as indicated by the IT coordinator: "Some key users did not know the workflow of their department and could not come up with the user requirements in time." One key user for Case 6C was the manager of the finance department, who knew the business process very well; however, she had serious computer phobia. So it took the project manager a much longer time than planned to persuade her to release the finance data to input them into the system. For applications with successful project and shakedown stages, key users' business knowledge and efforts in learning had positive effects on the outcomes of the implementations. The practice of Case 1A and Case 2A set good examples. The key users were the deputy managers of the functional department; they were eager to acquire the knowledge and techniques necessary for system operation, which contributed to the smooth project and shakedown process.

End-user training was thought necessary for successful implementation for all applications, and there was no difficulty getting end users involved in the training. One could get some idea from the comments of the training officer of company A: "So long as there is a 'demand' [from a senior manger], the end users have no choice but do their best to learn, or they may be dismissed."

IS Expertise of Vendor, Consultant, or IT Department

Most successful applications depended on external expertise, either vendor or consultant, with the IT department as a communication channel between external expertise and internal application. However, there were exceptions: 1A and 2A. Company A set up its own software branch company to support the IS implementation process. This approach turned out to be very successful. The CEO gave the following comments: "I studied many IS providers and consultants and even tried out some of them, but their solutions could not fit into our company, so I decided to hire an expert in the industry to expand our IT department into the software branch company to specially design IS system for us. Till now, it seems exciting." At the same time, there was a sad story for depending solely on the IT department without sufficient external expertise. The IT department of company B was asked to implement the application in case 3B based on the database of Oracle 9i; however, the project turned out to be a failure: 80% of its function was obsolete after it went live. In hiring external expertise, the capability of the vendor/consultant was an important factor in IS implementation success. When summarizing the lessons from 4B, the finance manager of company B said, "One of our most important mistakes was that we hired a vendor who was neither good at IT nor at finance management. The system itself is so terrible that, although the engineers and the consultants from the vender company tried their best, it failed inevitably." In summary, there is no absolute correct choice among the expertise of the IT department, the vendor, and the external consultant; the suitability may depend upon company characteristics.

Clear Goals

Seven applications had clear goals that were set in the initial stage and were adhered to throughout the implementations process. Many of the systems were targeted at saving cost to improve efficiency in order to face the fast-increasing challenge from the world market as China entered into WTO. Case 4B only had a three-page project plan, which did not include any deliverable benefits; it was almost forgotten during the implementation process. Although 3B clearly defined the goals of the project, the goals were regarded by the CEO as "being not able to add value to business" and was revised several times. That led to more than one year's delay, and the functions of the completed system differed greatly from the original one. So it is fair to conclude that clear goals in line with business strategy should be developed for IS implementation in China.

Business Process Redesign and Change Management

It was found that different companies employed different strategies in aligning business processes with those embedded in the system, which led to different results. Failure in redesigning business processes and matching them to IS systems resulted in not meeting user requirements (3B, 4B) and even could be wrong (4B). For all the successful applications, business process redesign (Davenport & Short, 1990) and change management (Benjamin & Levinson, 1993) was carried out, albeit to different extents. The more the business process was fine-tuned to meet the business goals of the systems, the more likely it was that the system could bring net benefits (1A, 8D) in the long term. Some companies recognized that optimizing the business process could bring them real value; as expressed by the sales manager of company A: "Our investment in business process optimization has made us more profitable and we would appreciate more effort in it." The CEO of company C also was very interested in adopting the business practice embedded in the BAAN system rather than adjusting the system to the company's current workflow.

Nevertheless, the successful companies were very cautious in adopting the best practice in IS applications that originated from developed countries. One of the main reasons that company A hired a software company to develop the IS system based on the technical structure of Western systems rather buying a mature Western county-based system promising best practice, such as SAP, was that they were afraid that what is good for U.S. and European countries could not work for them. For 3B, it turned out that the business processes embedded in the Oracle package did not fit company B. When asked about the lessons from the failure in 3B, the IT manager commented, "We made a wrong decision to change our management to the flow in the system suitable for market-oriented economy. It is too early for us or maybe not possible at all to follow the business rules of market-oriented economy."

This phenomenon suggests that while business process redesign and associated change management are necessary for successful IS implementation, companies in China should make management changes according to their own management characteristics and business goals rather than adopting the processes embedded in packages developed for Western business practices.[4]

Project Management and Database Management

Data from our case studies suggest that integrating a database poses a challenge for project management. Whether the database could be designed and managed as required by the IS is an important indicator for the success of the stage of project. The fatal factor for 3B was that the database design could not meet the production process. "Our warehouse is using the technique of 'back-flush' (a way of maintaining manufacturing warehouse), but the database does not support it, and we have to manually maintain the database. It is not only time consuming but also has resulted in wrong information." There was also difficulty getting necessary information. Whether the project team could overcome this difficulty is linked directly to the degree of completion and the requirement meeting. For example, the project team for 6C failed to get all the finance data for the system to set up, resulting in two months' delay and not meeting the requirement listed in the plan.

In summary, 10 factors (see Table 6) are critical for successful IS implementation in China. Among them, five are new China-specific CSF that emerged from this study, and the others are CSF that have been reported in the literature.

Table 6. A description of the CSF for IS implementation in China

CSF	Chartering	Project	Shakedown	Onward &Upward
Chinese-Specific CSF				
First-hand support	•	•	•	•
Internal communication	•	•	•	•
Accordance with government policy	•		•	•
Guanxi in steering committee	•	•		
Adjusting IS to changing business environment				•
CSF in Line With Literature				
Key user involvement		•	•	
IS expertise of IT department, vendor, consultant	•	•	•	•
Clear goals	•			
BPR and change management		•	•	
Project management		•		

Figure 3. IS implementation success model—a causal network of CSF

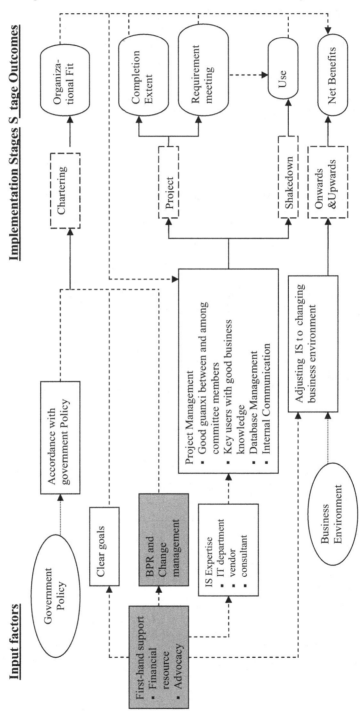

Toward an IS Success Model in Chinese Business

While recognizing that IS implementation is a complex phenomenon concerning many dimensions, it is our endeavor to propose a comprehensive framework based on the cross-case analysis (see Figure 3). The proposed model takes IS success as a multi-layer model. CSF are positioned on the left under Input Factors as the first layer. They are grouped into categories according to their content. For example, financial support and advocacy are arranged as components of first-hand support, and the items in circles are the contextual factors for the CSF. The next layer shows the implementation stages. Stage outcomes are depicted as the last layer. The dashed arrows signify the relationship between CSF; the rest of the arrows indicate input and output relationships.

Our IS success model highlights the overriding CSF as first-hand support and BPR and change management (in dark grey boxes in Figure 3). It also shows the virtuous relationships of necessity that should exist within CSF and stage outcomes (Akkermans & van Helden, 2002). For example, the arrow from first-hand support to change management implies that first-hand support is necessary for the smooth going of changing management; organizational fit, the outcome of the first stage, is a necessary success factor for the project management success as well as a predictor for the net benefit. The social-cultural contexts of CSF also are indicated in the proposed model. For instance, the business environment in the bottommost circle implies that it is necessary to upgrade the system in time to realize or to keep net benefit of the system due to the fast-changing business environment in China. The other context factor concerns government policy. We found "accordance with government policy" to affect directly the chartering stage of IS implementation projects. Such finding is especially useful and pertinent to understanding IS implementation in China, where the policy-related infrastructure and IT law are not as established as in Western countries.

Conclusion

The Intriguing Factors Identified

The key contribution arising from our case analysis lies in identifying the five CSF that are deemed intriguing in the Chinese context. They are first-hand support, guanxi between and among committee members, internal communication, accordance with government policy, and adjusting IS to changing business environment.

First-Hand Support

A striking finding is the significant role of persons in IS implementation; the first-hand influence may not be derived from official positions but rather from the personalities who are assuming the role of FH. Their availability, capability, and efforts are directly or indirectly necessary for enabling other CSF as well as the success in each stage. In some respects, it is similar to project champion discussed in the (Western) literature on IS implementation (Beath, 1991; Willcox & Sykes, 2000). For a given project, the champion often is not the project manager; rather, by virtue of seniority in the organization, the champion can rally support for the related project. FH in a Chinese firm, however, can go beyond advocacy and be directly influencing or even making decisions relating to the project directions as well as to details. This is a derivation from and a reflection of the personal rule of man in China rather than "universal or systemic rule by law" (Martinsons & Westwood, 1997, p. 221). In other words, the primary basis for trust and social order comes from persons rather than from any rules or regulations. The owner of the company defines the regulations and the rules, employing a far more subtle strategy (Tricker, 1988). Their subordinators, characterized as high power distance (Redding, 1990), just follow those rules and regulations to keep networks of relationships and social harmony, showing their deference, respect, obedience, and personal loyalty to the senior (Westwood & Chan, 1995).

Guanxi between and among Committee Members

In China, *guanxi* forms an invisible network that is popular and the most expedient way of getting things done. The subtlety is strongly highlighted by Hill (2003) that "there is a tacit acknowledgment that if you have the right *guanxi*, ... rules can be broken, or at least bent" (p. 87). Although being somewhat novel in IS implementation literature, this concept has drawn much interest in the field of organizational science for comprehending Chinese business culture (Chung & Hamilton, 2002; Farh, Tsui, Xin, & Cheng, 1998; Seligman, 1999) and provides a basis for trust (Farh et al., 1998). While it sometimes may be confused with the notion of social relationship, *guanxi* is distinct as a Chinese concept in that role obligation and friendship, where applicable, are instrumental in influencing behavior even in a business setting (Farh et al., 1998). Although relationship and social network are regarded as important in all cultures, it is generally unknown that in China social relationships may mean much more than what Westerners intend because they are so fundamental to the Chinese national character. As Vanhonacker (2004) put it, the cultivation of *guanxi* is an integral part of doing business: in the West, relationships grow out of deals; in China, deals grow out of relationships. As they have come to understand *guanxi* as an important mechanism for getting business done and building long-term business relationships in China, many Western businesses have tried to build *guanxi* to grease the wheels required to do business in China (Seligman, 1999).

Guanxi is not necessarily related to the norms of reciprocity as social relationship does. In the context of *guanxi,* relationships often are seen as ends in themselves rather than means for realizing individual goals. It is generally accepted that Confucianism-rooted, Chinese-characteristic, family-centered firms are shaping IS practice in China significantly. In other words, family is the traditional context of the enterprise, leading to informal organization structure and operation (Tricker, 1988). Although family connections may be replaced by institutional rules for employment decisions and although Chinese people are becoming more individually oriented with the ongoing modernization process, *guanxi* within a certain group is still an important source of identity and is unlikely to be replaced in the near future (Yang, 1986, 1988).

Internal Communication

Being conservative long has been regarded as a virtue in the bigger part of China's history, and the consequence of this on today's social norms cannot be underestimated. In the business arena, what this may translate into is a group of employees who seldom speaks their real opinions. Also, in a Chinese nature, high-context culture (Raman & Watson, 1994) in which most of the information is already in the persons involved and very little is in an explicit coded form, people are hesitant to do things while following a clear and strict schedule. Accordingly, they tend to see the project design and the related technical issues as a single entity and to store this entity in their minds rather than explicitly writing them out and making a clear plan. However, it is necessary for project teams to understand the business flows that IS may concern through talking with people from different functional sections. Such seemingly irreconcilable differences between the requirement for successful IS implementation and cultural habits probably can be addressed with explicit and extra efforts to create and foster an open-minded internal communication throughout the IS implementation process.

Accordance with Government Policy

The extensive efforts of China's government have been playing an import role in shaping IS application development. For example, under the strategy to self-dependently transform financial management, the government has sponsored several local software companies, has carried out a series of policies to set their products as the industry standard, and has mandated companies' acceptance. This effort has resulted in the fact that domestic software is in possession of the major financial application market, and the top ones are government-sponsored companies such as Kingdee and Usoft (CCID, 2002). The characteristics of Chinese government policy on IT, which is important for understanding its role, can be summarized as follows: "There is no clearly defined vision of the future of IT in China, nor is there a government institution capable of developing and implementing a comprehensive national IT plan. Authority is divided among agencies variously concerned with telecommunications, computers, information services, education, and R&D. This situation results both from Chinese political culture, which resists vesting power in a single institution, and from the government's deliberately allowing market forces to shape IT development rather than relying solely on central planning" (Kraemer & Dedrick, 1995, p. 64). Therefore, the Chinese government is playing and will continue to play a major role in guiding the economy. Accordingly, Chinese government policy accounts greatly for IS implementation success.

Adjusting IS to Changing Business Environment

The economy and business environment is underdeveloped, nonstandard, and in the process of changing quickly. Restricted by the closed economy before the industry reform in 1978, Chinese enterprises were characterized as having behind-hand and nonstandard management systems and business processes. The market also is planned rather than directed by the economy rules like those in developed countries. With the Chinese reform that started in 1978, companies have been changing to be self-responsible and market-oriented. Along with this progress of transforming, especially, Chinese enterprises were encouraged to improve their competitiveness through strategic use of information systems (Franz et al., 1991). IS managers, who are exposed to global market economy, actively have advocated and promoted IS for business functions, leading to the popularity of IS in China. For example, in 1996, with quite a number of discussions about enterprise resource planning systems appearing in influential presses, ERP became a buzzword overnight. However, economy reform in China has not yet been accomplished fully, and the old system is still deeply rooted (Dologite et al., 1998). Now the Chinese economy is growing at an ever faster pace. In this economy growing process, the utilization of computer-based information systems in business organizations also should be adjusted quickly in response to the changing business environment. One implication is that, due to the fact that the Chinese economy is in a special evolution stage within a mixed market and Chinese managers are utilizing IS in a different way than that of their counterparts in Western countries, it is timely to study IS practice in China.

Concluding Remarks

It generally is known that cultural and communication barriers are among the greatest challenges when conducting business in China. While the impact of cultural differences are undeniably being felt by many international companies, the literature has yet to suggest a clear distinction on how the Chinese are different from Western cultures in terms of implementing information system packages. Our case study provides a comprehensive understanding of IS implementation in China by identifying key factors for IS success. While some of the CSF correspond to the existing literature, others are Chinese-specific. It highlights that, in China, the success of IS implementation may be achieved using a different approach from that in Western, developed countries. In this light, Chinese enterprise executives should have a good understanding of the uniqueness of context in China while reflecting on the experiences of Western countries in order to enhance their business performance.

This chapter provides a basis for further studies in related areas on how IS practice is shaped by business culture and economic environment. Future research can be extended to validate the proposed IS success model via more case studies in selected industries and software packages in China. For practitioners, the findings of this study offer organizations a set of guidelines to adopt IS applications in order to improve their business performance.

References

Akkermans, H., & van Helden, K. (2002). Vicious and virtual circles in ERP implementation: A case study of interrelations between critical success factors. *European Journal of Information Systems, 11*(1), 35-46.

Alavi, M., & Joachimstahler, E. A. (1992). Revisiting DSS implementation research: A meta analysis of the literature and suggestions for researchers. *MIS Quarterly, 16*(1), 95-116.

Beath, C. A. (1991). Supporting the information technology champion. *MIS Quarterly, 15*(3), 355-372.

Benjamin, R. I., & Levinson, E. (1993). A framework for managing IT-enabled change. *Sloan Management Review, 34*(4), 23-33.

Bingi, P., Sharma, M. K., Maneesh, K., & Godla, J. K. (1999). Critical issues affecting an ERP implementation. *Information Systems Management, 16*(3), 7-14.

Bonoma, T. V. (1985). Case study research in marketing: Opportunities, problems, and a process. *Journal of Marketing Research*, *22*(2), 199-208.

Bullen, C. V., & Rockart, J. F. (1986). A primer on critical success factors. In J. F. Rockart & C. V. Bullen (Eds.), *The rise of managerial computing: The best of the center for information systems research* (pp. 384-423). Cambridge, MA: Sloan School of Management.

Butler, T., & Fitzgerald, B. (1999). Unpacking the systems development process: An Empirical application of the CSF concept in a research context. *Journal of Strategic Information Systems, 8*(4), 351-371.

Cameron, P. D., & Meyer, S. L. (1998, December). Rapid ERP implementation—A contradiction? *Management Accounting*, 58-60.

CCID: China Center for Information Industry Development. (2002). *Annual report on China's software market.* Beijing: CCID Consulting.

Chinese Economy and Trade Information Center. (2001). Report on Chinese industry information systems—The present status and plan for the tenth five-year plan [全国企业信息建设现状和"十五"规划情况调查报告, 国家经贸委经济信息中心].

Chung, W. K., & Hamilton, G. G. (2001). Social logic as business logic: Guanxi, trustworthiness, and the embeddedness of Chinese business practices. In R. Appelbaum, W. Felstiner, & V. Gessner (Eds.), *Rules and betworks: The legal culture of global business transactions* (pp. 325-346). Oxford, UK: Hart Publishing.

Cooper, R. B., & Zmud, R. W. (1990). Information technology implementation research: A technology diffusion approach. *Management Science*, *36*(2), 123-139.

Davenport, T. H., & Short, J. E. (1990). The new industrial engineering: Information technology and business process redesign. *Sloan Management Review*, *31*(4), 11-27.

DeLone, W. H., & McLean, E. R. (1992). Information systems success: The quest for the dependent variable. *Information Systems Research*, *3*(1), 60-95.

DeLone, W. H., & McLean, E. R. (2003). The DeLone and McLean model of information systems success: A ten-year update. *Journal of Management Information Systems*, *19*(4), 9-30.

Dologite, D. G., Fang, M. Q., Chen, Y., Mockler, R. J., & Chao, C. N. (1998). An information systems view of Chinese state enterprises. *The Journal of Strategic Information Systems*, *7*(2), 113-129.

Farh, J. L., Tsui, A. S., Xin, K., & Cheng, B. S. (1998). The influence of relational demography and guanxi: The Chinese case. *Organization Science*, *9*(4), 471-488.

Franz, C. R., & Robey, D. (1987). Strategies for research on information systems in organizations: A critical analysis of research purpose and time frame. In R. J. Boland Jr., & R. A. Hirshheim (Eds.), *Critical issues in information systems research* (pp. 227-251). New York: John Wiley and Sons.

Franz, C. R., Wynne, A. J., & Fu, J. H. (1991). Managing information systems to support functional business requirements in China. *International Journal of Information Management, 11*(3), 203-219.

Hammer, M. (1990, July-August). Reengineering work: Don't automate, obliterate. *Harvard BusinessReview, 68*(4), 104-112.

Hill, C. W. L. (2003). *International business: Competing in the global marketplace.* Boston: McGraw-Hill; Irwin.

Holland, C. P., & Light, B. (1999, May-June). Critical success factors model for ERP implementation. *IEEE Software, 16*(3), 30-36.

Hong, K. K., & Kim, Y. G. (2002). The critical success factors for ERP implementation: An organization fit perspective. *Information and Management. 40*(1), 25-40.

Ives, B., & Olson, M. H. (1984). User involvement and MIS success: A review of research. *Management Science, 30*(5), 586-603.

Kraemer, K. L., & Dedrick, J. (1995). From nationalism to pragmatism: IT policy in China. *IEEE Computer, 28,* 64-73.

Kwon, T. H., & Zmud, R. W. (1987). Unifying the fragmented models of information systems implementation. In R. J. Boland Jr. & R. A. Hirshheim (Eds.), *Critical issues in information systems research* (pp. 227-251). New York: John Wiley and Sons.

Lee, S., Goldstein, D. K., & Guinan, P. J. (1991). Informant bias in information systems design team research. In H. E. Nissen, H. K. Klein, & R. Hirschheim (Eds.), *Information systems research: Contemporary approaches and emergent traditions. Proceedings of the IFIP TC8/WG 8.2 Working Conference* (pp. 635-656). Amsterdam: Elsevier North-Holland.

Li, D. (1999). 探索中国信息化建设的道路 [Exploring the enterprise information systems in China]. *Computer Information Systems Practice, 10,* 4-6.

Lin, J. R., & Zhu, J. (2002). ERP在企业中的实与案例分析 [The implementation of ERP in enterprise and its case study]. *Industrial Engineering and Management, 2,* 54-55.

Lu, M., Qiu, Y., & Guimaraes, T. (1988). A status report of the use of computer-based information systems in PRC. *Information and Management, 15*(5), 237-242.

Markus, M. L., & Tanis, C. (2000). The enterprise systems experience—From adoption to success. In R. W. Zmud (Ed.), *Framing the domains of IT manage-*

ment: Projecting the future through the past (pp. 173-207). Cincinnati, OH: Pinnaflex Publishing.

Martinsons, M. G., & Westwood, R. I. (1997). Management information systems in the Chinese business culture: An explanatory theory. *Information and Management, 32*(5), 215-228.

Miles, M. B., & Huberman, A. M. (1994). *Qualitative data analysis: An expanded sourcebook* (2nd ed.). Thousand Oaks, CA: Sage Publications.

Nah, F., Lau, J., & Kuang, J. (2001). Critical factors for successful implementation of enterprise systems. *Business Process Management, 7*(3), 285-296.

Nandhakumar, J. (1996). Design for success? Critical success factors in executive information systems development. *European Journal of Information Systems, 5*(1), 62-72.

Olson, D. L. (2004). *Managerial issues of enterprise resource planning systems.* Boston: McGraw Hill; Irwin.

Parr, A., & Shanks, G. (2000). A model of ERP project implementation. *Journal of Information Technology, 15*(4), 289-304.

Parr, A. N., Shanks, G., & Darke, P. (1999). Identification of necessary factors for successful implementation of ERP systems. In O. Ngwenyama, L. D. Introna, M. D. Myers, & J. DeGross (Eds.), *New information technologies in organizational processes* (pp. 99-119). Boston: Kluwer Academic Publishers.

Patton, M. Q. (2002). *Qualitative evaluation and research methods* (3rd ed.). Thousand Oaks, CA: Sage Publications.

Pfeffer, J. (1981). *Power in organization.* Marshfield, MA: Pitman Publ. Co.

Pierce, J. L., & Delbecq, A. L. (1977). Organization structure, individual attributes, and innovation. *Academy of Management Review, 2*(1), 27-37.

Qi, E. S., Huo, Y. F., Chen, M. F., & Wang, Y. (2000). CIMS下的ERP管理模式---概念、论、及实施研究[The study on ERP under the CIMS environment from the aspects of conception, philosophy and implementation]. *Industry Engineering Journal, 3*(1), 11-13.

Raman, K. S., & Watson, R. T. (1994). National culture, information systems, and organizational implications. In P. C. Deans & K. R. Karwan (Eds.), *Global information systems and technology: Focus on the organization and its functional areas* (pp. 493-513). Hershey, PA: Idea Group Publishing.

Redding, S. G. (1990). *The spirit of Chinese capitalism.* New York: DeGruyter.

Reich, B. H., & Benbasat, I. (1990). An empirical investigation of factors influencing the success of customer-oriented strategic systems. *Information Systems Research, 1*(3), 325-347.

Reimers, K. (2002). Implementing ERP systems in China. *Proceedings of the 35th Hawaii International Conference on Systems Sciences* (pp. 3112-3121) [CD-ROM]. Los Alamitos, CA: IEEE Computer Society Press.

Rockart, J. F. (1979). Chief executives define their own data needs. *Harvard BusinessReview*, 57(2), 81-93.

Rockart, J. F. (1982). The changing role of the information systems executive: A critical success factors perspective. *Sloan Management Review*, 24(1), 3-13.

Seligman, S. D. (1999, September/October). Guanxi: Grease for the wheels of China. *China Business Review*, 26(5), 34-38.

Shanks, G., Parr, A., Hu, B., Corbitt, B., Thanasankit, T., & Seddon, P. B. (2000). Differences in critical success factors in ERP systems implementation in Australia and China: A cultural analysis. In H.R. Hansen, M. Bichler, & H. Mahrer (Eds.), *Proceedings of the European Conference on Information Systems* (pp. 537-544). Vienna: Wirtschaftsunivsitat Wien.

Soh, C. P. P., Yap, C. S., & Raman, K. S. (1992). Impact of consultants on computerization success in small businesses. *Information and Management*, 22(5), 309-319.

Sumner, M. (1999). Critical success factors in enterprise wide information management systems projects. In W. D. Haseman & D. L. Nazareth (Eds.), *Proceedings of the Fifth Americas Conference on Information Systems,* Milwaukee, Wisconsin (pp. 232-234) [CD-ROM].

Steers, R. M. (1977). *Organizational effectiveness: A behavioral view.* Santa Monica, CA: Goodyear.

Tate, P., & Maier, J. H. (1987, September). Dateline Beijing: The China syndrome. *Datamation, 1*, 33-42.

Thompson, V. A. (1969). *Bureaucracy and innovation.* Huntsville: University of Alabama Press.

Thong, J. Y. L., Yap, C. S., & Raman, K. S. (1996). Top management support, external expertise and information systems implementation in small businesses. *Information Systems Research*, 7(2), 248-267.

Tricker, R. I. (1988). Information resource management: A cross-cultural perspective. *Information and Management,* 15(1), 37-46.

Vanhonacker, W. R. (2004). Guanxi networks in China. *China Business Review,* 31(3), 48-53.

Walsham, G. (1993). *Interpreting information systems in organization.* New York: Wiley.

Westwood, R. I., & Chan, A. (1995). The transferability of training in the East Asian context. *Asia Pacific Business Review,* 2(1), 68-92.

Willcox, L. P., & Sykes, R. (2000). The role of CIO and the IT function in ERP. *Communications of the ACM, 43*(4), 33-38.

Williams, J. J., & Ramaprasad, A. (1996). A taxonomy of critical success factors. *European Journal of Information Systems, 5*(4), 250-260.

Yang, K. S. (1986). Chinese personality and its change. In M. H. Bond (Ed.), *The psychology of Chinese people* (pp. 106-160). Hong Kong: Oxford University Press.

Yang, K. S. (1988). Will societal modernization eventually eliminate cross-cultural psychological cultural differences? In M. H. Bond (Ed.), *The cross cultural challenge to social psychology* (pp. 67-85). London, UK: Sage Publications.

Yin, R. K. (1994). *Case study research: Design and methods* (2nd ed.). London: Sage.

Zhang, L., Lee, M., Zhang, Z., & Chan, J. (2002). A framework for enterprise resource planning systems implementation success in China. In T. Terano & M. D. Myers (Eds.), *Proceedings of the Pacific-Asia Conference on Information Systems,* Tokyo, Japan (pp. 688-701) [CD-ROM].

Zhang, Y. (2001). 企业家和中国企业信息化建设 [Entrepreneur and implementation IS in Chinese companies]. *Industry Technology Innovation, 3*, 4-6.

Zhu, C. Y., & Ma, G. H. (1999). *New horizon of management—ERP and supply chain management.* China: Electronics Audio and Video Press.

Zwass, V. (1998). *Foundations of information systems.* New York: Irwin.

Endnotes

[1] Critical success factors are defined as those few critical areas where things must go right for the business to flourish (Rockart, 1979). Williams and Ramaprasad (1996) distinguished among four types of criticality; the third type, "factors necessary for success", is adopted for this chapter.

[2] Translated from Chinese phrase 'Yi-Ba-Shou' [一把手], referring to the individual who has the greatest power and legal responsibility in an organization; this is to be differentiated from top management or the CEO. This term is widely understood in mainland China.

[3] "Guanxi" refers to the existence of direct, particular ties between an individual and others (Farh, Tsui, Xin, & Cheng, 1998).

[4] It is interesting to note that none of the companies considered radical business process reengineering advocated by Hammer (1990).

Chapter V
Developing Electronic Content for the Support of the European Cultural Inclusion:
From the Earlier "eEurope" Initiative toward the Future "i2010" Perspective

Ioannis P. Chochliouros, Hellenic Telecommunications Organization S.A. (OTE), Greece and University of Peloponnese, Greece

Ioannis Bougos, Hellenic Telecommunications Organization S.A. (OTE), Greece

Stergios P. Chochliouros, Independent Consultant, Greece

Anastasia S. Spiliopoulou, Hellenic Telecommunications Organization S.A. (OTE), Greece

Abstract

Various European initiatives and appropriate policies have been deployed (and are still in progress) to promote the creation and the distribution of new forms of electronically available content, aiming to support the "European cultural inclusion," which constitutes a high-priority societal issue. The present works investigates diverse

Copyright © 2007, Idea Group Inc. Copying or distributing in print or electronic forms without written permission of Idea Group Inc. is prohibited.

potential opportunities for realizing and offering innovative services, applications, and/or related facilities in the market through the proper use of modern electronic communications, especially for the promotion of cultural and social targets. Such options implicate vigorous participation both of state authorities and market industry players to launch dynamic business partnerships in parallel with efforts to improve quality of life and social cohesion in order to forward new ways of participating in society and to advance the European diversity and rich cultural heritage.

Introductory Framework: The Importance from the Development of the Global Information Society

The information society represents one of the most fundamental changes (Chochliouros & Spiliopoulou-Chochliourou, 2003c) of our time with enormous opportunities for the society as a whole. Any relevant development offers huge perspectives to improve the way in which citizens live and work (European Commission, 2003a). However, the scale and pace of change present decisive challenges for individuals, companies, and certain groups of citizens and regions (European Commission, 2001a), thus having a significant social importance within the wider context of all relevant policy measures applied. For the particular European framework, the European Commission and other responsible authorities very early have underlined the extreme importance of the social dimension of all possible initiatives that are able to affect to the shaping of any political, social, cultural, and civil dialogue, and so to provide assistance to various measures (either planned or in progress).

Among the fundamental priorities promoted by the contemporary European Union's (EU) policies were: (a) to increase awareness of the social implications of the information society; (b) to build an effective "Information Society dimension," *where appropriate*, into well-defined social policies and actions; and (c) to identify precise actions designed to maximize the contribution of the information society to promote digital inclusion, employment, and culture (Royal Danish Ministry for Foreign Affairs, 2000b). New (interoperable) technologies and their applications can convert the entire image of our society, thus providing appropriate tools for such a transformation in a global environment. In fact, the wider European model of knowledge, based on truly multicultural societies, a wealth of memory and identity, and a unique way of dealing with technologies (Chochliouros & Spiliopoulou-Chochliourou, 2005a), has many opportunities to prevail in the new world order. In other words, the European way of life (or, alternatively, the "European dream") can offer to the entire planet a powerful response to the overriding challenges of civilization and progress of cultural values. It can be expected (Rifkin, 2004, p. 3) that *the European Dream emphasizes community relationships over individual autonomy, cultural diversity over assimilation, quality of life over accumulation of wealth, sustainable development over unlimited material growth, deep play over unrelenting toil, universal human rights and the rights of nature over property rights, and global cooperation over the unilateral exercise of power.*

Since very early, knowledge and innovation have been estimated as the engines of growth for the European reality and, among those particular essential components, absolutely necessary to build a fully inclusive information society based on the widespread use of information and communication technology (ICT) in public services, enterprises, and residential consumers. The aim of our work is to investigate current European activities and specific targeted policies and to explain how commonly approved (and occasionally applied) measures can be the "prime" material for further promotion of various forms of electronic content, especially for the aim of the "European Cultural Inclusion" that composes a high-priority issue among the fundamental activities of the EU. This implies the possibility to promote the dispersion of various culturally related issues to the citizens and to promote either cultural heritage and values or the development/expansion of new interests and/or activities associated with various forms of literature and creative arts of the appropriate sectors. Moreover, we correlate these initiatives to the promotion and deployment of new forms of electronic content by examining parallel activities for improving both quality of life and social cohesion in order to support electronic inclusion for all citizens and to provide high-quality public services. These directions drastically affect the design of future policies and of relative investments in a great variety of activities in which diverse business partnerships can be created under appropriate terms with the active involvement of both private and public sectors. Especially for the latter (also including the wider state-related activities), enormous opportunities can appear through the proper development of new forms of content and by opening new ways of participating in society. Furthermore, this can constitute a real promotion of currently available electronic content via new (even interactive) forms of ICTs while simultaneously creating a practical incentive for the creation and distribution of new content especially designed and forwarded to extend cultural perspectives. ICTs and properly designed digital content that is able to be disposed through them are a powerful tool for preserving and promoting the European diversity and cultural heritage (DigiCULT Project, 2002). To this aim, we investigate the possible development of electronic government facilities and further challenges for extending cultural values in specific areas.

The European Approach: Creating Conditions and Finding Measures for Improving Quality of Life and Social Cohesion

Recent EU policies have set the ambitious objective for Europe "to become the most competitive and dynamic economy in the world" (European Commission, 2000, p. 2). So it has been recognized (European Commission, 2000) as an urgent and ambiguous need to exploit the opportunities of the new information economy and, in particular, the Internet (Lalopoulos, Chochliouros, & Spiliopoulou, 2005), considered a "multiplicity" of all possible underlying platforms and/or networks-infrastructures, fast developing services, and/or innovative facilities, and most importantly, various forms of content available electronically. To achieve this, the EU authorities (European Commission, 2000, p. 2) have draw up

a comprehensive eEurope Action Plan, using an open method of co-ordination based on the benchmarking of national initiatives, combined with recent initiatives, applied strategies and programmes, to bring the entire Europe on-line, thus covering a great variety of national, regional and local needs.

To this aim, the original *e*Europe program initially identified particular areas in which action at various levels should add value to the broader European reality. Up to the present time, the key target areas have been revised in various cases (European Commission, 2004a), while numerous reactions and responses have been taken into account toward realizing the correspondent social, economic, cultural, and technological goals (Council of the European Union, 2003).

The new digital and knowledge-based economy has a major impact on multiple issues, implicitly and explicitly related to a great variety of activities (European Commission, 2003a). Meanwhile, variable challenges appear as Europe has been in the driving seat of the world for a long time. Indeed, democracy has been created in Europe, science and technology have made significant progress and evolution, modern learning institutions have been founded in Europe (including more than 800 years of successful universities), while multicultural contacts have been performed for centuries inside and outside of Europe. Within this particular framework, EU authorities and member states have focused their attentions specifically on measures to ensure an adequate supply of high-quality digital content (especially offered through the Internet) that is able to cover diverse sectors' needs. Among the core objectives was to provide a favorable environment for private investment, to boost productivity, to modernize public services, and to strengthen and make known widely the cultural values and heritage of Europe (DigiCULT Project, 2002). Relevant improved possibilities based on a harmonized framework will foster investment and innovation, which will lead, in turn, to growth and increased competitiveness of the digital content industry and bring benefits to all partners involved (Royal Danish Ministry for Foreign Affairs, 2000a, 2000b).

Realizing the Target through Recent European Policies: Designing the Future

There are three main methods by which *e*Europe targets will be achieved:

Accelerating the Setting Up of an Appropriate Legal Environment

On a European level, a range of harmonized legislative proposals is being promoted, timely adopted, and properly implemented. As a response to convergence and to the changing market and technological conditions, a new pro-competitive regulatory framework, which reinforces competition and takes account of the increasing speed of developments in this sector, is being put forward (Chochliouros & Spiliopoulou-Chochliourou, 2003b). Moreover, the latest European legislation insists on special issues for regulating multiple audiovisual content services. Soon, telecommunications providers will be able to deliver broadcasting services in a quality equal to traditional TV (Stone, 2000), while at the same time, traditional content providers are entering the communications markets (European Parliament and Council of the European Union, 1989). Consumers will be able to access (i.e., watch or listen) to audiovisual content anytime, anywhere, and on all technical platforms (TV, computer, mobile phone, personal digital assistant [PDA], etc.). Thus, the core aim is to create a modern, market-oriented regulatory framework for the always-on converging digital economy and to stimulate the availability of online content through transparent and nondiscriminatory conditions. Simultaneously, citizen access to a variety of facilities will be guaranteed in a way that all citizens (both residential and corporate) will be able to reach various sources of digital data. Cultural diversity and pluralism will be stimulated through broadband development (Chochliouros & Spiliopoulou-Chochliourou, 2005a; Harvey, 1997). Access for all will be ensured, both on the supply and the demand sides, while online content will be offered through variable access facilities and infrastructures.

Supporting New Infrastructure and Services across Europe

Developments in this area depend mainly on the promotion of business initiatives; in particular, through private sector funding. Economic and innovation networks are global, and Europe needs to adopt a proactive view to support its growth and innovation (European Commission, 2004b). Services will be designed appropriately to respond to specific market needs and will be implemented properly in the market to create value. This also can be seen as fostering a new European generation of products and services in which content is present in all fronts touching on creativity, networking, and entrepreneurship (European Commission, 2003a). The challenge (and the priority) now is to create wider EU functioning markets without excluding less-favored regions; in fact, it is necessary to enable the active participation of geographically isolated citizens in social and democratic life in order to improve their living standards by bridging distance, and facilitating education and access to public services (European Commission, 2004d) on the basis of their particular economic and social structures. Appropriate investments have to be made in a way that does not distort competition and is technologically neutral. High-speed networks will open up new possibilities for collaborative learning and researching in many sectors; they will offer the necessary means so that information will be accessible and usable in various formats (text, images, video, audio, etc.) (European Commission, 2003b).

Applying the Open Method of Coordination and Benchmarking

This aims to ensure that actions are carried out efficiently, have the intended impact, and achieve the required high profile in all member states (European Commission, 2001c), conforming to general benchmarking practices and tools.

In order to reap both the economic and social benefits of technological progress, development will be based on the principles of equal opportunities, participation, and integration of all. This can be achieved if everybody has access to at least a basic set of the new services/applications offered. Citizen access can have different dimensions, such as availability, continuity, affordability, accessibility, and awareness. With the purpose of avoiding any probable obstacles, public policies can make the difference, for example, by promoting national information strategies and disposing attractive forms of content (Cap Gemini Ernst & Young, 2004). Aiming to provoke a creative response on vision and strategy for the future, several key issues can appear for further evolution in the EU context. These can be promoted by state authorities and supported either by market players and/or end users. Among the most important ones is the proper definition of the role of technology, acting as a "promoter" for development in parallel with cultural influences on both the design and implementation of modern applications (Rifkin, 2004; Streeten, 2000). Of specific importance becomes the role of public authorities as policymakers (also at national and local levels) considered in parallel with efforts to promote effective legislation (Chochliouros & Spiliopoulou-Chochliourou, 2003b) in order to fulfill current market and societal needs. In all possible approaches, it is necessary to investigate how modern technologies can make a difference overall (Chochliouros & Spiliopoulou-Chochliourou, 2003a) and what the strategic choices are facing critical areas such as, for example, knowledge, education, learning, innovation, and research and development (R&D).

Facing the previously outlined issues, the latest "i2010" (European Information Society in 2010) initiative will provide an integrated approach to information society and audiovisual media policies in the EU, covering regulation and R&D, and promoting an extended cultural diversity (European Commission, 2002a). Among its fundamental objectives is to combine all the necessary instruments in order to create a modern, market-oriented, digital information economy. It emphasizes electronic communications as a driver of inclusion and quality of life and will encourage fast growth built around the convergence at the levels of networks, services, and devices (European Commission, 2005a).

The Involvement of the State: Disposal of Content and New Forms of Electronic Government (E-Government) Facilities

Working and participation for all in the knowledge-based economy is among the high priorities of the EU. In fact, it has been made clear that achieving full employment would require a radical transformation of the economy and skills. Digital literacy is an essential element of the adaptability of the workforce and the employability of all citizens (Mehta & Shah, 1997), especially through the exploitation of the European public historical and cultural archives.

Europe is multilingual and multicultural. In the EU, a vast variety of languages is used extensively in services. At the presentation level (front office and Web pages on the Internet—the level at which citizens and enterprises are to interact with administrations), language is clearly a major factor in the effective delivery of modern services, applications, and facilities. This richness can be used as a strategic advantage of Europe, exploiting diversity within social groups, language groups, cultures, and tribes (Royal Danish Ministry for Foreign Affairs, 2000b). Diversity across Europe is a fact and a distinguishing feature from other trading blocks in the world (and is part of the Europe's ethos) (McGuigan, 2005). Modern technologies can be used to overcome the difficulties of access to content and can help to promote cultural pluralism and diversity. Many more outlets can be made available to users, who can choose freely between them and, thus, access a broader and more diverse range of content supply (Punie, 2004). Public spending can be more focused (European Commission, 2002b) in the cultural sector (culture and tourism). A digital environment can be offered, and publicly owned content can be made more readily available (European Parliament and Council of the European Union, 2003b). Recent European initiatives (European Commission, 2002b) have focused on competitiveness and sustainable development in the field of tourism, an activity that affects society in many different ways and has a profound impact on our social, cultural, and economic life. It relates to a wide range of areas; to name only a few: employment, regional development, education, environment, consumer protection, safety, new technology, transport, finance, taxation, and culture. Offering attractive and electronically available related content supports both business and cultural perspectives for evolution.

The new market (Dutta, Paua, & Lanvin, 2004) creates the potential to reduce social exclusion both through higher levels of growth and employment and by opening new ways of participating in society, such as, *for example*, creating new highways of knowledge and advancing the cultural heritage, based on the long intellectual tradition of Europe (European Commission, 2003a). Making European heritage and cultural creations available to a larger number of citizens will extend Europe's cultural values and diversity. The fact is that from recorded and documented history, Europeans are used to thinking and acting from a perspective of globalization. Ancient Greece perfected philosophical thinking for the betterment of the world; Rome legislated, not in confinement to strict borders but thinking of the entire humanity; European discoveries were meant to approximate peoples and cultures from all continents; the founding fathers of Enlightenment announced a new Age of Reason and Freedom for the modernization of all societies.

The benefits of the dispersion of knowledge must be accessible to all citizens, including those who have disabilities and those outside the labor market and the educational system; for the latter, major opportunities will be offered via the wider development of various cultural, artistic, literature, and other similar issues for further promotion of social inclusion (Cap Gemini Ernst & Young, 2004). Simultaneously, digital technologies provide the opportunity to more easily access and reuse the prosperity of information held and stored in the public sector (Millard, 2003). Public sector information (PSI) is a primary material for digital content products and services and will become an even more important content resource with the expected development of wireless content services. The electronic government can transform old public sector organizations and provide faster, more responsive services (Centeno, van Bavel, & Burgelman, 2004). It thus can increase efficiency, cut costs, and speed up standard administrative processes for citizens and business in multiple instances. The challenge for administrations is to adapt quickly to the new methods and to enable new innovative ways of working, including partnerships with the private sector (Cap Gemini Ernst & Young, 2004). In such a context, the promotion of cultural influence and cultural perspectives can contribute to the development of a great variety of applications (especially in the field of arts and tourism, as such information usually is gathered and published by public sector bodies at different levels of government). Learning how to manage and exploit all relevant data produced and stored creates a very high level of public value, which is probably an unavoidable step toward a future user-centered government (European Parliament and Council of the European Union, 2003b).

The European Commission, in cooperation with member states, the private sector, and regional authorities, will promote appropriate e-services in order to support usage and disposal of user-friendly public information (Millard, 2003). Such services will be deployed and built on interoperable interfaces, will use broadband communication, and will be accessible from all types of digital terminals (Chochliouros & Spiliopoulou-Chochliourou, 2005a). However, a specific framework for the conditions governing reuse of public sector documents is needed in order to ensure fair, proportionate, and nondiscriminatory terms of usage. In fact, ensuring that such conditions are clear and publicly available is a prerequisite for the development of an EU-wide information market (European Parliament and Council of the European Union, 2003b). As government services and important public information become increasingly available online (Punie, Burgelman, & Bogdanowicz, 2002), it will ensure that access to government Web sites becomes a significantly important issue (Department of Economic and Social Affairs of the UN, 2003). Public sector Web sites and their contents have to be accessible in order to ensure that all citizens can access information and take full advantage of the underlying potential (Council of the European Union, 2002b). Making public all generally available documents held by the relevant (state/regional/local) authorities, organizations, and institutions is a fundamental instrument for extending the right to knowledge, which is a basic principle of contemporary democracy (Millard, 2003). Services offered for such purposes should be simple to use, transparent, and useful (Clements et al., 2003). They should inspire trust and confidence, and create value (e.g., via adaptation and translation).

Creating and Distributing European Content: Options for Extending Cultural Perspectives

The importance of digitizing Europe's cultural heritage already was recognized by the European Commission. This gave rise in 2001 to the Lund Principles and the corresponding Lund action plan (*eEurope* Action Plan, 2001), and to the creation of a National Representatives Group on digitization (Council of the European Union, 2002c). The imperative respect for cultural diversity underpins and sustains the Europe of culture in accordance with the principle of subsidiarity.

A key challenge to the European content industries is to exploit fully the opportunities created by the advent of digital technologies. Support for digitization of production and distribution of digital content is, therefore, more than simply essential (European Commission, 2003b). Content has been raised as the main ally of ICTs in the coming years. Among content, it is worth mentioning commercial content, free content, peer-to-peer content, content developed by communities of interest, public content, and self-creative content. A deliberate strategy to address these possibilities will overcome the dichotomy between industrial and free productions (Chochliouros & Spiliopoulou-Chochliourou, 2005b). Some of the drivers for content production can be entertainment, education (lifelong learning, training, public [digital] libraries, open universities), health, and the sciences. Some of the challenges ahead are the consideration of marketing as a key issue to sell content; the necessity to "involve" everybody, including older and disabled people; fostering new business models; and developing interfaces to make information broadly available (in language and format).

Europe has a strong base on which it can build a dynamic digital content industry—a long-established print publishing sector and extensive cultural heritage and linguistic variety that can be exploited as well as a significant, growing audiovisual sector (European Commission, 2001b). Digital content plays an important role in this evolution (European Parliament and Council of the European Union, 2005). Content production also has given rise to rapid job creation in recent years (MacLean & Davis, 1999) and continues to do so. Content industries thus can create significant added value by exploiting and networking the European cultural diversity (Mundy, 2000). (These industries are nowadays a fast-growing segment of the wider EU information economy). In fact, diversity is the engine of creativity, innovation, growth, and variety, which can trigger additional wealth (Peacock & Rizzo, 1994). Entrepreneurs, investors, and young people are the first among those who will contribute to the creation of this valuable approach. Europe's internal diversity (her unique base of world spoken languages and of outreach cultures) is a potential lever of global business networks and of creative responses to the new trade challenges that remain in short supply of crafty multicultural skills (Jameson & Miyoshi, 1998).

Furthermore, a matter of high importance focuses on the abilities of those industries to ensure an adequate quality in the content for the new (electronic) media by combining particular features such as artistic freedom and power, creativeness, innovation, and pluralism (European Parliament and Council of the European Union, 1989). This is a basic policy challenge, as audiovisual and multimedia content are fundamental driving forces for the success of novel technologies in general and of broadband in particular (European Commission, 2002a). The audiovisual sector essentially has a cultural dimension and constitutes not only an expression of creativity (particularly of identities) and a fundamental means of promoting democracy but also an economic activity of growing importance that still is destined to evolve both quantitatively and qualitatively (Council of the European Union, 2002a). Relevant positive impacts in cultural terms can contribute to a deeper understanding of Europe's cultural multiplicity and richness and to a wider acceptance of the European integration process.

Moreover, Europe has more than 100,000 cultural institutions (museums, libraries, and archives) employing more than a million people (Fuegi & Jennings, 2004). They contain a wealth of information, which can be made more accessible and more effectively exploited. The degree of access to this information determines how far people can experience their cultural heritage and benefit from it in their work or studies. By digitizing their collections and making them available online, libraries, archives, and museums can reach out to citizens and make it easier for them to access material from the past (Bower & Roberts, 1995). In any case, the digital heritage will become (as far as possible) and remain accessible to the public, while the digital heritage materials, especially those in the public domain, shall be free of "unreasonable" restrictions (Mannerheim, 2000). Among the suggested actions is the uptake of new technologies for the creation of new forms of (electronic) content; the digitization of materials, ensuring lasting accessibility; and the development of modern services. In addition, digitized material can be a key asset for educational purposes and can enrich European cultural life (Pung, Clarke, & Patten, 2004). Other issues, such as socioeconomic disparities between regions and groups of citizens, also will be addressed. In terms of electronic inclusion, a multi-channel approach has to be taken into account in order to render the services available to citizens and enterprises (Punie et al., 2002) through several different communication means (kiosks, Web-TV, mobile connectivity, etc.). Other important objectives include stronger support and closer cooperation between educational communities and the content industry, with consequent mobilization of material and immaterial resources (Throsby, 2001).

A parallel major issue focuses on the creation of a "virtual European library," aiming to make Europe's cultural and scientific record accessible to all (European Commission, 2005b). It deals with the digitization, online accessibility, and digital preservation of the rich European cultural values. It combines multicultural and multilingual environments with technological advances and new business models. Digital libraries are "organized collections" of digital content that are made available to the public. They consist of material that has been digitized, such as digital copies of books and other physical material from libraries and archives; alternatively, they can be based on information originally produced in digital format. This is increasingly the case in the area of scientific information in which digital publications and enormous quantities of information are stored in digital repositories. Three main strands now are adopted to realize widespread and easy access to information: (a) online accessibility (i.e., a prerequisite for maximizing the benefits that citizens and companies can draw from the information); (b) digitization of analogue collections for their wider use in the globalization context of information society; and (c) preservation and storage to ensure that future generations can access the digital material and to prevent precious content from being lost.

Special attention will be given to support the creation of a framework supportive of the commercial exploitation of PSI and the development of multilingual services, thus encouraging the expansion, distribution, and support of European audiovisual works and of multimedia applications/products, and promoting the dissemination of live cultural events over the Internet (European Commission, 2004c). The Internet offers positive benefits by empowering consumers, lowering the barriers to the creation and distribution of content, strengthening education, and offering wider access to even richer sources of digital information. Currently, there is a clear need for coordination within the safer Internet field, both on the national and the European level. The coverage of safer Internet use (European Parliament and Council of the European Union, 1999) will be extended, primarily with the aim of improving the protection of children and minors, to new online technologies (including mobile and broadband content), online games, peer-to-peer file transfer, text and enhanced messages, and all forms of real-time communications (e.g., chat rooms, instant messages). Moreover, intensified action will be performed to ensure that areas of illegal and harmful content and conduct of concern are covered, with an emphasis on crimes against children (such as child pornography and trafficking in children) and on racism, violence, ethnic and/or xenophobic ideas. The involvement of all relevant actors (especially a greater number of content providers in the different sectors) is a crucial factor for further success. To this aim, promotion of industry self-regulation (Shoniregun, Chochliouros, Laperche, Logvynovskiy, & Spiliopoulou-Chochliourou, 2004) and content-monitoring schemes, development of filtering tools and rating systems provided by the industry, and increased awareness of industry services (as well as fostering international cooperation between all parties concerned) plays a fundamental role in consolidating a safer environment and contributing to removing probable obstacles.

Greater coordination of digitization programs across Europe will be assured to guarantee broader access to Europe's common heritage (European Parliament and Council of the European Union, 2003a). A key challenge is the need to improve digitization techniques in order to make digitization more cost-efficient and affordable. Cultural and intellectual assets of our society, which are created, usable, and available in digital form and which form the memory of tomorrow, are dependent on rapidly changing technologies and on fragile media, and are widely distributed geographically; these assets are, therefore, at great risk of being irremediably lost unless positive measures are taken to preserve them and to keep them available for the future.

However, before implementing any relevant effort, several factors have to be examined and/or analysed very carefully in order to avoid inconveniences and to promote effective implementation, offering real benefits. Some among these may be as follows:

1. The digitalisation of cultural goods is not always "even" or "analogue" for all cases, especially if taking into account the latest enlargement of the EU with 10 new member states. Existing differences between the European countries at various levels (such as the penetration of [broadband] technologies and the maturity of markets to support liberalization, competition, and new commercial practices) may limit industry's ability to develop global European products.

2. Any "lack" of clarity and homogeneity in rules on access and exploitation of PSI can affect further development (European Commission, 1998). As already explained in a previous section, PSI may be important in many cases in a way to facilitate development and to provide opportunities for growth through modern business schemes (Cap Gemini Ernst & Young, 2004).

3. The full implementation of modern regulatory measures and practices, to fulfill the multiple requirements of a competitive and liberalized market, is not always easy. This also implies the need to support basic issues such as matters for sufficient security and protection of content through electronic communications, matters relevant to intellectual property rights (IPR) (European Parliament and Council of the European Union, 2001), digital rights management (DRM), access to content (Chochliouros & Spiliopoulou-Chochliourou, 2005b), and so forth.

4. Insufficient linguistic and cultural customization of digital content are quite important issues. Such customization can help European companies to establish a global presence and exploit new markets, thus extending their business activities (European Commission, 2005a).

5. Appropriate cooperational schemes between educational (and cultural) institutions, communities, and the content industry can facilitate offering of modern applications (and/or products) through the development of appropriate business partnerships between market players (Cap Gemini Ernst & Young, 2004).

6. The "nature" of content is changing rapidly due to the expansion of modern electronic communications, thus promoting convergence effects. Consequently, content has to be enriched with a variety of features in a way to be accessible by various networks and/or equipments in the context of combined services/applications offered. This makes absolutely necessary the inclusion of diverse additional characteristics in order to follow up miscellaneous requirements and preferences (Rifkin, 2004).

7. Market responses from content distribution and availability vary enormously. Content now has become an essential ingredient in stimulating online sales.

At the same time, developments in particular areas, such as in mobile Internet access and the increasing importance of mobile e-commerce, mean that content providers will have to adapt their products to new access devices (Council of the European Union, 2003).

Overview and Conclusion

Information technologies have the potential to make Europe's cultural and scientific heritage visible and available for present and future use (European Commission, 2004d). In the context of recent initiatives, their role in promoting culture needs to be explored and further encouraged. Once appropriately digitized, Europe's cultural heritage can be a driver of networked traffic. It will be a rich source of raw material to be reused for added value services and products in sectors such as education, tourism, and entertainment (Peacock, 1994). If properly preserved, the material can be used time and time again. Furthermore, digitization efforts will have considerable spinoffs for firms developing new technologies.

Europe has encouraged various practical policies and applied measures to exploit the potential of the digital economy, especially to deliver growth and innovative online public services. It has so implemented proactive plans from the earlier *e*Europe policies to the latest i2010 initiative, which constitutes a core issue of the renewed competitiveness strategy. The strategy set aimed at preparing the transition of the EU to a knowledge-based economy and society and increased digitization and use of Internet access (especially in public service and in cultural institutions). Reinforcing social, economic, and territorial cohesion by making relevant services (and products) more accessible, including in regions lagging behind, is an economic, social, ethical, and political imperative.

Convergence of media and information facilities, networks, and devices provide unique opportunities for the manipulation of digitally available content: for firms to modernize their business processes and deliver a wide range of services; for consumers to experience a range of new media-content services; and for governments to offer efficient, modern, interactive public services online. European society and the economy as a whole are increasingly dependent on digital information, while its archiving will be essential in the future to provide a comprehensive view of European development and collections. In a global competitive environment, where relevant (business) actions take place, it will be important to strengthen the market position of European interactive media content producers and to support distribution and marketing of content produced and disposed (Punie, 2004). Furthermore, turning Europe's historic and cultural heritage into digital content will make it usable for European citizens for their studies, work, or leisure, and will give innovators, artists, and entrepreneurs the raw material that they need. However, access for all has to be ensured, both on the supply and the demand sides, through proper EU regulatory measures and practices. New services, applications, and content will create new markets, reduce costs, and provide the means to increase productivity and, hence, growth and employment throughout the economy.

In such a context, digital libraries can contribute significantly to contemporary European policies; in particular, to the areas of the information society, multilingualism, and cultural promotion. Private involvement and public/private partnerships are a key element in achieving this goal. The need to continue to develop methods and guidelines for long-term preservation of these records, documents, collections, and archives is absolutely essential for safeguarding the heritage of Europe (Council of the European Union, 2002c). Other specific initiatives can help the way forward (e.g., a European search engine, a Content Clearing House, or an EU Cultural Portal). In fact, the public sector collects, produces, reproduces, and disseminates a wide range of information in many areas of activities (e.g., social, economic, geographical, weather, tourist, business, patent, and educational information) that have to be made widely available at all levels through a diverse range of multi-modal electronic communications applications and mainly the Internet (Cap Gemini Ernst & Young, 2004). These can facilitate the creation of European community-wide information products based on public sector documents in order to enhance an effective cross-border use of relevant data by private companies for added value services in the market(s) (Centeno et al., 2004). The common dimensions and mutual knowledge of cultures in Europe in a society based on knowledge, freedom, democracy, solidarity, and respect for diversity, are essential components of citizens' support for and participation in European integration.

Acknowledgments

The major author of the present work, Dr. Ioannis P. Chochliouros, would like to express his profound gratitude to the co-authors for their valuable contributions for the full completion of the exposed work. Furthermore, both Dr. Ioannis P. Chochliouros and Dr. Stegios P. Chochliouros would like to dedicate the present effort to the memory of their father Panagiotis, who was always an active inspiration for their activities.

References

Bower, J., & Roberts, A. (1995). *Developments in museum and cultural heritage information standards.* The International Committee for Documentation of the International Council of Museums (ICOM-CIDOC). Retrieved July 14, 2005, from http://www.cidoc.icom.org/stand1.htm

Cap Gemini Ernst & Young. (2004). *Report on online availability of public services. How is Europe processing?* Brussels, Belgium: DG Information Society of the European Commission.

Centeno, C., van Bavel, R., & Burgelman, J.-C. (Institute for Prospective Technological Studies [ITPS]). (2004). *eGovernment in the EU in the next decade: The vision and key challenges (technical report EUR 21376 EN).* Brussels, Belgium: Directorate General Joint Research Center (JRC) of the European Commission.

Chochliouros, I. P., & Spiliopoulou-Chochliourou, A. S. (2003a). Perspectives for achieving competition and development in the European information and communications technologies (ICT) markets. *The Journal of the Communications Network: TCN, 2*(3), 42-50.

Chochliouros, I. P., & Spiliopoulou-Chochliourou, A. S. (2003b). Innovative horizons for Europe: The new European telecom framework for the development of modern electronic networks and services. *The Journal of the Communications Network: TCN, 2*(4), 53-62.

Chochliouros, I. P., & Spiliopoulou-Chochliourou, A. S. (2003c). The challenge from the development of innovative broadband access services and infrastructures. In EURESCOM & VDE Verlag (Eds.), *EURESCOM SUMMIT 2003—Evolution of broadband services—Satisfying user and market needs* (pp. 221-229). Heidelberg, Germany: European Institute for Research and Strategic Studies in Telecommunications GmbH (EURESCOM).

Chochliouros, I. P., & Spiliopoulou-Chochliourou, A. S. (2005a). Broadband access in the European Union: An enabler for technical progress, business renewal and social development. *The International Journal of Infonomics: IJI, 1*, 5-21.

Chochliouros, I. P., & Spiliopoulou-Chochliourou, A. S. (2005b). Content regulatory aspects in broadcasting activities and in modern electronic communications. *The Journal of the Communications Network: TCN, 4*(2), 60-67.

Clements, B., Maghiros, I., Beslay, L., Centeno, C., Punie, Y., & Rodriguez, C. (2003). *Security and privacy for the citizen in the post-September 11 digital age: A prospective overview. Report to the European Parliament Committee on citizens' freedoms and rights, justice and home affairs (LIBE), IPTS-JRC, [EUR 2083 EN].* Brussels, Belgium: European Parliament.

Council of the European Union. (2002a). *Council resolution of 21 January 2002, on the development of the audiovisual sector [OJ C32, 05.02.2002]* (pp. 4-6). Brussels, Belgium: European Commission.

Council of the European Union. (2002b). *Council resolution of 25 March 2002, on the eEurope action plan 2002: Accessibility of public websites and their content [OJ C86, 10.04.2002].* (pp. 2-3). Brussels, Belgium: European Commission.

Council of the European Union. (2002c). *Council resolution of 25 June 2002, on preserving tomorrow's memory—Preserving digital content for future generations [OJ C162, 06.07.2002]* (pp. 2-3). Brussels, Belgium: European Commission.

Council of the European Union. (2003). *Resolution of 18 February 2003 on the implementation of the eEurope 2005 action plan [Official Journal (OJ) C48, 28.02.2003].* (pp. 2-9). Brussels, Belgium: European Council.

Department of Economic and Social Affairs of the UN. (2003). *E-government at the crossroads, world public sector report.* New York: United Nations.

DigiCULT Project. (2002). *Technological landscapes for tomorrow's cultural economy. Unlocking the value of cultural heritage. Full report.* Brussels, Belgium: European Commission, Directorate General for the Information Society. Retrieved June 15, 2005, from http://www.digicult.info/pages/report.php

Dutta, S., Paua, F., & Lanvin, B. (eds.). (2004). *The global information technology report 2003-2004: Readiness for the networked world.* New York: Oxford University Press.

eEurope Action Plan. (2001). *Action plan on coordination of digitization programmes and policies: Implementation framework for digitization coordination across Europe. Follow up of experts meeting, Lund, Sweden, 4 April 2001.* Brussels, Belgium: European Commission. Retrieved August 12, 2005, from http://www.cordis.lu/ist/digicult/lund_principles.htm

European Commission. (1998). *Green paper on public sector information: A key resource for Europe [COM(1998) 585, 20.01.1999].* Brussels, Belgium: European Commission.

European Commission. (2000). *Communication on eEurope 2002—An information society for all. Draft action plan, prepared by the European Commission for the European Council in Feira, 19-20 June 2000 [COM(2000) 330 final, 24.05.2000].* Brussels, Belgium: European Commission.

European Commission. (2001a). *Communication on impacts and priorities [COM(2001) 140 final, 13.03.2001].* Brussels, Belgium: European Commission.

European Commission. (2001b). *Communication on certain legal aspects related to cinematographic and other audiovisual works [COM(2001) 534 final, 26.09.2001, Official Journal (OJ) C43, 26.9.2002, pp.06-17]*. Brussels, Belgium: European Commission.

European Commission. (2001c). *Communication on the impact of the e-economy on European enterprises: Economic analysis and policy implications [COM(2001) 711 final, 29.11.2001]*. Brussels, Belgium: European Commission.

European Commission. (2002a). *Fourth report from the European commission to the council, the European parliament, the European economical and social committee and the committee of the regions, on the application of directive 89/552/EEC ("television without frontiers") [COM(2002) 778 final, 01.06.2003]*. Brussels, Belgium: European Commission.

European Commission. (2002b). *European tourism forum 2002—Brussels, 10th December 2002—Agenda 21—Sustainability in the European tourism sector—Background document*. Brussels, Belgium: European Commission. Retrieved August 15, 2005, from http://europa.eu.int/comm/enterprise/services/tourism/tourism_forum/outcome.htm

European Commission. (2003a). *Communication on electronic communications: The road to the knowledge economy [COM(2003) 65 final, 11.02.2003]*. Brussels, Belgium: European Commission.

European Commission. (2003b). *Investing in networks and knowledge for growth and jobs—Final report to the European council [COM(2003) 690 final/2, 21.11.2003]*. Brussels, Belgium: European Commission.

European Commission. (2004a). *Communication on delivering Lisbon—Reforms of the enlarged union [COM(2004) 29 final/2, 20.02.2004]*. Brussels, Belgium: European Commission.

European Commission. (2004b). *Communication on connecting Europe at high speed [COM(2004) 61, 03.02.2004]*. Brussels, Belgium: European Commission.

European Commission. (2004c). *Proposal for a recommendation of the European parliament and of the council, on the protection of minors and human dignity and the right of reply in relation to the competitiveness of the European audiovisual and information services industry [COM(2004) 341 final, 30.04.2004]*. Brussels, Belgium: European Commission.

European Commission. (2004d). *Communication on challenges on the European information society beyond 2005 [COM(2004) 757 final, 19.11.2004]*. Brussels, Belgium: European Commission.

European Commission. (2005a). *Communication on electronic communications: i2010—A European information society for growth and employment [COM(2005) 229 final, 01.06.2005]*. Brussels, Belgium: European Commission.

European Commission. (2005b). *Communication on i2010: Digital libraries [COM(2005) 465 final, 30.09.2005]*. Brussels, Belgium: European Commission.

European Parliament and Council of the European Union. (1989). *Directive 89/552/EEC of 3 October 1989 on the coordination of certain provisions laid down by law, regulation or administrative action in member states concerning the pursuit of television broadcasting activities (OJ L298, 17.10.1998, p.23) [original text as amended by the latest European Parliament and Council Directive 97/36/EC (OJ L202, 30.07.1997, pp.60-70)]*. Brussels, Belgium: European Parliament and Council of the European Union.

European Parliament and Council of the European Union. (1999). *Decision No.276/1999/EC of 25 January 1999, adopting a multiannual community action plan on promoting safer use of the Internet by combating illegal and harmful content on global networks (OJ L33, 06.02.1999)*. (pp. 1-11). Brussels, Belgium: European Parliament and Council of the European Union.

European Parliament and Council of the European Union. (2001). *Directive 2001/29/EC on the harmonization of certain aspects of copyright and related rights in the information society (OJ L167, 22.06.2001)*. (pp. 10-19). Brussels, Belgium: European Parliament and Council of the European Union.

European Parliament and Council of the European Union. (2003a). *Decision No.1151/2003/EC of 16 June 2003, amending Decision No 276/1999/EC adopting a multiannual community action plan on promoting safer use of the Internet by combating illegal and harmful content on global networks (OJ L162, 01.07.2003)*. (pp. 1-4). Brussels, Belgium: European Parliament and Council of the European Union.

European Parliament and Council of the European Union. (2003b). *Directive 2003/98/EC of 17 November 2003, on the re-use of public sector information (OJ L345, 31.12.2003)*. (pp. 90-96). Brussels, Belgium: European Parliament and Council of the European Union.

European Parliament and Council of the European Union. (2005). *Decision No456/2005/EC of 9 March 2005, establishing a multiannual community programme to make digital content in Europe more accessible, usable and exploitable (OJ L79, 24.03.2005)*. (pp. 1-8). Brussels, Belgium: European Parliament and Council of the European Union.

Fuegi, D., & Jennings, M. (2004). *International library statistics: Trends and commentary based on the libecon data*. Retrieved May 10, 2005, from http://www.libecon.org/pdf/International LibraryStatistic.pdf

Harvey, F. (1997). National cultural differences in theory and practice: Evaluating Hofstede's national cultural framework. *Information Technology & People, 10*(2), 132-146.

Jameson, F., & Miyoshi, M., (Eds.). (1998). *The cultures of globalization*. Durham: Duke UP.

Lalopoulos, G. K., Chochliouros, I. P., & Spiliopoulou-Chochliourou, A. S. (2005). Challenges and perspectives for Web-based applications in organizations. In M. Pagani (Ed.), *The encyclopedia of multimedia technology and networking* (pp. 82-88). Hershey, PA: Idea Group Reference.

MacLean, M., & Davis, B. H. (Eds.). (1999). *Time & bits: Managing digital continuity*. Los Angeles: Getty Trust Publications.

Mannerheim, J. (2000, August 13-18). The WWW and our digital heritage—The new preservation tasks of the library community. In *Proceedings of the 66th IFLA Council and General Conference,* Jerusalem, Israel. Retrieved July 21, 2005, from http://www.ifla.org/IV/ifla66/papers/158-157e.htm

McGuigan, J. (2005). Culture and the public sphere. *European Journal of Cultural Studies, 8*, 427-443.

Mehta, K. T., & Shah, V. (1997). Information revolution: Impact of technology on global workforce. *Journal of International Information Management, 6*(2), 85-94.

Millard, J. (2003). *e-public services in Europe: Past, present and future. Research findings and new challenges. Final paper.* Retrieved May 10, 2005, from ftp://ftp.cordis.lu/pub/ist/docs/epublic-services.pdf

Mundy, S. (2000). *Cultural policy. A short guide*. Strasbourg, France: Council for Cultural Co-operation, Council of Europe Publishing.

Peacock, A. (1994). *A future for the past: The political economy of heritage*. Edinburgh, UK: David Hume Institute.

Peacock, A., & Rizzo, I. (Eds.). (1994). *Cultural economics and cultural policies*. Dordrecht: Kluwer Academic Publishers.

Pung, C., Clarke, A., & Patten, L. (2004). Measuring the economic impact of the British Library. *New Review of Academic Librarianship, 10*(1), 79-102.

Punie, Y. (2004). Convergence and divergence in the new media landscape. *Proceedings of the Seminar on Threats to Pluralism—The Need for Measures at the European Level,* Organized by the European Parliament Committee on Citizens' Freedom and Rights, Justice and Home Affairs. Brussels, Belgium.

Punie, Y., Burgelman, J-C., & Bogdanowicz, M. (2002). The future of online media industries. Scenarios for 2005 and beyond. *The IPTS Report, 64,* 35-42.

Rifkin, J. (2004). *The European dream*. Cambridge: Policy Press.

Royal Danish Ministry for Foreign Affairs. (2000a). *Building a global community: Globalisation and the common good.* Copenhagen, Denmark: Royal Danish Ministry for Foreign Affairs.

Royal Danish Ministry for Foreign Affairs. (2000b). *The power of culture: The cultural dimension in development.* Copenhagen, Denmark: Royal Danish Ministry for Foreign Affairs.

Shoniregun, C. A., Chochliouros, I. P., Laperche, B., Logvynovskiy, O., & Spiliopoulou-Chochliourou, A. S. (Eds.). (2004). *Questioning the boundary issues of Internet security.* London: e-Centre for Infonomics.

Stone, A. (2000). Interactive TV's really big picture. *BusinessWeak online.* Retrieved August 10, 2005, from http://www.businessweek.com/bwdaily/dnflash/sep2000/nf2000097_655.htm

Streeten, P. (2000). Culture and sustainable development: Another perspective. In J. D. Wolfensohn, L. Dini, G. F. Bonetti, I. Johnson, & J. Martin-Brown (Eds.), *The international bank for reconstruction and development/THE WORLD BANK (2000): Culture counts. Financing, resources, and the economics of culture in sustainable development* (pp. 41-48). Florence, Italy: Washington, DC: World Bank.

Throsby, D. (2001). *Economics and culture.* Cambridge, UK: University of Cambridge Press.

Chapter VI

Understanding the Cultural Roots of India's Technology Development from Homi Bhabha's Post-Colonial Perspective

Ramesh Subramanian
Quinnipiac University, USA

Abstract

This chapter introduces Homi K. Bhabha's post-colonial social theory of interstitial perspective and then discusses the application or overlay of the constructs that emanate from that to the roots of India's technology and (subsequent) IT development and its complementary effect in shaping Indian Information Systems professionals. The chapter spotlights various events and persona in India's history, including the current crop of IT professionals emerging from the subcontinent. It then overlays Homi Bhabha's constructs to verify that the constructs do, indeed, apply to India's developments in the IT arena. The chapter thus aims to offer a cultural and social-theory viewpoint with philosophical underpinnings to explain the roots and current happenings in the field of IT in India.

Introduction

Currently, Indian IT companies provide a plethora of services across the value chain to global IT markets. These companies are beneficiaries of various trends, some global and some local. Global trends include the rise and acceptance of globalization and outsourcing (especially IT outsourcing) by companies and governmental agencies in industrialized nations. Local trends include the availability of large numbers of skilled IT professionals in India, their proficiency in the English language, and the generally lower salaries in India. The 2005 McKinsey-Nasscom report indicates that India currently accounts for 65% of the global offshore IT industry and 46% of the global BPO (Business Process Outsourcing) industry. The report further predicts that India IT and BPO industries will continue to grow more than 25% annually and generate revenues ranging from US$60-80 billion by the year 2010 (Nasscom, 2005). Given this trend, it is not surprising that several recent studies have focused on the strategic and practical aspects of outsourcing to India (and countries like India), the growth of Indian BPO firms, the merits of offshoring to India, and so forth. Studies by Dedrick and Kraemer (2005), Parthasarathy (2003), and Kapur and Ramamurthi (2001) are representative of such research.

The increasing presence of India in the global IT arena has resulted in an increasing number of Indian IT professionals engaged either in providing service at customer locations or in the process of transferring knowledge from the customer's location to Indian locations (so that various business processes can be carried out of India). Further, there has been a steady growth of Indian-led companies, especially in the U.S., that act as via-media between U.S. companies offshoring their IS/IT operations and Indian vendor companies located in India. These, in turn, have increased awareness in North America and Europe of India and its large, technically-proficient, inexpensive, English-speaking IT workforce. There is a general acceptance among companies in North America and Europe as well as their customers that customer service most likely would be provided by an Indian IT professional working in India. The Indian IT professional's presence thus is becoming ubiquitous in the global IT arena.

There is general agreement in Western media, echoed by the leaders of the IT industry in the U.S., on the high level of technical and professional capability of Indian IT professionals. It is apparent even to a casual observer that these professionals are globe trotters, successfully engaged in complex IT projects in a variety of problem domains in organizations spread around the globe. What has not received much analysis in popular media or academic literature is the Indian IT professional's psyche—the apparent ease with which he or she is able to straddle different cultures, fit seamlessly into alien corporate cultures, work on international teams, learn new languages, change accents, and generally get along, while maintaining a vigorous personal and cultural agenda.

Given this increased visibility of India and the Indian IT professional in today's corporate environment, it would be useful to learn more about the factors that have shaped the development of IT in India and the emergence of the globally functioning Indian IT professional. Such a study could focus on questions such as the following:

- How has the sociopolitical culture of India evolving over the last two centuries affected India's growth in technology, especially information technology?
- How and to what extent has that culture influenced Indian intelligentsia, Indian politicians, and Indian IT professionals, who, in turn, cyclically influence technology development?
- What are the current trends and what are likely to emerge in the future in this arena?

This chapter, therefore, is an attempt to look into these questions in detail. We address these questions by focusing on the cultural, psychological, and sociological underpinnings that have influenced the development of technology in colonial and post-colonial India, using a post-colonial social theory framework provided by Homi Bhabha, a noted post-colonial theorist. We hope that our analysis will provide a context for understanding India's IT developments, the historical factors that have played a role in those developments, and the sociocultural aspects that have shaped the present-day Indian IT professional. This chapter is thus an attempt to offer a new colonial-cultural perspective on the roots of IS and IT developments in India and to understand the Indian IT professional. This approach is unique in that it overlays historical developments with post-colonial theory.

What is Post-Colonial Theory?

Post-colonialism deals with many issues for societies that have undergone colonialism: the dilemmas of developing a national identity in the wake of colonial rule; the ways in which writers from colonized countries attempt to articulate and even celebrate their cultural identities and reclaim them from the colonizers. ... Post-colonialism is about dealing with the legacy of colonialism. Perhaps somewhat surprisingly the most prominent form this has taken to date has been in the cultural realm, especially in respect of identity politics and literary studies. Thus, the most common way the term has been used is in reference to a genre of writing and cultural politics, usually by the authors from the country which were previously colonized. All post-colonialist theorists admit that colonialism continues to affect the former colonies after political independence. (Wikipedia, 2006)

Homi K. Bhabha is a well-known post-colonial theorist. Bhabha's works (1990, 1994) offer a deep look at the finer nuances of the colonial experience and the ways in which it shapes the colonizer as well as the colonized subject. Bhabha's discourse offers a varied tapestry of positive values of the colonial experience, which seems to apply more accurately to the experience of India and to explain some of the developments of India in the technology spectrum.

After studying Bhabha's post-colonial constructs, it is apparent that at least some of the constructs can be applied directly to explain the phenomenon of India's IT development and the Indian IT professional. In the following, we briefly discuss Bhabha's constructs that we find relevant to this chapter. After that, we trace the relevant sections of India's history from the late 18th century to the latter part of the 20th century. We use historical vignettes to trace this history and then overlay the vignettes with Bhabha's constructs with a view to explaining the cultural history behind India's continuing technology development, especially information technology development.

Homi Bhabha's Interstitial Perspective

In his two important works, *Nation and Narration* (1990) and *Location of Culture* (1994), Bhabha argues against the tendency to treat third-world nations as if they were a homogeneous block. Instead, Bhabha argues that nations and cultures arise from the hybrid interactions of various counteracting national and cultural constituencies (Perloff, 1998). He refers to this as the interstitial perspective, because in this outlook, nations, the concept of nationness, and culture evolve in the interstitial spaces between and overlapping two cultures—the culture of the colonizer and the culture of the colonized subject.

It is in the emergence of the interstices—the overlap and displacement of domains of difference—that the intersubjective and collective experiences of nationness, community interest, or cultural value are negotiated. ... Terms of cultural engagement, whether antagonistic or affiliative, are produced performatively. The representation of difference must not be hastily read as the reflection of pre-given ethnic or cultural traits set in the fixed tablet of tradition. (Bhabha, 1994, p. 2)

Expanding on this basic notion, Bhabha then goes on to explain several constructs that inform this discussion of culture emanating from the interstitial spaces of the colonial experience. For this discussion we focus on four of these constructs: stereotype, mimicry, sly civility, and hybridity. As we will see, these constructs impact and help in the formation of a certain cultural trait in both the colonizer and the colonized subject.

Bhabha's interpretation of these constructs is based on his identification of the existence of a certain ambivalence in colonial dominance. This enunciation of the ambivalence in colonial dominance is considered one of Bhabha's most important contributions to post-colonial studies and is made clear as we look more closely into the constructs themselves.

Stereotype, Mimicry, Hybridity, and Sly Civility: The Colonizer and the Colonized

One of the instruments that the colonizer uses in ensuring continued dominance of the subject is by using the notion of fixity, or stereotype. By using a fixed notion of the kind of person the subject is or the culture (or lack thereof) that the subject represents, and by repeating this frequently, the colonizer is able to maintain his or her superiority over the colonized. However, Bhabha argues that the same colonizer also exposes his or her ambivalence in stereotyping, which he refers to as *splitting*. This split is explained by the following:

[T]he (Black) native is both savage and yet the most obedient and dignified of servants; ... he is mystical, primitive, simple-minded yet the most worldly and accomplished liar, and manipulator of social forces. (Bhabha, 1994, p. 82)

Bhabha explains this as the moment of ambivalence of the colonial fantasy; that is, on the one hand, the colonized can be reformed, whereas on the other, the colonized can never be reformed and, hence, has to be governed. The more important implication is that this ambivalence can be used by the native to subvert the power relationship between the colonizer and the native.

According to Lacan, the objective of mimicry is to camouflage rather than to harmonize with the background. "It is not a question of harmonizing with the background, but against a mottled background, of becoming mottled" (Bhabha, 1994, p. 85).

According to Edward Said, another post-colonial thinker, one of the major conflicts in colonial discourse is between the forces of domination, stability, and stasis vs. changes that occur over time and mimicry. Expanding on this, Bhabha states that colonial mimicry is the desire for a reformed, recognizable subject, while at the same time making sure that the change is not disruptive enough (meaning that it should not be a complete change to parallel the colonizer himself or herself).

Parallel to this is the notion of hybridity or in-betweenness, which is developed by the subject to counter the effects of colonization or dominance. The subject develops a hybrid personality that mimics the role that is required to be performed by him or her. According to Rajan (1998), Bhabha's notion of hybridity suggests that "hybridity is the permanent subversion … that there will always be hybridity because there will always be a program, always an excess of the actual over the programmed, and always an indeterminate area of spillage and proliferation arising from the program and the excess. Hybridity is normal because resistance is unavoidable."

Bhabha also introduces the notion of sly civility that is ultimately subversive in a passive-resistant sort of way. It is exhibited by the colonized subject when exhorted into doing something by the colonizer.

To illustrate, Bhabha gives the example of a conversation between native catechists who were trying to convert some Indian peasants to Christianity in the early 19th century. The peasants, who were asked to comment on the Bible, maintained a sly (but ultimately subvertive) civility by stating that the words in it were beautiful, but "your priests are a non-vegetarian class. We cannot believe that anybody who eats meat can transmit the word of God."

Bhabha's argument is that at least in the case of the English colonial experience, the colonizer was filled with a certain ambivalence with regard to the exercise of power vs. leniency and the desire to control and rule in order to change vs. the desire to maintain a stasis. Subsequently, this ambivalence was used and exploited by the native for his or her own benefit.

Having briefly discussed the interstitial perspective of Bhabha, we now change track and peer into the history of some specific events pertaining to India from colonial times to post-independence.

Applying Homi Bhabha's Concepts to the History of India's Technology Development: Vignettes from History

We present next some vignettes from the history of India from the late 18th century to the mid-20th century. These vignettes are connected either directly or indirectly to technology development in India. In some cases, the connection to technology, in general, and information technology, in particular, is apparent. In other cases, they are not that apparent but provide a sociocultural context in which technology developments took place. While we refer to technology in a general sense, it is to be noted that much of the technologies were clearly related to India's need to develop electronics and computing technologies for the purposes of assisting and enhancing national development programs.

Vignette 1: Colonial Education Policy in India

India officially came under British rule in 1757, when Robert Clive led the forces of the East India Company and defeated the army of Siraj-ud-daulah, the Nawab (Muslim equivalent of the King) of Bengal (SSCNet, n.d.). The Charter Act of 1813 decreed that English would be taught in the Indian education system. The idea was to introduce European knowledge to the Indians to counterbalance Indian customs and traits that were considered inferior. The British rulers hoped to enforce the notion of moral superiority of the Western colonists through the use of the English language. Language became an instrument of subjugation. In 1835, Lord William Bentninck issued a New Education Policy, which decreed that English should be the official language of the courts, diplomacy, and education (Doyle, 1998). (Until then, Persian was accepted as the official language of diplomacy, even though there was a multitude of languages spoken by the people of the Indian subcontinent). This meant that any Indian who aspired to a job in the business, legal, and administrative sectors needed to possess knowledge of the English language. This led aspiring Indians to take up the English language, aided by the establishment of examining universities in Calcutta, Bombay, and Madras in 1854. These universities were modeled after London University. In 1857, the University of Calcutta officially was formed under the University Act of 1857 during the administration of Lord Canning. The basic objectives of the university were to "encourage Her Majesty's subjects of all classes and denominations, 'in the pursuit of a regular and liberal course of education,' and at the same time ascertain by means of examination the proficiency of all persons in different branches of Literature, Science and Art. Such persons would be rewarded by Academic Degrees as evidence of their attainments" (Banerjee, 1957, p. 127). The British rulers were very interested in mapping and surveying the Indian subcontinent with a view to determining its natural characteristics and resources. To further this end, technical programs were started even before the University Act. The earliest technical institute outside of Europe was established in Madras (the Madras Survey School) in 1794 (Muthiah, 2003). The idea ostensibly was to train the locals, who then would assist the British surveyors in mapping India. The colonial education policies proved to be a boon to aspiring Indians, many of whom rapidly gained proficiency, especially in the sciences. We will look at several of these Indian technology luminaries in later sections. The most prominent among them is arguably C. V. Raman, who eventually became the first Asian to be awarded the Noble Prize in Physics in 1930. Raman received his BA and MA degrees in physics from the Presidency College, Madras, and later conducted his research at the Indian Association for the Cultivation of Science (IACS), Calcutta, and at Calcutta University as the Palit Chair in Physics. It is worth noting that Raman as well as the founder of IACS and the vice-chancellor of Calcutta University were all Indian natives educated under the colonial education system (INSA, n.d.).

Analysis

The introduction of the English language coupled with Western scientific education brought about unintended consequences to the British. It heralded a renaissance period among Indians of all classes and religions. It enabled Indians belonging to various classes and religions and speaking different languages to be educated in Western science and technology and to communicate with each other. This communication eventually led to collective opposition of the British rule, culminating in independence to India. This period also saw the emergence of Indian scientists and technologists who used British education in India as a stepping stone to pursue further higher education in Cambridge and Oxford universities. Notable Indian scientists of the time were Sir C. V. Raman and J. C. Bose, who went on to develop and lead important educational and scientific institutions in India. These and other individuals developed a hybrid personality that allowed them to co-exist comfortably in traditional Indian environments as well as Western scientific environments. We see the existence of colonial ambivalence in the British rulers who, one the one hand, wanted to subjugate the Indians and prove their intellectual superiority by introducing them to Western education and then allowing the Indians eventually to espouse liberal ideas such as individual freedom and self-identity, which then enabled them to organize and fight for independence. The British rulers were hoping that their Indian subjects would maintain stasis through mimicry, but instead, they engendered a generation of intelligentsia who exhibited great facility in a hybrid existence.

Vignette 2: Jawaharlal Nehru and the Socialist Mindset

While M. K. (Mahatma) Gandhi is widely regarded as the father of independent India, Jawaharlal Nehru, the first Prime Minister of India, can be considered the real architect of modern technology-proficient India. Nehru was born to a privileged class. His father, Motilal Nehru, was a successful lawyer during the British rule. As most children from wealthy, privileged families at the time of British rule, Jawaharlal was educated in England Harrow and later at Cambridge, where he studied natural sciences. During this time, he developed a fascination for technology and its varied uses. After his return to India from England in 1912, he was energized into action by the Freedom movement led by Gandhi and joined him in India's freedom struggle. He also soon became one of the persuasive and eloquent leaders of the Congress party.

With independence achieved in August 1947, Nehru became the first prime minister of India. At that time, poverty was the main problem in India. Nehru wanted India to become strong and self-reliant. Nehru had been imprisoned for several stretches of time during the freedom struggle against British rule. While in prison, Nehru started reading and getting interested in Marxist philosophy. He also liked the Attlee Consensus, which was defined by Clement Richard Attlee of the Britain Labor Party as "a mixed economy developing toward socialism. … The doctrines of abundance, of full employment, and of social security require the transfer to public ownership of certain major economic forces and the planned control in the public interest of many other economic activities" (Yergin & Stanislaw, 1998). His focus was on self-reliance and alleviation of poverty in India. He liked the Soviet model of a planned economy and five-year plans and promoted those ideas with vigor. His idea was to integrate technology and industrialization with central planning. In doing this, Nehru became a strong cheerleader for technology-aided governmental planning and policymaking (Yergin & Stanislaw, 1998). Computers and computing clearly would play a large part of this process. Computing in India was heralded through the pioneering vision of P. C. Mahalanobis, a physicist turned statistician.

Analysis

Nehru is arguably the most hybridized individuals of India's colonial and post-colonial history. In his formative years, he followed his family's traditions of being the perfect colonial subject by immersing himself in Western thought, scientific education, and liberal Western views. His liberal views coupled with strong national views attracted him to Gandhi and the independence movement. The socialist movement and Marxist ideologies transformed him into a renaissance man, one who saw India as a country steeped in backward social norms and ideologies, whose growth could be achieved only through self-reliance in technology. Nehru was considered by the British as urbane and Western-oriented, which helped him during negotiations with the British. In this, he also practiced sly civility—behind his suave and urbane façade, Nehru held a fierce vision of a technologically proficient modern India.

Technological and Social Renaissance in Colonial and Post-Colonial India

The following four vignettes focus on the period just prior to independence and extend to the period a few decades after India's independence. They are used to showcase the growth of technology in post-colonial India and the main visionaries and drivers of those technologies—the scientists, engineers, and technocrats. We first show all the vignettes and then follow with a combined analysis using Bhabha's constructs.

Vignette 3: Mahalanobis and the Advent of Social Computing in India

Prasanta Chandra Mahalanobis was born to a well-to-do, progressive family in Bengal in 1893. His family was progressive because his grandfather had started a Hindu renaissance and reform movement called Bhramo Samaj, which introduced monotheistic concepts as well as ideas from Christianity, Judaism, and Islam into the interpretation of Hindu religious texts such as the *Upanishads*. Mahalanobis studied physics in Cambridge from 1913 to 1915. He also completed a *Tripos* in physics. A Tripos is a university examination in Cambridge University, England, the passing of which qualifies a candidate partly or wholly for admission to an Honors Degree (University of Cambridge, 2004). Just before Mahalanobis left for India for a vacation, he was introduced to a few volumes of *Biometrika* by his tutor W.H. Macaulay. *Biometrika* is a journal of statistics that publishes original theories and their applicability in practice. Mahalanobis was captivated by the articles he read, which turned his interest toward statistics.

Mahalanobis did not return to Cambridge, even though he had a research fellowship there. Instead, he joined the Presidency College in Calcutta as a professor of physics. His interest in statistics continued. His early mentor was Acharya N. B. Seal, Principal of Calcutta University. Seal asked Mahalanobis to help him in a statistical study of examination results at Calcutta University. He did several statistical studies over the next few years and set up a Statistical Laboratory in the Presidency College in the 1920s. In 1932, he founded the Indian Statistical Institute. First, the British government in India merely wanted Mahalanobis to train personnel in statistics. But Mahalanobis constantly argued for a greater role of statistics in national surveys and national planning efforts. In 1943, the great famine of Bengal struck. Mahalanobis was asked by the erstwhile government to conduct a statistical survey of the yield of paddy crops (Indian Statistical Institute, n.d.).

This study attracted the attention of Nehru to Mahalanobis. The use and importance of statistics to undertake national surveys for central planning became apparent to Nehru, who himself was educated in the sciences. After independence, Nehru asked Mahalanobis to prepare a mathematical model for India's first Five Year Plan.

Vignette 4: The Technological Vision of the Entrepreneur J. N. Tata

The story of India's development in technology, including information technology, will not be complete without the story of Jamshedji Nusserwanji Tata (1839-1904), one of the Indian leaders who met at the formative meeting of the Indian National Congress in 1885 in Bombay. Tata Enterprises, which encompasses vast holdings in iron and steel, power utilities, and textiles, was founded by Tata, who accumulated a fortune from his textile mill at Nagpur. The Tata family, starting from J. N. Tata onwards, has played an astonishingly important role in the technological development of India.

The story of Tata Enterprises and the Tata family has been chronicled by Russi M. Lala (1981). Lala points out that in the 1880s, "the (Indian) intelligentsia realized that a lot of the wealth of India was being funneled to England, and the need to patronize Indian goods was important" (Lala, 1981). Tata, as owner of a textile mill in Nagpur, constantly had to tussle with the British, who tried their best to stifle Indian entrepreneurship and competition.

According to Lala, Tata's textile mill, named the Empress Mills to honor the day Queen Victoria was proclaimed Empress of India, opened on January 1, 1877. Yet, Tata really shared Indians' craving for self-rule from the British and also was critical of the impact of British rule on the Indian economy. Lala also chronicles that Tata was not only present at the first session of the Indian National Congress in Bombay but also supported it financially and remained its member all his life (Lala, 1981).

Tata saw the need to develop India's technical prowess much before India gained independence. Toward the end of the 19th century, Tata became convinced that the future progress of India depended crucially on research in science and engineering. His astounding foresight can be seen in his important role in founding three of India's greatest technological institutions: the Indian Institute of Science (IISc), the Bhabha Atomic Research Institute (BARC), and the Tata Institute of Fundamental Research (TIFR). All three institutes played a very important role in developing technology for governance, once India became independent.

The founding of the Indian Institute of Science (IISc) through the persistent efforts of Tata in 1911 is probably the most important milestone in the story of India's homegrown technological development. The British rulers were at first disdainful of Tata's proposal. It is noted in the literature that Lord Curzon, viceroy-designate during the time, upon being presented with the proposal, apparently asked, "Where are the students qualified enough to enter such a university?" Tata eventually won the battle of wills and the institute started functioning in 1911 (Financial Express, 2004). From those ideological beginnings, the IISc has become one of the top technological institutes in the world today. It is regularly listed as one of the institutions that trains the top computing professionals in India.

Vignette 5: Homi Bhabha, J. R. D. Tata, and the Establishment of TIFR

Homi Jehangir Bhabha (not to be confused with Homi K. Bhabha, the post-colonial theorist) was one of the greatest scientists of modern India. He was born to a wealthy, aristocratic family in Bombay in 1909. He showed signs of genius at a very young age and went on to study mechanical engineering in Cambridge. However, he was fascinated by the work of his mathematics teacher Paul Dirac and started developing a keen interest in mathematics and theoretical physics. He started his research in 1930 in the Cavendish Laboratory in Cambridge and received his PhD in theoretical physics for his work in cosmic ray and elementary particle theory in 1935. By that time, he was already internationally famous in the field of theoretical physics (Fotadar, 2000). He stayed in Cambridge until 1939 and then left for India for a vacation. While in India, news broke out that World War II had started, and he was unable to return to Cambridge (Mausam, 2000). This proved to be especially beneficial to Indian science. In 1940, C. V. Raman, a Nobel Laureate in Physics, invited Bhabha to join the faculty of the Indian Institute of Science, where he joined as Reader in Physics.

Bhabha, aware of the importance of translating theory into experiments, began putting his knowledge into the practical study of cosmic rays. Realizing the need for an institute fully devoted to fundamental research, he helped with funds from J. R. D Tata (a member of the Tata family and then the head of Tata Group) and established the Tata Institute of Fundamental Research (TIFR) in Bombay in 1945 (Fotadar, 2000).

Vignette 6: Vikram Sarabhai, Rocketry and Space in Modern India

After independence in 1947, India focused all its energy in nation building, focused on economic and industrial development, fully understanding the key role of science and technology. Indian rocketry was born, thanks to the technological vision of Prime Minster Jawaharlal Nehru. Prof. Vikram Sarabhai took the challenge of realizing this dream (Bharat Rakshak, n.d.).

Vikram Sarabhai also had his science education in Cambridge and earned his Tripos in 1940. Upon his return to India, he worked under C. V. Raman at the IISc until 1945, when he returned to Cambridge to finish his PhD in 1947. In 1957, Russia launched Sputnik-I, which fired the imagination of Vikram Sarabhai (Parthasarathy, 2003).

With Nehru's support, INCOSPAR (the Indian Committee for Space Research) led by Sarabhai was set up in the late 1950s.

Analysis

From the previous, it is very apparent that India's leaders recognized from the beginning that the technologies discussed earlier would all converge in ways that would enable the government to monitor the state of the nation and to take adequate actions. The combination of high technology and computers focused on national development and self-reliance was the essence of their vision.

It is also very important to note the sociocultural background of Nehru as well as the scientists and entrepreneurs who became closely associated with him during India's formative years. An analysis of the profiles of the scientists Homi Bhabha, Vikram Sarabhai, and Mahalanobis, as well as the entreprenueurs who actively supported their endeavors (namely, J. N. Tata and J. R. D. Tata) reveals certain interesting commonalities.

They were all considered renaissance men of their times. They were all from successful families with good economic and educational backgrounds. Their families had become successful under British rule and were occupying various positions of importance under the British government. Almost all of them were educated in England, mostly at Cambridge University, in the sciences. All of them chose to return to India, fueled by the dream of Indian independence and the opportunity to contribute to nation building.

All of them shared a similar outlook toward religion. They believed that India was shackled by its Hindu religious norms and chose to follow, at various levels, forms of reform movements. Nehru often was outspoken in his views that Indians should rid themselves of the shackles of religion, which he felt was a main cause of India's backwardness.

It is also very interesting to note that almost all of the scientific and educational institutions that played such an important developmental role in India's formative years were started through active funding by private enterprise, mainly through entrepreneurs such as J. N. Tata and J. R. D. Tata. J. N. Tata and Motilal Nehru (father of Jawaharlal Nehru) were involved in the formative stages of India's freedom movement.

Vignette 7: Post-Independent India and IT Stagnation

Author's note: This and the next vignette are adapted from Subramanian (2005).

In 1947, India gained independence from British rule. The first prime minister of India as well as senior leaders, politicians, and scientists believed strongly that India's path to development should be achieved through self-reliance. Central planning, similar to that in the Soviet Union, was the approved form of governance and planning. The reasons for this were India's long years of subjugation by the British and manipulation by multi-national organizations that often supplied India with substandard or obsolete technologies. Indian leaders wanted to prove to the world that India could succeed on its own through self-reliance. Imports, especially technology imports, were regulated strictly under a *permit raj*, or rule of the permits. Imports of computers and other IT equipment were made excruciatingly difficult. Indian technocrats were forced to develop indigenous technologies, which often took long to develop and were below international standards. But this period was also a boon to engineers, especially software engineers, who learned the art and science of building computer technology from scratch. Software engineers learned to write extremely efficient code to eke out the best performance from scarce computing resources. The UNIX operating system became the standard in India's banking industry, thanks to the availability of nonproprietary strains of the Unix O/S that were available to Indian programmers. In the mid-1970s, Indian entrepreneurs started software service firms that adopted the modus operandi of sending Indian programmers to locations in the U.S. and UK to work in-house at client locations. These software engineers and the Indian companies that employed them thus gained valuable knowledge about international software services and standards. Indian IT professionals also realized that they comfortably could morph from one situation to another, from low-tech, underdeveloped infrastructures to high-tech, developed infrastructures. This dual existence and hybridity eventually would be recognized and made use of by firms in the Western world. This also provided the foundation for India's growth in the IT arena in the last decades of the 20th century.

Vignette 8: Economic Liberalization and the Rise of IT Entrepreneurs

In 1984, Prime Minister Indira Gandhi, daughter of Jawaharlal Nehru, was assassinated. Her son, Rajiv Gandhi was elected to become prime minister of India. Rajiv Gandhi, like his mother Indira Gandhi and grandfather Jawaharlal Nehru, was educated in the elite schools of India and UK. But unlike his mother and grandfather, Rajiv Gandhi was a hybridized individual, who believed that India should join the rest of the globe in trade and started a policy of economic liberalization and openness. This move also was propelled in part by India's weak balance of the payment situation at the time when Rajiv became the prime minister. Rajiv Gandhi favored computerization of all governmental functions. Even before becoming the prime minister, he had personally supported the government's technology arm, the National Informatics Center, in its bid to oversee computerization of the Asian Games held in India in 1982. After he became prime minister, economic liberalization took a solid hold. Import duties on computer equipment were eased substantially, and software exports were given high importance.

This started a trend of rapid development of the IT industry in India. Software houses were started by entrepreneurs, whose primary business model was to train and send Indian IT professionals to work on-site in North America and Europe. Some of these professionals left their parent companies to join firms in the U.S. and

UK. Some of them eventually played very important roles in formation of dot-com companies during the 1990s. Many started new companies in the US, and several others occupied very high positions in IT startups, no doubt due to their technical proficiency and their ability to function well in rapidly changing and foreign environments. The year-2000 (Y2K) problem added further impetus to this trend. Indian software service companies provided many more software engineers to the West in order to address the problem. Thus, India started to gain ground in the software service arena.

Software services became even more important after the global economy started a downward trend in 2000. Many companies in the U.S. and UK started outsourcing their software operations to India. Anything that could be done outside of the U.S. or UK was sent to India to be processed by software engineers working there. These software engineers often were paid much lower salaries than their counterparts in the U.S. and UK, and thus, it was argued, it made sense to outsource software service operations to Indian IT firms. N. R. Narayana Murthy, the Chairman of Infosys, a top IT company in India, is said to have noted, "Common wisdom says that India and America can never work together. When it is day for you, it is night for me. But uncommon wisdom says that in a day we get two shifts, (with) each person working one shift. That is a great advantage" (Sadagopan, 2004). Thus, the Indian IT professional became even more ubiquitous in global IT environments.

India's growth in IT is not without challenges. The first challenge is the great animosity and backlash that the perceived loss of jobs has created in the Western world. Indian software and BPO workers recently have had to endure verbal abuse and other threats, especially in the U.S. Politicians are in the process of enacting anti-outsourcing laws. The second challenge is that in India, the rapid growth has caused a severe shortage of trained IT professionals. In some cases, the growth in the number of very young, highly paid software and BPO workers has caused social upheavals and animosities among various segments of society. The night-only nature of BPO jobs has caused increased stress and breakdowns in families. Many cases of health problems and intra-family strife have been reported in the media. It would be interesting to see if Bhabha's constructs (i.e., hybridity, mimicry and sly civility) can combine with the overall resilience of the average Indian psyche to counteract and overcome these challenges of IT growth in India.

Conclusion: Validating Homi Bhabha and Current Challenges

It can be seen from the previous vignettes that Indian leaders had begun to actively prepare for the end of British rule by the turn of the 19th century. These were hybridized individuals who straddled the British or Western world and the Indian culture. They all seemed to comfortably embrace British values and British education and were at home in England and in India. They all had thriving intellectual associations with scientific and intellectual luminaries in Britain and in continental Europe. However, they were focused clearly on the issue of India's independence and cleverly (i.e., through a sly civility) used their associations to establish and nurture a strong technology infrastructure in the early stages of India's independence movement. They were very proficient in performing the mimicry that Bhabha describes. Indian leaders and intelligentsia were rooted very deeply in the notion of an independent India. In order to achieve that status, they became what the colonizer most fears: a person who is transformed to match the colonizer's ideals and who, therefore, seeks out and demands Western concepts such as liberation and freedom. It is also apparent that the colonized subjects practiced the act of sly civility with ease. J. N. Tata fashioned himself to be an industrialist and even named his mill after the Empress of England, while, at the same time, joined the Indian National Congress and planned for an independent India that was technologically self-sufficient. These trends have continued to this day. Indian IT professionals and technologists continue their characteristic of being extremely hybrid individuals who exist comfortably in the interstitial spaces between cultures and nations. Thus, based on our analysis of the history of India, it is safe to conclude that the constructs proposed by Homi Bhabha does provide one convincing explanation for the route and progress of technology and IT development in India. Further research is required on other colonized countries in order to see if parallels exist between them and India and whether Homi Bhabha's constructs could be used to explain their current state of IT development.

As India continues its impressive growth in the IT sector, this analysis also provides a vehicle for understanding more clearly the Indian psyche, the Indian IT professional, and challenges faced by him or her in today's increasingly global work environments.

References

Banerjee, P. N. et al. (1957). *Hundred years of the University of Calcutta*. Calcutta: University of Calcutta Press.

Bhabha, H. K. (1990). *Nation and narration*. Routledge.

Bhabha, H. K. (1994). *The location of culture*. Routledge.

Bharat, R. (n.d.). *History of the Indian space program—2*. Retrieved October 15, 2004, from http://www.bharat-rakshak.com/SPACE/space-history2.html

Caslin, S. (n.d.). Going native. *The Imperial Archive: Key concepts in post-colonial studies*. Retrieved November 29, 2005, from http://www.qub.ac.uk/en/imperial/key-concepts/Going-native.htm

Dedrick, J., & Kraemer, K. L. (1993). Information technology in India: The quest for self reliance. *Asian Survey, 33*(5), 463-492.

Doyle, T. (1998). Western education in nineteenth-century India. *The Imperial Archive Project*. Retrieved November 29, 2005, from http://www.qub.ac.uk/en/imperial/india/educate.htm.

Financial Express. (2004, August 15). The Jamshedji who trounced Lord Curzon. *Tata.com*. Retrieved October 15, 2004, from http://www.tata.com/tata_sons/media/20040815.htm.

Fotadar, U. (2000). Homi J. Bhabha: The Indian titan. *Sulekha.com*. Retrieved October 15, 2004, from http://www.sulekha.com/expressions/articledesc.asp?cid=86596

Indian National Science Academy. (n.d.). *The torch bearers of Indian renaissance*. Retrieved November 29, 2005 from www.iisc.ernet.in/insa/ch2.pdf

Indian Statistical Institute. (n.d.). *The road traversed: A brief history of the Indian Statistical Institute*. Retrieved October 15, 2004, from http://www.isical.ac.in/history.html

Kapur, D., & Ramamurthi, R. (2001). India's emerging competitiveness in tradable services. In *Proceedings of the Academy of Management Conference*.

Lala, R. M. (1981). *The creation of wealth: A tata story*. IBH Publishing Co.

Library of Congress. (1995). India—"Origins of the congress and the Muslim league" section. *Library of Congress Country Studies*. Retrieved August 8, 2005, from http://lcweb2.loc.gov/cgi-bin/query/r?frd/cstdy:@field(DOCID+in0025)

Mausam. (2000). Building scientific institutions in India: Saha and Bhabha—A report. *Mausam* (1997134). Retrieved October 15, 2004, from http://216.239.41.104/search?q=cache:CFqksnC7Z6EJ:www.cs.washington.edu/homes/mausam/hukka.ps+cavendish+lab+homi+bhabha&hl=en&start=7

Muthiah, S. (2003, November 24). Pioneers in technical education. *The Hindu*. Retrieved November 24, 2005, from http://www.hindu.com/thehindu/mp/2003/11/24/stories/2003112400230300.htm

NASSCOM. (2005). *The National Association of Software and Service Companies, India (NASSCOM)*. Retrieved February 24, 2006, from www.nasscom.org/download/Mckinsey_study_2005_Executive_summary.pdf

Parthasarathy, R. (2003, April 3). Vikram Sarabhai (1919-1971): Architect of Indian space programme. *The Hindu*. Retrieved October 15, 2004, from http://www.hindu.com/thehindu/seta/2003/04/03/stories/2003040300100300.htm

Perlhoff, M. (1998). *Cultural liminality/aesthetic closure?: The "interstitial perspective" of Homi Bhabha.* Retrieved October 25, 2004, from http://wings.buffalo.edu/epc/authors/perloff/bhabha.html

Rajan, B. (1998). Modern Philogogy v.95, n4 (May 1998): 490-500. *Reviews of the Location of Culture.* Retrieved October 25, 2004, from http://prelectur.stanford.edu/lecturers/bhabha/reviews.html#re

SSCnet. (n.d.). Battle of plassey. *History Politics.* Retrieved November 29, 2005, from http://www.sscnet.ucla.edu/southasia/History/British/Plassey.html

Subramanian, R. (2005). *India and information technology: An historical and phenomenological perspective.* (Submitted for publication, currently under review).

University of Cambridge. (2004). *Glossary of Cambridge terminology.* Author. Retrieved November 30, 2005, from http://www.cam.ac.uk/cambuniv/pubs/works/appendix1.html

Wikipedia. (2006). Post-colonialism. *Wikipedia: The Free Encyclopedia.* Retrieved February 22, 2006, from http://en.wikipedia.org/w/index.php?title=Post-colonialism&oldid=40610890

Yergin, D., & Stanislaw, J. (1998). *Commanding heights.* Free Press.

Chapter VII

Digital Culture and Sharing:
Theory and Practice of a Brazilian Cultural Public Policy

Saulo Faria Almeida Barretto, Research Institute for Information Technology, Brazil

Renata Piazalunga, Research Institute for Information Technology, Brazil

Dalton Martins, Research Institute for Information Technology, Brazil

Claudio Prado, Research Institute for Information Technology, Brazil

Célio Turino, Ministry of Culture, Brazil

Abstract

In this chapter, we present the digital culture project that has been developed in Brazil and supported by the Brazilian government, which determined that the process of digital inclusion should be seen fundamentally from a cultural perspective. The chapter started with the principle of sharing of information as the basis of the information society, thus provoking a reanalysis of the productive scenario in which users actively participated in the productive scenario instead of acting as mere consumers of the technically innovative process being carried out by the companies. We discussed the concept of cultural property in the context of this new society and presented the Brazilian vision of this scenario.

Copyright © 2007, Idea Group Inc. Copying or distributing in print or electronic forms without written permission of Idea Group Inc. is prohibited.

Introduction

The question is whether sharing is the beginning or the end of the process of re-evaluating culture in the context of the information society. Therefore, one should analyze the idea of a repositioning of the values related to the productive processes as a result of the influence of this concept. Is it possible to identify a movement of transformation in the concepts of work, value and property as the human productive process becomes collective, shared, and cooperative? From this investigation, implications arise such as the fact that human potential could become decentralized instead of being synthesized and tied to specific activities according to the logic of industrialization (e.g., confining a worker to a single machine carrying out a single action). In other words, all human potentiality becomes seen as an active element in the construction of the processes. It thus becomes possible to envision human activities being determined by each player as a function of his or her level of interaction, interest, motivation, and so forth, with the world that surrounds him or her. In other words, it is cultural action as an element of value and of a consequent generation of autonomy.

This has been the position adopted by the Brazilian Government's Ministry of Culture, which considered that the process of digital inclusion should be seen fundamentally from a cultural perspective. In this way, the Ministry of Culture appropriated not only the means of production but also the way in which production occurred, strengthening sharing as the principle of digital inclusion in Brazil. Such appropriation aimed to create conditions for the establishment of autonomy to the citizens, who thereby gained an active role in the scenario of digital production and not merely a role as spectators and/or users of a process in development. In this context, two structural pillars have been seen as fundamental to the exercise of protagonism: the specification of a multimedia kit and the usage of free software. Both pillars performed essential roles in this process. Through the language of text, audio, video graphic images, and software, cyberspace has the potential to promote, among other things, an ample notion of citizenship, culture, education, and political activism.

Based on the aforementioned, this chapter presents the main digital culture actions developed in Brazil, with the support of the Brazilian government, as part of a program called Living Culture (*Cultura Viva*). These included the concept of cultural hotspots; a description of the multimedia kit and its relevance to the project; the training program for content producers; the creation of conditions for autonomy and dissemination of the project's philosophy; the models of intellectual property adopted; the environments created to motivate collaboration, publication and sharing among the members of the project's network; and the results obtained up to now.

Fundamental Principle: Sharing

The principles of exchange, sharing, and cooperation in the conception, design, and implementation of ideas have accompanied human evolution as a strategy and methodology for processes of technological innovation from primordial times to the present age of digital communication networks. Using as a basis the sharing of ideas and the building of knowledge, the social validation of this information has established networks of cultural propagation and intellectual appropriation, which demonstrate their potential through processes of cooperation and transfer of information through information technology.

The network of relationships opens channels in which information flows and enables us to interrelate socially. This is the key element allowing us to understand how the sharing of information and the cognitive processes derived from it occur. A network consolidates a cognitive dimension, given that it is constituted as a living, dynamic element, reconfiguring itself constantly through the various actions that its players exert on "that which they receive from others" (Lévy, 1993, p. 56). These actions, the basis of the process of cultural production and technological innovation, essentially are fed by sharing information disseminated by the most diverse channels of communication and, principally, by subsidizing the building of an interacting collective intelligence.

The semantic structure of the network, a cartography in constant evolution from the connections between the points in the network, is connected intimately to the technologies that endorse symbolic production and a base support for information sharing. Examples are so vast that they include signs painted on prehistoric walls to peer-to-peer file-sharing systems. We thus can understand culture as a means of exchange, where the relationships established and defined on the network are based on a policy of rights resulting from the exchanges carried out (Myers, 2005). Through policies of rights, moderation levels are defined as well as the permission of use and cultural dissemination, thereby restricting possibilities for articulating and sharing information in a network.

To speak of the sharing of information in the information society involves reflecting on how this process occurs. It is important to reflect on the determining role of technology, giving full potential to the process and the motivations that lead to the sharing of information on the Web through its cultural dimension.

As Castells (1999) said, what characterizes the current technological revolution is the application of knowledge and information for the generation of knowledge and mechanisms for the processing/communication of information. The result is a very close relationship between the social processes of creation and manipulation of symbols (the culture of the society) and the ability to produce and distribute goods and services (the productive forces) (Castells, 1999).

Here we are faced with what Pierre Lévy describes as the information economy (Perret, 2002), where capital consists of the ideas in circulation, making sense of their acquisition from the moment when communities interact and people think together. This information economy establishes itself through a triad of values formed by technology as a structural support for the construction of ideas (e.g., cables, computers, printers), by the cultural aspect represented by knowledge registered as a sign, and by the social aspect of the relationship between people, which guarantee the circulation of the capital of ideas.

Reanalyzing the Productive Scenario: Appropriation of the Means of Production

It was hackers, influenced by the counterculture of the 1960s, the creation of the Internet and its principal technological standards, together with research institutions, universities, businesses, and the U.S. military that, in turn, influenced the conception of the primary technology upon which the Internet and the first systems of communication and information sharing were based. From this varied composition of players and their cultural influences emerged and continue to emerge various practices in the use of information technology in the field of information sharing, ranging from models that restrict the dissemination of information through protection mechanisms that use patents and intellectual property rights, such as in the cathedral model, to models that promote the availability of information and its dissemination through the public domain (Raymond, 1999), such as in the case of the bazaar model.

The basic difference between the cathedral and bazaar models consists in how the topology of the network used in the construction of knowledge based on information sharing is structured. The cathedral posits a monolithic model, a large bureaucratic structure that supplies a productive model limited by patents and intellectual property rights. Therefore, the production and sharing of information is limited to the scope of the structure. The topology of the network is restricted to the limits of the cathedral with no possibility of emergent growth. Whereas, the bazaar takes as its founding principle that it is the network and not a monolithic structure. As the network, it is the most efficient structure for the production of information through the free exchange of information about products, leading to greater efficiency in the process of technological innovation. Therefore, the network topology structured on the bazaar model foresees a free association of knots, allowing the construction of various possibilities of exchange and cooperation through the information that circulates through the network and the free appropriation of this information by the knots in the network.

Another fundamental point for consideration in terms of the bazaar model is that networked cooperation allows a contributing member of the network to obtain important comments and analysis about his or her product, allowing him or her, in turn, to improve, modify, and reinterpret it, a factor that encourages network participation and cooperation for reasons other than financial.

We also can detect applications of the bazaar model in what can be referred to as the democratization process in technological innovation, occurring in the relationship between businesses and their community of users. This process takes into consideration that many users of certain products and services have the competence, interest, and need to innovate for themselves (von Hippel, 2005). These groups of users organize themselves into communities that have certain interests in common and a technical knowledge or interest in obtaining such knowledge that will enable them to suggest alterations in product design in the characteristics to be developed or incorporated into the products.

The basic idea behind the democratization process in technological innovation, a principle in tune with the bazaar model, consists of the fact that the innovation structure of the company linked to a certain product is on its own in terms of its ability to incorporate and conceive technological innovation to serve specific market demands and not just average trends. This process connotes a topological opening up in the network of technological innovation beyond the limits of the corporations, thus allowing the user community to participate in the conception process of new projects instead of just offering opinions about what should or should not be incorporated into a particular product. Here, the emergence of a model of innovation centered on the user and based on his or her needs and his or her ability to articulate and produce in a network becomes much more apparent. Quoting von Hippel (2005, p. 110), "Innovation that is distributed openly is an attack on the structure of the social division of labour."

This topological opening up of the network therefore allows the recreation of the relationship of exchange between users and producers through a new relationship established between innovators and innovation. This new relationship incorporates and considers users as active members of the productive scenario instead of mere consumers of the process resulting from the technical innovation carried out by the company. We therefore have a model that sees the technological innovation process from the point of view of new social organizations, taking into account the bazaar model as an extension of innovative possibilities.

What we can see is that this trend validates the initial statement by Castells (1999, 237) that "there is as a result, a very close relationship between the social processes of creation and manipulation of symbols (the culture of the society) and the ability to produce and distribute goods and services (the productive forces)." This closeness, in the context of digital culture, is due to the intensification of the sharing and availability of information through information technology that leads to new possibilities of networked management and production. This intensification is owed to the recognition of and the creation of conditions for the bazaar model to evolve and integrate into an innovative management practice within a state organism for cultural management, such as has been done by the Ministry of Culture.

Nevertheless, production on the net and technological innovation appear to be mediated by relationships of rights and intellectual property in a world of possibilities and of interpretations of the role of culture as property and the relationships of exchange that lead to the act of cultural production itself.

Culture as Property

The central question in the discussion of cultural production as intellectual property revolves around economic and financial questions surrounding the forms of its dissemination, production, and consumption, and around the models for financing cultural production. The reigning argument of the industry is that the free circulation of cultural production would lead to a financial condition such that the market would be unable to remunerate the author of the work, thus denying him or her the opportunity to continue to develop his or her work based on research and artistic experimentation.

Copyright originated as a way to regulate the copying of books in the Renaissance with the invention of the printing press (Liang, 2004). Perceiving the potential of a new industry—the printing industry—some form of regulation was needed to protect a new business from book copying. The idea of copyright appeared, which from then on became the regulatory model for the industry of cultural dissemination applied in the most varied areas of the media and information technology.

The model remained relatively stable until the cost of cultural dissemination (now shared among multiple players through the knots of the Internet) fell to an almost negligible amount, leading to intense pressure to share information in all kinds of formats: text, music, and video files; in short, the whole range of languages used in cultural production that requires the technology to be digitalized.

The fact is that the reduction in the cost of implementation on a large scale on the Internet and its resulting popularization led to intense sharing of resources on the net, causing pressure on the part of the culture industry in terms of its rights over those resources now freely circulating over the networks. A classic example of this pressure was the *Napster* case, which allowed that "in the analogical world, it costs money to make a copy of something. In the digital world, it costs money preventing copies from being made" (Shirky, 2001, 5).

This social change has supported a series of models for licensing intellectual property rights, allowing the creation of structural conditions for new models of copyrighting and its limitations. The basic idea behind these new models is to provide a legal structure that supports the fact that the technological and social structure of emerging networks and of personal computers is able to provide resources so that users undergo a transformation from being consumers of culture to producers. Moreover, they supply a legal structure, allowing the creator of a work of culture to choose what level of access, what kind of use, and to what extent of cultural dissemination the work is to be allowed. Essentially, the creator makes the decision to release or not to release some, or even all, of his or her rights.

This discussion about licenses began in the world of software development in which users were closer to innovations in information technology and first felt the emergence of the networks of cooperation, which, as we can see today, have emerged in a cultural field that is much bigger and more wide-ranging in terms of media cooperation and production. The Free Software Foundation[1] and the products derived from its actions and initial production of free software had a fundamental role in the first demonstrations of the feasibility of cooperative production and of information sharing as a possible dynamic in network management. Software applications such as Linux, Apache, and Mozilla, among many others, are proving the value of alternative models to copyrighting as a mechanism for protecting intellectual property rights.

The demonstration of the feasibility and the verification of the emergence of a new paradigm in other sectors apart from software allows us to see that digital culture is based profoundly on a transformation of a sociocultural character, revealing new forms of production and new ways of relationships and exchange. Today, we can broaden the discussion on cooperation and the sharing of information through countless models and examples coming from the world of software, enabling us to think up and create new models that use digital culture as a tool for cultural management in the 21st century. As such, we can bypass traditional models of intellectual property rights that, in a certain fashion, restrict the creative capacity of the emerging networks, increasingly seen as truly organic systems of cultural conception.

A Brazilian Vision of this Scenario

Brazil is a country of multiple visions and social divisions, many ethnic groups, and an intense sense of regional identity in terms of the language, production, and feeling of culture. At the same time as one looks at Brazil and finds creativity and a spirit of solidarity, one is also faced with inequality, social injustice, workers without work, families without homes, young people with no prospects for the future, native peoples without rights, and a people without a state. Even so, the country manages to maintain its spirit of solidarity. The people of Brazil are inventive, entrepreneurial, and joyous. It is within these multiplicities and regional differences that Brazil is placed and from where it wishes to build its own vision of digital culture as a tool for cultural management and as a form of cultural and political action.

A strong supporter within Brazilian politics is the Minister of Culture, Gilberto Gil, musician and important member of the Tropicalista counterculture movement of the 1960s, who has adopted the concepts of digital culture and the new emerging models as official policy. In one of his speeches, he stated his position as being in tune with these principles: "I, Gilberto Gil, Brazilian citizen and citizen of the world, Minister of Culture of Brazil, work in music and in the ministry and in the all the dimensions of my existence, inspired by the ethics of the *hacker*, and concerned with the questions of my world and of the age, such as the question of digital inclusion, the question of free software and the question of the regulation and the development of the production and the dissemination of audiovisual contents, whichever the medium, whatever the purpose" (Gil, 2005).

In addition, there is in Brazil a political context propitious to the implementation of projects in the area of digital culture through the construction of public policies as well as an intense debate, both academic and institutional, about the noninstitutionalized nature present in all classes, which has been recognized by federal government officials, as has also been made clear by the minister of culture: "We in Brazil already have vast accumulated experience in the field of free software and digital inclusion, with hundreds of projects and prototypes, and even deep academic debate, and this far-ranging mobilization of people, intellects and creativeness has flowed into the government itself, which has embraced the cause and made digital culture one of its strategic public policies" (Gil, 2005).

In 2004, the Minister for Culture inaugurated the *Cultura Viva* (Living Culture) Program, which includes digital culture in its brief. The *Cultura Viva* program was conceived as an organic network for creativity and cultural management. Its principal structuring action is the *Ponto de Cultura* (cultural hotspots), whose name arose from the inaugural address of Culture Minister Gilberto Gil when he suggested that "an anthropological *do-in* (Japanese massage) should be made, a massaging of the nation's vital points." Instead, therefore, in determining local actions and conducts, the *Cultura Viva* program aims to encourage creativity, releasing the potential of that which already exists and creating conditions for an exchange of experiences between agents of this network. The objective is to avoid an institutionalized and stratified structure, overbearing in its form of management and control, as is common in government bureaucracy, and encourage a strengthening of social networks. In this context, the cultural hotspots is the reference point for the structuring of the cultural and social network, and the digital culture is the element that consolidates and broadens the exchange between the configured network, providing through the technology available the mechanisms for exchange, discovery, and continuity of the actions of the hotspots, giving these an autonomy that goes beyond the bounds of government support, initially based on the consolidation of the process.

With the *Cultura Viva* program, the Brazilian Ministry of Culture has given support to initiatives and methodologies of technology reappropriation and development of the social technology of communities of free software developers such as MetaReciclagem (www.metarecicagem.org), Arca (http://arca.ime.usp.br/), Mídia Tática Brasil (www.midiatatica.org), and SubMídia (www.submidia.radiolivre.org) grouped around the *Instituto de Pesquisas em Tecnologia da Informação* (Research Institute for Information Technology)—IPTI (www.ipti.org.br), which is the institution responsible for the coordination of the project's digital culture activities.

The Brazilian Program for Digital Culture

Before giving a detailed presentation of the Brazilian program for digital culture, it is important to define what exactly is a *Ponto de Cultura* (cultural hotspot). It is a space for interaction between cultural players already existing prior to the *Cultura Viva* project. They are groups immersed in their local realities, spread all over Brazil, producing culture and media in various forms and linked with their local communities, therefore possessing legitimacy within their work focus and axis of insertion. They produce culture, ranging from costume design, theater, dance, and music to digital inclusion. The idea behind the *Cultura Viva* project is to link the cultural hotspots virtually through a digital network, creating infrastructure for cooperation and the sharing of information for cultural and political action, while using digital culture as a tool for cultural management.

A cultural hotspot is defined by the Brazilian Ministry of Culture as an amplifier of the cultural expression of its community, as a place where one can make music, videos, community cinema or television and that has a digital recording studio available with the capacity to record, edit, and upload what has been recorded onto the Internet (Gil, 2004). This idea makes feasible laboratories for experimentation and multimedia interaction, which supply infrastructure to broaden cultural expression. These laboratories, however, are not aimed at being a meeting place for the local community, attracting interest merely for access to technology, but rather a laboratory inserted into a cultural space and open to all its history, thereby expanding the story of its creation, improvisation and cultural experimentation.

It is important to highlight the organizational and managerial structure that provides the conditions for effective and ample cultural cooperation to occur. It is in this moment that a new conception of digital culture (as a potential qualitative jump) permitted the creation of a management structure that would mix the cathedral and bazaar model, as cited earlier. The Ministry of Culture is the aggregating element guiding the public policies of cultural management; therefore, its role within the project is to serve as an agglutinating reference for the most diverse cultural entities spread throughout the country. It is these entities that accompany the edicts of the Ministry, its regulatory forms, and its investment abilities in cultural development. As such, they envision the Ministry as a well-positioned leader capable of bringing together culture. Taken from this point, the project innovates and permits investment funds to come from the Ministry, while allowing it to have its own demands, local experiences of management, mechanisms of innovation, research, and decentralized development. It is the network and its technological structures that permit us to create new conditions for management and the recognition that the methodology of the project is emergent and in constant evolution.

As a way of making the infrastructure available for equipping the cultural hotspot with the necessary resources to broaden cultural expression and to create a laboratory environment capable of supporting the multiple languages of the digital culture, a multimedia kit was designated, currently being set up in various cultural hotspots selected by the Ministry of Culture through public tender.

The objective of the kit is to provide a local infrastructure within the culture hotspot to facilitate the production and sharing of culture among users connected to the network. The kit is centered on a server for multimedia applications. Connected to the multimedia server is an application server supporting thin client workstations, providing a simple solution for resource sharing using low-cost computers. All the kit's computers and its productive equipment are operated entirely by Free Software and are connected to the Internet.

However, it should be remembered that the key to the transformation into digital culture is the emergence of networks allowing the sharing of information. We are therefore talking of the social dimension in which this sharing occurs. This dimension is not merely the implantation of the kits in the culture hotspots nor is it due to the Internet connection. It is the channel of expression that allows any citizen, company, or community (in short, anybody connected to the Internet) to be a member of the hotspots network and, therefore, to interact with an intense chain of cultural production. What enables this social dimension to occur are the systems of cooperation that have been customized for the Project; that is, *Conversê* (www.converse.org.br), *Estúdio Livre* (www.estudiolivre.org), and the *Encontros Regionais do Conhecimento Livre* (Regional Encounters for Free Communication).

Conversê is a site for open and collective publishing, customized for the project using a Drupal platform (www.drupal.org) with the aim of weaving a network for conversation and the circulation of information between the culture hotspots. This environment has various characteristics that are additional to collective publishing, such as the ability to create chat communities around specific interests, configure the way the system will allow for those chats of greater personal interest to be followed, and allow the exchange of private messages among users. Through this publishing environment and the multimedia kit, users will interconnect online and engage in an infinite number of possibilities for conversation, articulation, construction of knowledge, and technical support in the use of the equipment, software, or similar doubts, always participating in the spirit of community, living and exchanging a diversity of experiences. The project also aims to connect with other communities in addition to the members of the hotspots network, thereby broadening the potential for conversation, dissemination, and the circulation of ideas. It will be as if a huge bazaar were installed on the Internet, connecting cultural hotspots and their subjective views with the reality of an emerging network of hotspots.

Estúdio Livre is an environment developed by TikiWiki (www.tikiwiki.org) with the objective to develop a platform for sharing cultural content produced by the hotspots through the production of free media. It is in this environment that the hotspots will be able to share video, music, text, documents, and other files, creating structural conditions for a real cultural remix from diverse productions emerging from different localities, realities, and hotspots of view: a veritable repository of thinking and of innovative cultural production and broadening of cultural expressions using the digital culture.

Having created the structural conditions from the kit's hardware, free software, and the cooperating systems, the Project organizes local encounters called *Encontros Regionais de Conhecimento Livre* (Regional Encounters for Free Knowledge) that work as install fests (installation parties widely used by the Linux community as a way of promoting the system and as a form of technical support for beginners) based on the open source culture. These encounters are held in each of the regions in which the cultural hotspots are located. At them, members of the communities get together with the cultural hotspots to exchange their experiences in the digital culture area for the construction of temporary laboratories, providing the conditions for the hotspots to experience real situations for using the kits as a production and cooperation tool. The appropriation of the hacker ethic, the hacker way of building knowledge together with the communities, and the creation of relationships is what will be taken from the workshops and continued through *Conversê* chats and through the sharing of information through *Estúdio Livre*.

It is in this way that the *Cultura Viva* project appropriates for itself the concept of the information economy and proposes the implementation of its three founding pillars through the technology of the kits, the logical infrastructure of operation provided by the free software, and the saving and use of previous knowledge from the conversations and the social dimension of interaction through the support systems and regional encounters. Our model of infrastructure consisting of the three pillars of the economy of knowledge is illustrated in Figure 1.

The physical infrastructure is the actual kit, which permits the development of basic digital technology and computational operations. Above this structure, between

Figure 1. Infrastructure of the points network

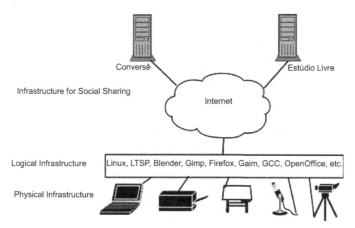

the infra logic, are the open source softwares that permit us to operationalize the computers and accessories in a way that enables us to take advantage of its effective computational potential. In addition to this, we have an infra of sharing, which enables pairs to meet in the network, develop meaningful conversations, exchange experiences, and engage in the collective construction and collaboration of new ideas and cultural products.

With this infrastructure, the necessary conditions are in place for the emergence of the cultural hotspots' Network, with great possibility of its appropriation by it members as a form of cultural and political action.

Results

Some of the data collected throughout the duration of the project allows us to evaluate the dimension of the program and its insertion within the communities involved. We initiate our analysis with the Free Knowledge Workshops that are taking place throughout various Brazilian states with the participation of almost all of the 250 points involved in the program.

In total, 1,211 people participated in the Free Knowledge Workshops held in 13 states (Figure 2). Members from the cultural hotspots and local community members located close to the hotspots participate in these workshops, come into contact with the technology and experience in the ways these technologies are being used within each hotspot. These workshops and the intense participation of the local communities is what permits the first direct contact with the techniques of network collaboration

Figure 2. Workshop participants listed by state

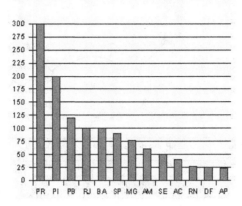

and supplies the social material necessary to initiate conversation and collaboration at the local level, which later will unfold on the Internet. It is worth emphasizing that even though the workshops are organized by state, many hotspots members have made trips in order to participate in workshops taking place in other states, creating greater conditions for the effective exchange within the hotspots network.

The effects of these workshops within an environment of interaction and sharing on the Internet are now more manifest. The cultural hotspots system of collective publication is owned by 1,916 users, 56 chat rooms, and 676 themes currently in discussion. These numbers bring us to believe that in relation to the number of workshop participants, there was a great increase by about 40% in the number of Web system users. Many users bring their communities and invite friends and acquaintances. The greater number of themes in discussion, approximately three users for each theme, also demonstrates a diversity of conversations and the ample use of the system by its users. In addition, there is *Estúdio Livre*, what we see as the system of collaborative production as it comes to be used more frequently. To date, 633 users have produced 159 songs and 121 videos and have released and made available about 105 images and 317 written pages in Wiki.

Final Remarks

The *Cultura Viva* project, as conceived and executed from the emergence of organized network communities, developing actions of cultural and political essence, are constructing and disseminating a strategy of action based on the principles of a digital culture. By incorporating this strategy, the Brazilian Ministry of Culture fostered social and political conditions for the creation of a network of cultural hotspots through the use of free technologies inspired by the hacker ethic and based on digital culture as an element of new possibilities in addition to artistic expression.

This project was initiated at the end of 2004 and nowadays includes approximately 250 Cultural Hotspots, receiving resources from the federal government to fund their cultural activities, training in the use of free multimedia tools for production and/or documentation of their cultural works, and participating collaboratively in a network of sharing information and knowledge. Shortly, 250 more cultural hotspots, previously selected through a public tender, will join this network.

In this way, the project intends to delve into a hidden Brazil and stimulate the cultural hotspots to expose themselves, exhibit themselves, and show that sharing their cultural identities is the best possible way to preserve their differences and cultural richness.

References

Castells, M. (1999). *A sociedade em rede*. São Paulo, Brazil: Paz e Terra.

Gil, G. (2005). Talk of the ministry of culture, Gilberto Gil, at Universidade de São Paulo (USP). Retrieved October 21, 2005, from http://www.cultura.gov.br/noticias/discursos

Gil, G. (2004). Talk of the ministry of culture, Gilberto Gil, about the national programme for culture, education and citizenship—*Cultura viva*, during a meeting with artists in Berlin. Retrieved October 21, 2005, from http://www.cultura.gov.br/noticias/discursos/

IBGE. (2000). *Censo demográfico 2000*. Retrieved October 21, 2005, from www.ibge.gov.br

Lévy, P. (1993). *As tecnologias da inteligência*. São Paulo, Brazil: Editora 34.

Liang, L. (2004). *Guide to open content licenses*. MA: Piet Zwart Intitute.

Myers, F. (2005). Some properties of culture and persons. In R. A. Ghosh (Ed.), *CODE: Collaborative ownership and the digital economy* (pp. 45-60). Boston: MIT Press.

Perret, R. (2002). Inteligência coletiva Pierre Lévy e a sociedade ciberdemocrática. *Observatório da Imprensa*. Retrieved October 21, 2005, from http://www.observatoriodaimprensa.com.br/artigos/eno110920021p.htm

Raymond, E. (1999). *The cathedral & the bazaar*. MA: Oreilly & Assoc.

Shirky, C. (2001). Prestando atenção ao Napster. In A. Oram (Ed.), *Peer-to-peer: O poder transformador das redes ponto a ponto* (pp. 23-40). São Paulo, Brazil: Berkeley Ed.

von Hippel, E. (2005). *Democratizing innovation*. Boston: MIT Press.

Endnotes

[1] http://www.fsf.org, retrieved February 21, 2006.
[2] http://creativecommons.org, retrieved February 21, 2006.

Section III

Information Resources Development Challenges

Chapter VIII

Dialogue Act Modeling:
An Approach to Capturing and Specifying Communicational Requirements for Web-Based Information Systems

Ying Liang
University of Paisley, UK

Abstract

Web-based information systems (WBIS) aim to support e-business using IT, the World Wide Web, and the Internet. This chapter focuses on the Web site part of WBIS and argues why an easy-to-use and interactive Web site is critical to the success of WBIS. A dialogue act modeling approach is presented for capturing and specifying user needs for easy-to-use Web site of WBIS by WBIS analysis; for example, what users want to see on the computer screen and in which way they want to work with the system interactively. It calls such needs communicational requirements, in addition to functional and nonfunctional requirements, and builds a dialogue act model to specify them. The author hopes that development of the Web site of WBIS will be considered not only an issue in WBIS design but also an issue in WBIS analysis in WBIS development.

Introduction

Web-based information system (WBIS) is a new type of information system that uses information technology, the World Wide Web (WWW), and the Internet to support e-business and information source management worldwide. They provide a new way of managing, manipulating, exchanging, sharing, and supplying global information and services online. They enable customers and companies worldwide to communicate with each other through the Internet and to demand and supply business information resources and services around the world without meeting each other. Appearance and use of WBIS in business have changed people's lives because it brings a new culture into business. People living with this new culture do not have to buy things in local shops; instead, they can buy things in global e-shops using the Internet. Suzuki (1997) in general defines culture as the response pattern shared by a specific group of people, which is shaped through interaction with the environment. Furthermore, Ratner (2003) specifically defines global culture as what is common to all human beings and the response pattern as how people interact with the Web site in the context of the Web site. Based on the Suzuki and Ratner definitions (Ratner, 2003, p. 48), WBIS culture is defined in this chapter as follows:

WBIS culture is what is common to all users of WBIS and that the "response pattern" is how the users will interact with the Web site of WBIS in the context of WBIS.

Such response patterns can be perceived as an interactive communication procedure of getting the goal of a user's task by using WBIS within a business context. A well-accepted WBIS should allow the user to control the interactive communication procedure and to decide how to interact with the Web site of WBIS. It is agreed that in order to have an easy-to-use Web site, WBIS basically needs to be user-centered and interactive as a computerized business tool linking customers to companies or linking companies to companies in e-business and information source management. They must be attractive to online users and compatible with other similar systems. Usability of WBIS relies greatly on the developer's understanding of the user's need for the Web site of WBIS (e.g., what users want to see on the computer screen and in what way they want to work with the system interactively). Unfortunately, understanding such a need currently is ignored in WBIS analysis because usability is often thought of as an issue in design rather than in analysis. For example, Cato (2001) emphasizes that the developer needs to pay greater attention to user interface and interaction design in WBIS design if he or she wants a system to be effective. However, in principle, understanding a user's need is the task of system analysis but not the task of system design. This means that the user requirements for effective interactive communication with the Web site must be captured in WBIS analysis. This is not supported by current modeling approaches, and new modeling approaches are needed for this purpose. Actually, a new modeling approach called *dialogue act modeling* (Liang, 2004, 2005a, 2005b) has been created specially by the author for capturing and specifying such user need and will be introduced in this chapter. The objective of the chapter is to help the WBIS developer (in particular, WBIS analysts) to understand the following:

- Different cultures of WBIS and traditional IS.
- Role of the Web site in the WBIS culture and its impact on user-centered and interactive WBIS.
- Need of considering usability of the WBIS Web site in WBIS analysis.
- Need of understanding the user need for easy-to-use Web site in WBIS analysis.
- The dialogue act modeling approach and how to use it to capture and specify such user need in WBIS analysis.

WBIS Culture and its Impact on WBIS Development

Observation has shown many differences between WBIS and traditional IS (Deshpande et al., 2002) and, unlike traditional IS, WBIS provides a Web site as the communication medium to interact with the user over the Internet. This drives the two systems to run in different environments with different cultures (i.e., what is common to the user and that the response pattern is how the user will interact with the system in its context are different). Figure 1(a) and 1(b) show that the response pattern of traditional IS culture is how the user will interact with IS directly in the context of IS, whereas the response pattern of WBIS is how the user will interact with IS through the Web site of WBIS in the context of WBIS.

The WBIS culture requires WBIS analysis to involve users much more than traditional IS analysis and requires WBIS development to make WBIS with the following specific characteristics:

- **WBIS should have a Web site and Web pages:** The user of WBIS uses the Web site to get information and services on the Internet, and WBIS uses Web pages to display the result on the computer screen. The Web site must enable the user to complete his or her tasks effectively within the minimum time. Thus, developing a Web site and Web pages in consultation with users is vital to Web site success (Lawrance, Gorbitt, Tidwell, Fisher, & Lawrance, 1998).

Figure 1(a). Traditional IS culture

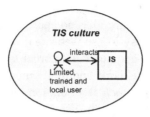

Figure 1(b). WBIS culture

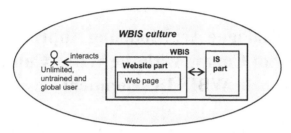

- **WBIS have an unlimited number of varied users:** Users of WBIS are unlimited and live in different places of the world (Conallen, 2003). This makes it impossible to train all of them in training sessions. Untrained users may have fear and anxiety when using WBIS. Thus, easy-to-use Web sites of WBIS is critical to such a user.
- **WBIS often have global users:** WBIS in principle is used by anyone, anywhere in the world. Global users normally have different cultural backgrounds and use different languages (Ratner, 2003). It has been found that the possibility of information systems failing through culture shock increases because of business globalization and creation of WBIS connecting the world. When it becomes apparent that some users do not behave as expected, the problem of what to do next becomes crucially important in Web site development (Szewczak & Snodgrass, 2002).
- **WBIS have a volatile development process:** This makes requirements analysis and specification difficult (Barry Y Lang, 2001).

These characteristics of WBIS make WBIS development more difficult than traditional IS development, because the problem of the development is no longer constructing the technical infrastructure of the system but is creating a good Web site that is easy-to-use and meets the user's needs (Lazar, 2001). Analysis and design of WBIS are inevitable to face this problem. A lot of effort already has been made to solve it in WBIS design (Burdman, 1999; Corry, Frick, & Harsen, 1997; Fuccella, 1997; Lazar, 2001; Lynch & Horton, 1999), but not much has been done for this in WBIS analysis. Thus, the WBIS analyst has no approach available to capture and understand the user's needs for easy-to-use Web site in WBIS analysis, although it is the task of analysis in software engineering. In order to help to improve the situation, this chapter will address this issue and represent a new modeling approach called *dialogue act modeling approach* as a solution to this problem. The author hopes it is useful to capture and specify the user's need for the Web site of WBIS in WBIS analysis.

The Dialogue Act Modeling Approach: A Way of Coping with the WBIS Culture in WBIS Development

The dialogue act modeling approach presented in this chapter aims to cope with the WBIS culture in WBIS development. It provides a new modeling technique for capturing and specifying interactive communications between the user and WBIS in WBIS analysis with the concern about usability of the Web site of WBIS, because easy-to-use and the usefulness of the technology have the most influence on the user of computerized systems in general (Davis, 1989), and they also are applied to the user of a Web site (Lederer, Maupin, Sena, & Zhuang, 1998). IS needs to be developed around the user's needs (Norman & Draper, 1986), which becomes even more critical in WBIS development because traditional IS without Web can control where a user can go, but WBIS cannot do so, as users of WBIS want to be active when they are on the Web, and they even can take paths that were never thought or intended by the developer (Nielsen, 2000). They intend to control the paths toward the goal of their tasks on the Web and to complete their tasks in a minimum amount of time with a minimal amount of frustration. Such a user need should be understood in order to avoid the failure of WBIS, because some of e-business Web sites have failed completely and have been shut down because of poor usability (Ratner, 2003). Users will not purchase things from a Web site that is complicated and frustrating to use (Lazar, 2001), and they are extremely goal-driven on the Web (Nielsen, 2000). Difficult-to-use Web sites of WBIS can cause computer anxiety when users use the computer, which is thought of as one of the most prevalent emotions experienced by users (Downton, 1993; Negron, 1995). However, it can be remedied through positive computing experiences (Cambre & Cook, 1987) with an easy-to-use Web site.

How to make easy-to-use Web sites for WBIS is a big challenge of WBIS development, because it is quite hard to produce a useful product that is easy to learn and use (Cato, 2001). It needs effective development approaches and techniques such as modeling techniques that enable WBIS developers to consider the user's attitude and motivation and to allow users to build interactions in the way they like and create their own power over interaction. Traditional IS development approaches and techniques are not really proper or adequate to WBIS development because of the dynamic and evolving nature of WBIS (Lang, 2002; Zelnic, 1998). The dialogue act modeling approach was created, therefore, for overcoming the lack and by considering the following:

- How to create a WBIS specification that will lead to a success of user-centered and interactive WBIS in development.
- What kind of modeling techniques can help to promote usability of the Web site of WBIS.
- How to make users involved in Web site analysis.

Experts in this field have attempted to create a new discipline, Web Engineering, for research and establishment of sound scientific, engineering, and management principles and systematic approaches to successful development and maintenance of high quality WBIS (Murugesan, Deshpande, Hansen, & Ginige, 2001). Engineering a Web application is to diversify problems to application domain analysis, navigational structures, and user interface design (Conallen, 2003). The dialogue act modeling approach falls into the application domain analysis specifically in Web Engineering, as illustrated in Figure 2. It will analyze the Web site part of WBIS through dialogue act modeling and analyze the IS part of WBIS by object modeling and behavior modeling (Booch, Rumbaugh, & Jacobson, 2005). The analysis process was established based on the one defined by Sommerville (2004). The details of the approach are described as different parts in the following.

Part One: Pragmatic View and Descriptive View Used by the Approach

The pragmatic view and the descriptive view are the two typical modeling views used in traditional IS modeling in the past:

- **Pragmatic view:** Used for observing the pragmatic aspects of IS (e.g., business properties such as customers) as part of reality within the business context. The action workflow approach (Denning & Medina-Mora, 1995) used this view. It has much focus on pragmatic concepts but little focus on semantics of the

Figure 2. Dialogue act modeling approach for WBIS requirements analysis and specification

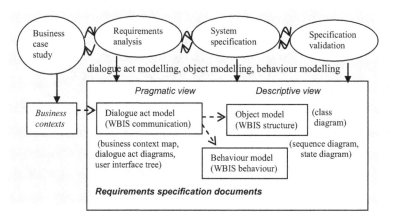

system (Agerfalk, 2002; Erickson & Kellogg, 2000; Eriksen, 2002; Holm & Ljungberg, 1996).

- **Descriptive view:** Used for observing the semantic aspects of IS (e.g., business processes and entities) as an image of reality. Data modeling (Chen, 1976), process modeling (DeMacro, 1978), and object modeling (Booch, 1994; Coad & Yourdon, 1991; Jacobson, 1992; Liang, 2003; Rumbaugh, Blaha, Premerlani, Eddy, & Lorensen, 1991) used this view. In general, it has much focus on business processes but little focus on business properties such as customers.

Nearly every existing modeling approach used one of these two modeling views only but not both in IS modeling because they aimed to focus on either pragmatics or semantics of a system during analysis. However, the dialogue act modeling approach presented in this chapter uses both views, because it aims to focus on both pragmatics and semantics of the system in WBIS analysis. It uses the pragmatic view in dialogue act modeling and the descriptive view in object modeling and behavior modeling.

Part Two: Dialogue Act Modeling for Building Dialogue Act Model for the Web Site of WBIS

To cope with the WBIS culture in WBIS analysis, a modeling technique is needed for capturing and specifying the interaction pattern (i.e., interactive communication) required by the user. Such user requirements are called *communicational requirements* in this approach. However, none of current modeling approaches (Conallen, 2003) has provided this needed technique, because they do not focus on communicational requirements. Although some lifecycle models for Web development such as ones presented by Burdman (1999), Fleming (1998), and Lynch and Horton (1999) have addressed the specific need of Web sites, they have limited user input to such requirements. In terms of these approaches, Web developers only can guess or imagine what users find easy or confusing in interactive communication, but they cannot capture what the user actually wants. To overcome this problem, the dialogue act modeling approach provides a dialogue act modeling technique that creates a dialogue act model for the Web site part of WBIS by WBIS analysis (see Figure 3).

This modeling technique focuses on communicational requirements and the pragmatic aspects of WBIS. It describes business participants as speakers and hearers of dialogues, performers of dialogue acts, and seekers of business information in the dialogue act model.

(a) Communicational Requirements

In traditional IS analysis, user requirements are either functional requirements (i.e., what the system needs to do) or nonfunctional requirements (i.e., what constraints are on the system, such as performance and security constraints). However, in WBIS analysis, users can make additional requirements for the Web site part of WBIS. They may provide their own opinions and judgments on layout and display of a Web site and request to develop the Web site of WBIS according to their needs (e.g., easy-to-understand and follow) rather than according to the developer's need (e.g., easy-to-implement). To distinguish them from functional and nonfunctional require-

Figure 3. Dialogue act modeling in WBIS analysis

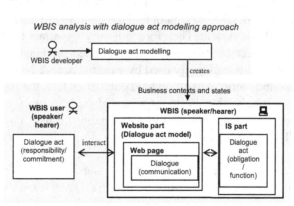

ments, they are called *communicational requirements* by this approach. Modeling communicational requirements can draw a developer's attention to the user's need for the Web site of WBIS at the early stage of development and avoid the WBIS changes caused by a difficult-to-use Web site of WBIS after development. The following user concerns about the WBIS Web site are considered communicational requirements of WBIS:

- **Business contexts:** They are the ranges of e-business and information resource management. In WBIS, they are initiatives and entrances required by the user for starting interaction with the Web site. They can be business activities such as "car for sale" or business indexes such as "cars" (see Figure 6). Different business contexts may cover similar things or even another business context. For example, the car company may want a business context "car for sale," and its customers may want a business context "cars" that includes "car for sale."
- **Dialogues:** They are the conversations preferable to the user in a business context. The user wants to use them to communicate with the system interactively on the Web site and will consider them visible feedbacks from the system in a business context. For example, after car buyers enter into the business context "cars," the system displays the dialogue "car list" on the screen as the feedback (result) of the dialogue act "offer car" demanded by car buyers (see Figure 7). Commitments of the user or the system such as "offer car" can be identified based on the dialogues.

The benefits of capturing communicational requirements in WBIS analysis are expected to be the following:

- **The Web site of WBIS developed based on communicational requirements should be user-centered and not complicated and frustrating to use:** The research on the use of the Web has found that errors occur frequently mainly because users do not know or cannot find the right way to interact when communicating with the system. Therefore, it is very important to make the Web experience as predictable as possible and to create a predictable Web site that is understood quickly and easily used by a user (Lazar & Norcio, 2000). Although Lazar and Norico (2000) pointed out that, in fact, the Web experience is inherently unpredictable due to the nature of the Web itself, we still can try to predict a user's Web experience by understanding the user's way of interacting with WBIS from the user's point of view. This at least can reduce errors caused by a user's misunderstanding of interaction. A successful Web site of WBIS must offer contents needed by users, and the content must be found easily by interactive communication. Otherwise, it will leave users frustrated and possibly unable to achieve their goals (Preece et al., 1994). This requires the developer to understand communicational requirements deeply.

- **Recognition of feedbacks displayed on the computer screen as dialogue:** Dix, Finlay, Abowd, and Beale (2004) emphasize that interactivity is at the heart of all modern interfaces and is important at many levels. Interaction between user and computer is affected by social and organizational factors. Awareness of the factors can help to limit any negative effects on the interaction. Barfield (2004) said, "The system's role in the interaction is to provide the user with information about the interaction. This information, this communication between the system and the user, is feedback and it is what helps the user build up the model that the designer wants to build up. ... Feedback is new media means keeping the user informed about what is going on" (p. 195). He defined three feedbacks:
  - **Feedback after:** The user should know that the system heard their request and is doing it. When the request is done, the user should be made aware of this and see the results.
  - **Feedback before:** The system should tell the user what will be going on if the user takes a particular action. The user must be told what actions he or she can initiate, and it should be clear how he or she can initiate them.
  - **Feedback during:** The user should be notified about what is happening, when it has started happening, how it is progressing, and when it is expected to finish, during generation of feedback.

Capture of communicational requirements helps to identify the feedbacks expected by the user for the business context and decide what the system needs to do for them. It has been known that some modeling approaches and techniques for traditional IS analysis, such as use case modeling, object modeling, and behavior modeling with UML (the Unified Modeling Language) (Booch et al., 2005), can be adopted in order to capture and specify functional and nonfunctional requirements for the IS part of WBIS (Conallen, 2003). However, none of them can help to capture and specify communicational requirements for the Web site part of WBIS. Therefore, we investigated a wide range of approaches and found that the Speech Act Theory (Austin, 1962) in the social science could help do this because dialogue means speech to act. According to the theory, a dialogue between a user and a system means one or more dialogue acts in an e-business society. This can help to analyze the role of the user and the system as well as user commitments and system obligations. In WBIS design and implementation, a system obligation may mean a function of WBIS, and a user commitment may mean a precondition of functions. Navigation structures and user interfaces should be designed and implemented based on communicational requirements.

(b) Speech Act Theory and Interactive Communication Analysis

Although there are many theoretical frameworks developed by researchers in social science for conversation analysis, the language/action perspective has been most influential in human computer interaction and applied as a social approach for analysis and design of computer-mediated conversations for people interacting with each other (Preece et al., 1994). For example, Winograd (1988) showed an approach that used this perspective to view languages as a means by which people act. This perspective often refers to the speech act theory defined initially by Austin (1962) for describing the phenomenon in a social society that people use speech to act; for example, demanding or promising something. The theory was expanded by Searle (1969) with the definition of a speech act with four different subacts:

1. **Utterance acts** with uttering words.
2. **Prepositional acts** with referring and predicating.
3. **Illocutionary acts** with stating, questioning, commanding, and promising.
4. **Perlocutionary acts** with causing an effect on hearers.

The expanded theory considers an utterance as the speech within each turn (person A says something, then person B says something, then person A says something again, etc.) Often, the utterances of the conversation can be grouped into pairs: a question and an answer or a statement and an agreement. An utterance should be understandable to the listener and should be sufficiently unambiguous for the listener to understand (Dix et al., 2004). This means that the speaker should be aware of the model of understanding the listener and vice versa.

The speech act theory explains how people in a society use language for talking about events in the external world as observers and also for the communication act within the world as actors in the society (Agerfalk & Erisson, 2004). In computer society, IS was even defined as language systems in general used to perform communication acts (Goldkuhl & Lyytinen, 1982). Current modeling approaches based on this theory include COMMODIOUS (Holm & Ljungberg, 1996), conversion-for-action schema (Winograd & Flores, 1987), DEMO (Dietz, 2001), and action-oriented conceptual modeling (Agerfalk & Erisson, 2004). Application of these approaches in IS analysis has proved that this theory can support significant understanding of the pragmatic aspects of the system. This success has encouraged us to use this theory in WBIS analysis, because WBIS also uses a language on a Web site for interactive communication and other things such as demanding and promising something.

However, current modeling approaches observe interaction between users and IS in the developer's perspective and regard it as input/output of the system, such as data flows, and communication acts as data transformations internal through different media such as a computer screen. Such observation is not sufficient in WBIS modeling because WBIS in general not only deals with data transformations but also provides business information and customer services. The interactive communication between the user and WBIS is more than data flows because it can mean other things, such as information flows (e.g., car list), organization flows (e.g., sales department), and service flows (e.g., buy car). The dialogue act modeling approach treats the interactive communication as dialogues. It defines a sequence of dialogues as a dialogue flow and a speech act as a dialogue act. In addition, it defines the four subacts that refer to the ones defined by Searle (1969):

1. **Utterance act** is production and communication of physical written messages such as "buy car" displayed on the computer screen.
2. **Prepositional act** is performed by an object such as "car" and its attributes.
3. **Illocutionary act** is performed by a business service (activity) such as "sell car."
4. **Perlocutionary act** such as "buy car" is performed by the hearer. It has the effect on the business context.

The problem with the language/action perspective in practice is that conversations may be vague and may result in misunderstandings and promises failure. But Flores (1988) suggested that people would be better at communicating with each other if the types of commitments were made explicit to all parties involved during conversations. This means that if the modeling approach can make commitments of the user and the system explicitly during interaction, misunderstanding can be reduced. The dialogue act modeling approach thus aims to capture and specify such commitments and make them visible on the WBIS Web site for avoiding misunderstanding and for increasing usability.

(c) Dialogue Act Model and Dialogue Act Diagram

The approach provided a dialogue act diagram (see Table 1) along with the dialogue act modeling technique as a notation representing the dialogue act model. In WBIS analysis, this diagram is produced while building the dialogue act model for the Web site part of WBIS. It provides a communication medium in order for the analyst and the user to decide communicational requirements for the Web site of WBIS.

Table 1. Notation and definition of dialogue act diagram

Element	Diagrammatic Notation	Definition
Start dialogue with a business context	Speaker → Business context → Hearer	First directed communication from speaker to hearer(s). It must be displayed on computer screen as the entrance into the business context in WBIS.
Dialogue	Speaker → Dialogue → Hearer	Directed communication from speaker to hearer(s). It can be displayed on the computer screen to mean a demand/promise for information/service.
Actor (speaker/hearer)	Name	Speaker/hearer of a dialogue. A performer of a dialogue act.
Dialogue flow	Dialogue ↓	Connection of dialogues happening in the business context.
Dialogue act	Dialogue act (oval)	Activity performed by hearer(s) as a result of a dialogue. It is regarded as a precondition of the next dialogue.
Resource flow	⇒ Resource ⇒	Resource sent through actors within the business context.
State of business context	State	Effect on the business context.

(d) Impact of Dialogue Act Model on User Interface Design in WBIS Design

Communication requirements have an impact on design of WBIS user interface (Web site), which often is regarded as a whole of the system by the user (Dix et al., 2004). The user's metal model of IS has a critical impact on the user's ability to use systems effectively (Szewczak & Snodgrass, 2002). Different users may have different ideas (user model) on how to reach the goal of their tasks within a business context, and ideally, the WBIS analyst should consider different alternatives of communicational requirements in dialogue act modeling. Then, the analyst should let the user compare the alternatives and choose the ones most appropriate for him or her. However, the problem with system modeling is that user modeling techniques have been developed in order to instantiate user models as part of computer systems, but as yet, there are no usable conceptual user modeling techniques (Preece et al., 1994). Most user modeling relies on checklists of user characteristics but not on identification of user responsibilities and commitments in use of the system. The dialogue act modeling aims to support user modeling by visualizing different scenarios of interaction explicitly using dialogue act diagrams. The user and the developer then can discuss the model together in front of the diagrams.

Figure 4. Dialogue act modeling for the Web site part of WBIS

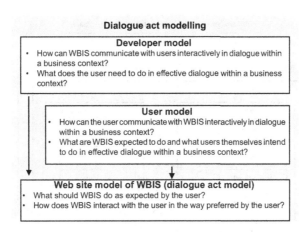

Dialogue act modeling also intends to provide a basis for designing user interface elements such as dialogue boxes displayed on the computer screen. Johnson (2000) has explored the problem with the current graphic user interface design in his book: "A particular annoying design error one often sees in this is dialog boxes that provide no way out other than a direction that users do not really want to go in. Often dialog boxes only seem to trap users and none of the choices is what the user wants" (p. 316). Such dialog boxes force users to stop thinking about their task and, instead, force users to find out how to use the system. A way of solving this problem is to display the dialogue act diagram on the Web site to assist the user in building up an appropriate scenario of interaction and dialogue box, as shown in Figure 4, when the user does not know what they exactly want and expect.

It has been recognized that developers need to be aware of a user's cognitive needs and cultural conditioning (Preece et al., 1994). If the user interface of WBIS is built based on the developer model but not based on the user model, then the user will have great difficulties in learning and using the system, which can result in a lot of frustration, time-wasting, and error-making. The dialogue act modeling is to build a Web site model acceptable to both users and developers.

Part Three: Object Modeling/Behavior Modeling for Building Object Model/Behavior Model for the IS Part of WBIS

In general, WBIS is currently designed and implemented using object-oriented technology in WBIS development with the descriptive view (Conallen, 2003). Therefore, in WBIS analysis, once the dialogue act model is created as the specification of

Figure 5. Object modeling and behavior modeling in WBIS analysis

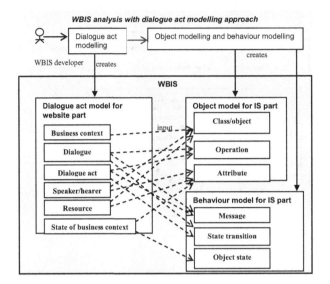

the pragmatic aspects of a WBIS, it can be used as input to the object model and the behavior model that are the specification of the semantic aspects of the same system, as illustrated in Figure 5.

Object modeling focuses on business entities (business objects) involved in business contexts and components (system objects) for design and implementation of the system. It is to build an object model for the IS part of WBIS that shows the static structure of the system. A class diagram in UML (Booch et al., 2005) is used to represent this model through this mdoeling (see the example in Figure 9).

Behavior modeling focuses on behavior of business entities and components in the object model. It is to build a behavior model for the IS part of WBIS that shows interactions between objects as well as object states in the system. Sequence diagrams and statecharts in UML (Booch et al., 2005; Harel, 1987) are used as the notation to represent this model; the sequence diagram shows the interactions between objects with messages; the statechart describes the states of individual objects in the system (see examples in Figures 10 and 11).

Part Four: The Analysis Process

The dialogue act modeling approach provides an analysis process for modeling WBIS in WBIS development. The process will be demonstrated using examples.

The Analysis Process with the Approach

The analysis process with the dialogue act modeling approach consists of four stages.

Stage One: Build a Dialogue Act Model for the Web Site of WBIS With the Pragmatic View

This stage includes the following two steps:

- **Step 1:** Elicit business contexts such as "cars" and "car search" and stakeholders (users) such as "customer" and "car company," who may use the system in future. Consider the following for identifying users and capturing business contexts:
    o   For whom is this system being developed?
    o   Who are the intended users?
    o   What roles do these users play? (Different roles imply different criteria and needs within the business context.)
    o   What are the business goals in the user perspective?
    o   What do these users expect to do with or get within the business context?

    Business contexts and users (called *actors* of the business context) and their connection are explicitly described in a business context map provided by the approach (see the example shown in Figure 6).

Figure 6. Business context map

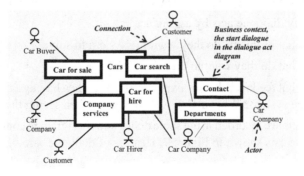

Analysis in this step focuses on the business contexts and the user need for the Web site of the system. It should show the business contexts that really interest the user.

- **Step 2:** For each of the business contexts, describe what dialogues with the Web site are expected by actors when they work in this business context online. A user-centered language should be used to name dialogues and dialogue acts in this step, as it is critical to making an easy-to-use Web site for WBIS. It must be the language (words and images) that the user can understand simply and quickly (Cato, 2001). In this case, a business language containing business terminologies may be an appropriate user-centered language, because business concepts (e.g., buy and sell) and properties (e.g., customer) described in such language should be acceptable and understandable to users worldwide who speak different languages but share a common business language in global business.

    a. Identify dialogues between the Web site and actors by asking actors:
        - What things/terms/phrases do they want to see on the Web site?
        - What conversations do they want to take with the system?
        - What feedbacks do they expect to see on the Web site?
        - What do they like or dislike in the current way of working? What are their preferable ways of working? How should the Web site be fitted into those ways in their view?
        - What skills and knowledge do they have? Are they motivated to learn and use the Web site, and how much do they like to learn?
        - Communication requirements must be captured from all the actors in this step, because they can have different points of view on dialogues.

        It is important for the company to determine the dialogues they hope to take with the user in the business context and vice versa. Dialogues are connected by dialogue flows in a dialogue act diagram, as shown in Figure 7, which was produced for the business context "cars" in Figure 6. Actors are described as the speakers/hearers of the dialogues.

    b. Identify dialogue acts by asking actors:
        - What problems do they want the system to solve?
        - What do they want to do along the dialogues?
        - What feedback do they expect to receive when having a dialogue with the system? What action will they take when seeing the feedback?

Dialogue acts are recorded in the same dialogue act diagram in connection with the dialogues, as shown in Figure 7. Hearers of dialogues are the performers of these dialogue acts. Each dialogue may be connected with many dialogue acts if it has different hearers in the business context.

c. Show states of the business context (effects of performance of the dialogue acts)

The dialogue acts have an effect on the business context, and they can change the states of it. For example, the "sell car" dialogue act made the state of the business context "cars" changed from "offered" to "ordering," as illustrated in Figure 7.

Table 2 summarizes what Figure 7 means to the Web site of the system in the "cars" business context. Figure 8 is a user interface tree that shows how this dialogue act model can be the input to the design and help design user interfaces and navigation structures in the user's perspective.

Figure 7. Dialogue act diagram for the business context "cars"

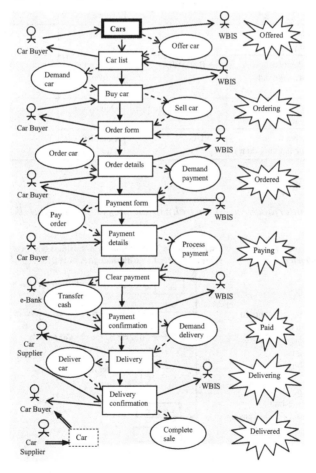

180 Liang

Table 2. The business context "cars"

{precondition} Dialogue (meaning)	Speaker-Hearer	Dialogue act performed by hearer	Resource	State of business context
{} **Cars** (want to find a car)	Car Buyer-WBIS	Offer car		Offered
{Offer car} Car list (promise to offer cars)	WBIS-Buyer	Demand car		
{Demand car} Buy car (want to own a car)	Car Buyer-WBIS	Sell car		Ordering
{Sell car} Order form (demand order details)	WBIS- Car Buyer	Order car		
{Order car} Order details (promise to buy car)	Car Buyer-WBIS	Demand payment		Ordered
{Demand payment} Payment form (want to get pay)	WBIS- Car Buyer	Pay order		Paying
{Pay order} Payment details (promise to pay)	Car Buyer-WBIS	Process payment		
{Process payment} Clear payment (want to get cash)	WBIS-e-Bank	Transfer money		
{Transfer cash} Payment confirmation (state payment accepted)	e-Bank-WBIS	Demand delivery		Paid
{Demand delivery} Delivery (want to deliver car)	WBIS-Car Supplier	Deliver car		Delivering
{Deliver car} Delivery confirmation (state delivery done)	WBIS	Complete car sale	Car	Delivered

Figure 8. User interface tree—hierarchy of business contexts and dialogues

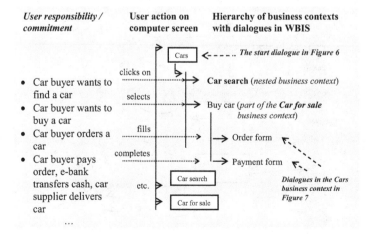

Figure 9. A class diagram for the cars business context

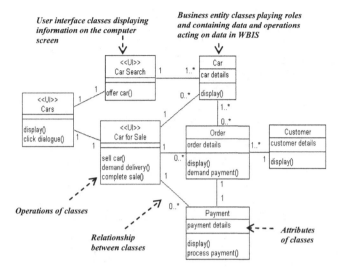

Stage Two: Build an Object Model for the IS Part of WBIS With the Descriptive View

This stage is to use the object modeling technique to build an object model for the IS part of WBIS based on the dialogue act model produced at the first stage. A class diagram in UML (Booch et al., 2005) is used to represent the object model at this stage. Figure 9 is the object model of a Web-based car sale and rental system and shows the static structure of the system. It includes classes such as "car" with attributes such as "car details" and operations such as "display car details." Classes can be identified from the dialogue act model, as illustrated in Figure 5. For example, in Figure 9, the business contexts (e.g., cars) in Figure 7 were mapped into user interface classes (UI); business entities and resources involved in the dialogues (e.g., car) were mapped into business classes; dialogues (e.g., car list) and dialogue acts (e.g., offer car) were mapped into operations of the classes (e.g., display() and offer car()); and properties of business entities (e.g., car details) were mapped into attributes of the business classes.

Classes have relationships with each other in the object model. They also can be identified from the dialogue act model. For example, Figure 9 shows a relationship between the UI class "car for sale" and the entity class "car" that was identified from the start dialogue "cars" and the 'buy car" dialogue in Figure 7.

Figure 10. Sequence diagram for the business context "cars"

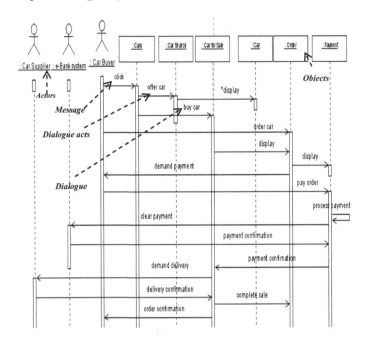

Stage Three: Build a Behavior Model for the IS Part of WBIS with the Descriptive View

This stage is to use the behavior modeling technique to build a behavior model for the IS part of WBIS based on the dialogue act model produced at the first stage and the object model generated at the second stage. Sequence diagrams and statecharts in UML (Booch et al., 2005; Harel, 1987) are used to represent various aspects of this model.

The objects of the classes in the object model have to send messages to each other in order to achieve the goal of a user's task in collaboration. This aspect of the system is represented in the sequence diagram (see the example in Figure 10). A message can be sent as a demand of operations in the objects. For example, in Figure 10, the "offer car" message is sent by the ":Cars" object to the ":Car Search" object for demanding the "offer car" operation in the ":Car Search" object.

Messages passing between objects can be identified from the dialogue act model, as indicated in Figure 5. A dialogue/dialogue act is a message. The speaker and the hearer of the dialogue are the sender and the receiver of the message. A dialogue act is an operation demanded by the sender of the message. For example, in Figure

Figure 11. State diagram for the objects of the "car" class

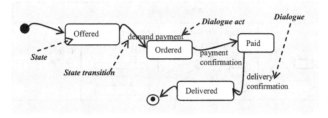

10, the "offer car" dialogue act in Figure 7 was mapped into a message sent by the ":Cars" object to the ":Car Search" object.

In WBIS, an object can be in different states at different times, and its behavior depends on these states. Events occurring in the system can trigger the transitions of the states. As shown in Figure 5, states and behavior of the objects can be identified from the dialogues, dialogue acts, and states of business contexts in the dialogue act model. For example, in Figure 11, states of the ":Car" object were identified from the states of the business context "cars," and state transitions and the behavior of the object were identified from the dialogues and dialogue acts in Figure 7.

Stage Four: Validate the Three Models (System Specification) Against Communicational Requirements

This stage is to validate the three produced models against communicational requirements. The models are modified by repeating the previous stages if they do not meet the requirements. This means that this analysis process can be iterative in WBIS analysis.

Future Trends

The dialogue act modeling approach presented in this chapter aims to cope with the WBIS culture, in particular to capture and specify communicational requirements for the Web site part of WBIS by WBIS analysis. In addition, visualization of interaction, evaluation of usability of WBIS in analysis, and modeling culture aspects of WBIS also will be important to WBIS analysis and critical to the success of WBIS. They should be addressed in the future research on the application domain analysis in Web engineering.

Figure 12. Web site structure for visualization of interaction

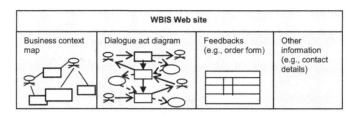

Visualization of Interaction

Our Web experience has shown that the current Web sites often hide the Web site structure from the user, and the user easily lost his or her path to reach the goal of the tasks in using the Web sites. Therefore, it will be useful if the path of the Web site of WBIS can be visualized. This can be done by displaying the business context map on the Web pages on the Web site of a WBIS, as a road map, for helping the user to find the goal of his or her tasks quickly. The dialogue act diagram produced for a business context then is displayed when the user selects the goal on the map. Feedbacks and other information also are displayed corresponding to the dialogues. The current business context and dialogue with its dialogue act(s) are highlighted in red as a milestone, telling where the user currently is on the path in interaction and what is going on (see the Web site structure in Figure 12). By looking at the dialogue act diagram, the user can understand how to get the goal of his or her task through them without confusion, frustration, and anxiety, because the user now can foresee what he or she will go through in interaction with the Web site before starting a business context. The user can control the dialogues and see what has been done and what is yet to be done in the business context.

Ideally, different groups of users with different interests and cultural backgrounds can participate in the production of business context maps and dialogue act diagrams for the Web site of a WBIS in analysis and show their ideas on interactions. Then they can compare, negotiate, and choose the business context map and the dialogue act diagrams that mostly suit their needs. This will reduce the culture conflict among users having different cultures and between the user and WBIS. In general, visualization of interaction is hoped to bring the following benefits into use of WBIS:

- It can help the user to avoid confusing and misunderstanding the business context and the interaction. In particular, it can help users who were not involved in development to understand and use business contexts, dialogues, and dialogue acts correctly and effectively by looking at the Web site structure and displayed diagrams.

- It transfers the power and control to the user in interaction. The user can work in the way he or she likes by using his or her favorite business context map and dialogue act diagrams in interaction.
- It provides a virtual guide on the Web site, which leads the user to achieve his or her goal step-by-step through dialogues. This is particularly useful when the system is new to users.
- It provides a basis for constructing a framework for developing the common interaction pattern for the WBIS culture, like the design patterns used in object-oriented development (Buschmann, Meunier, Rohert, Sommerlad, & Stal, 1996; Gamma, Helm, Johnson, & Vlissides, 1995; Liang, 2000).

Usability Evaluation in WBIS Analysis

Changes made at late stages of software development inevitably will lead to a huge increase of development cost in software engineering. Many changes required by the user are due to the late evaluation and testing of the software system in the software lifecycle. The research on software engineering has found that the cost of changes can be reduced significantly if errors are found and corrected earlier in the software lifecycle. This is the same in the WBIS development. Usability evaluation should be carried out as early as possible in order to reduce the cost of changes. However, in current Web site development, the errors and problems that led to poor usability of the Web site are often discovered at the very late stage of development, because the current development approaches do not encourage the earlier usability evaluation. New modeling approaches like the dialogue act modeling approach should be provided for capturing and specifying communicational requirements in WBIS analysis so that the developer enables the testing of usability according to them at the analysis stage.

Modeling Culture Aspects of WBIS

Culture now becomes a crucial issue in Web site design because the Internet has grown globally (Ratner, 2003). However, current modeling approaches used in WBIS analysis do not model the culture aspects of WBIS specifically, although they can significantly affect Web site modeling and design in WBIS development. The Web culture in the future should be acceptable and understandable to global users who have different culture backgrounds and live in different regions. It will provide a basis for capturing, representing, and interpreting interaction with the Web site of WBIS in the global user's perspective. The cultural aspects of WBIS will have to be modeled in analysis because of this need. These aspects can be the terminologies and languages used for recording and publishing information, culture impacts on the Web site, psychology concerns, concerns of different user groups (e.g., groups in age, gender, local social committee), and so forth. Issues on how to capture and specify the culture aspects of WBIS and in what perspectives should be addressed in future research. Cultural rules and business rules possessed by users in their social lives in different regions should be targeted when modeling these aspects. Specific groups of users, such as children, the elderly, disabled people, and so forth, also may need to be considered and consulted due to their different needs and culture rules.

Conclusion

WBIS is different from traditional IS because of its different cultures shown in Figures 1(a) and 1(b). WBIS consists of a Web site in addition to the IS part, but traditional IS only has the IS part. This difference lets WBIS have its own characteristics, such as Web pages, unlimited and untrained and global users. It also makes new challenges and problems for WBIS development; for example, how to find the user need for the Web site part of WBIS. This chapter addressed these issues by considering how to enhance the ability of WBIS analysis to cope with the WBIS culture and to meet the need of developing user-centered and interactive Web sites for WBIS.

This chapter explained why it is important to win users by providing a user-centered and interactive Web site for WBIS and why it is critical to capture and specify the user needs for such Web sites (they were called communicational requirements in this chapter) in WBIS analysis. It claimed that a well-accepted Web site of WBIS should satisfy the communicational requirements for effective interaction with WBIS, and therefore, the capture and specification of communicational requirements is an important task of WBIS analysis. It pointed out that communicational requirements are the new type of user requirements to WBIS with Web, and none of the existing modeling approaches had considered them in WBIS analysis. This chapter, therefore, promoted a new modeling approach called *dialogue act modeling approach* that was created, in particular, for WBIS analysis (see Figure 2) with a focus on communicational requirements. This approach is also hoped to be useful in WBIS design, because it can produce the interface tree (see Figure 8) as input to the user interface design.

This approach provides a new modeling technique called *dialogue act modeling* for capturing and specifying communicational requirements for the Web site part of WBIS with the pragmatic view. It also uses the existing object modeling and behavior modeling techniques (Booch et al., 2005) in analyzing and modeling the IS part of the same system with the descriptive view in WBIS analysis. The dialogue act model built by dialogue act modeling is input to the object model and the behavior model built by object modeling and behavior modeling (see Figure 2 and Table 3).

Table 3. Dialogue act modeling for WBIS requirements analysis and specification

Observation	Pragmatic View	Descriptive View	
Model	Dialogue act model	Object model	Behavior model
Requirements Analysis	Pragmatic aspects of WBIS within the business context	Data semantics of WBIS	Function Semantics of WBIS
System Specification	Actors and their responsibilities/ commitments within the business context Interaction between actors and WBIS Dialogue to act in the e-business society States of business contexts	Classes (business entities) Attributes of classes. Relationships between classes	Operations of classes (business processes) Object interactions Object behavior

This approach is different from other modeling approaches in IS analysis, because (a) it emphasizes analysis and specification of communicational requirements, while the other ones do not consider such requirements; (b) it enables the developer to observe WBIS from a user-oriented view, while the other ones observe WBIS from a developer-oriented view; and (c) it starts to develop easy-to-use and interactive Web sites for WBIS from analysis, while the other ones start to do this from design in the WBIS development.

It is hoped that the dialogue act modeling technique also can be used together with other modeling techniques, such as component modeling techniques in component-based software engineering (Heineman et al., 2005; Szypersky, 2002), because dialogues and dialogue acts in the dialogue act model can be used to create components in the component-based systems. It is also expected that the dialogue act model built by this approach can assist in writing user guides for use of WBIS and drawing users' attentions to their own responsibilities and commitments in the use of WBIS. In the future, a software tool will be produced for assisting WBIS developers in the use of this approach to develop new WBIS using computers.

References

Agerfalk, P. J. (2002). Messages are signs of actions: From Langefors to speech acts and beyond. In J. Barjis (Ed.), *The 7th International Workshop on the Language-Action Perspective on Communication Modelling* (pp. 81-100). Delft, The Netherlands: University of Technology.

Agerfalk, P. J., & Erisson, O. (2004). Action-oriented conceptual modelling. *European Journal of Information Systems, 13*, 80-92.

Austin, J. L. (1962). *How to do things with words*. Oxford, UK: Oxford University Press.

Barfield, L. (2004). *Design for new media interaction design for multimedia and the Web*. Harlow, UK: Pearson Education Limited.

Barry, C., & Lang, M. (2001). A survey of multimedia and Web development techniques and methodology usage. *IEEE Multimedia, 8*(2), 52-60.

Booch, G. (1994). *Object-oriented analysis and design with applications* (2nd ed.). Reading, MA: Addison-Wesley.

Booch, G., Rumbaugh, J., & Jacobson, I. (2005). *The unified modeling language: User guide* (2nd ed.). Reading, MA: Addison-Wesley.

Burdman, J. (1999). *Collaborative Web development*. Reading, MA: Addison-Wesley.

Buschmann, F., Meunier, R., Rohert, H., Sommerlad, P., & Stal, M. (1996). *Pattern-oriented software architecture—A system of patterns*. Chichester, UK: John Wiley & Sons.

Cambre, M. A., & Cook, D. L. (1987). Measurement and remediation of computer anxiety. *Educational Technology, 27*(12), 15-20.

Cato, J. (2001). *User-centered Web design*. Harlow, UK: Addison-Wesley.

Chen, P. (1976). The entity relationship model—Towards a unified view of data. *ACM Transaction on Database Systems, 1*(1), 9-36.

Coad, P., & Yourdon, E. (1991). *Object-oriented analysis*, Englewood Cliffs, NJ: Prentice Hall.

Conallen, J. (2003). *Building Web applications with UML* (2nd ed.). Harlow: Addison-Wesley.

Corry, M., Frick, T., & Harsen, L. (1997). User-centered design and usability testing of a Web site: An illustrative case study. *Educational Technology Research and Development, 45*(4), 65-76.

Davis, F. (1989). Perceived usefulness, perceived ease of use, and user acceptance of information technology. *MIS Quarterly, 13*(3), 319-340.

DeMarco, T. (1978). *Structural analysis and system specification*. New York: Yourdon Press.

Denning, P. J., & Medina-Mora, R. (1995). Completing the loops. *Interfaces, 25*(3), 42-57.

Deshpande, Y., et al. (2002). Web engineering. *Journal of Web Engineering, 1*(1), 3-17.

Dietz, J. L. G. (2001). DEMO: Towards a discipline of organisation engineering. *European Journal of Operational Research, 128*(2), 351-363.

Dix, A., Finlay, J., Abowd, G. D., & Beale, R. (2004). *Human computer interface* (3rd ed.). Harlow, UK: Pearson Education.

Downton, A. (1993). *Engineering the human-computer interface.* London: McGraw-Hill International.

Erickson, T., & Kellogg, W. A. (2000). Social translucence: An approach to designing systems that support social processes. *ACM Transactions on Computer-Human Interaction, 7*(1), 59-83.

Eriksen, S. (2002). Designing for accountability. In O. W. Bertelsen (Ed.), *Human-computer interaction* (pp. 177-186). Aarhus, Denmark.

Fleming, J. (1998). *Web navigation: Designing the user experience.* Sebastopol, CA: O'Reilly and Associates.

Flores, F., Graves, M., Hartfield, B., & Winograd, T. (1988). Computer systems and the design of organizational interaction. *ACM Transactions of Office Information Systems, 6*(2), 153-172.

Fuccella, J. (1997). Using user-centered design methods to create and design usable Web sites. In K. Smart (Ed.), *Systems documentation* (pp. 69-77). New York: ACM Press.

Gamma, E., Helm, R., Johnson, R., & Vlissides, J. (1995). *Design patterns: Elements of reusable object-oriented software.* Wokingham, UK: Addison-Wesley.

Goldkuhl, G., & Lyytinen, K. (1982). A language action view of information systems. In M. Ginzberg & C. Ross (Eds.), *Information systems* (pp. 13-29). Ann Arbor, MI.

Harel, D. (1987). Statecharts: A visual formalism for complex systems. *Science of Computer Programming, 8*(3), 231-274.

Heineman, G., Crnkovic, I., Schmidt, H. W., Stafford, J. A., Szyperski, C., & Wallnau, K. (Eds.). (2005). *Component-based software engineering.* Berlin: Springer-Verlag.

Holm, P., & Ljungberg, J. (1996). Multi-discourse conversations. In J. D. Coelho, T. Jelassi, W. König, H. Krcmar, R. O'Callaghan, & M. Sääksjärvi (Eds.), *Information systems* (pp. 835-848). Lisbon.

Jacobson, B. (1992). *Object-oriented software engineering—A use case driven approach.* Harlow, UK: Addison-Wesley.

Johnson, J. (2000). *GUI bloopers dont's and do's for software developers and Web designers.* San Francisco: Morgan Kaufmann.

Lang, W. (2002). Hypermedia systems development: Do we really need new methods? In E. Coyd (Ed.), *The informing science and IT education conference* (pp. 883-891). Cork, Ireland.

Lawrance, E., Gorbitt B., Tidwell A., Fisher J., & Lawrance, J. (1998). *Internet commence digital models for business*. Chichester, UK: John Wiley & Sons.

Lazar, J. (2001). *User-centered Web development*. Sudbury, MA: Johes and Bartlett.

Lazar, J., & Norcio, A. (2000). System and training design for end-user error. In S. Clark, & B. Lehaney (Eds.), *Human-centered methods in information systems: Current research and practice* (pp. 76-90). Hershey, PA: Idea Group Publishing.

Lederer, A., Maupin, D., Sena, M., & Zhuang, Y. (1998). The role of ease of use, usefulness, and attitude in the prediction of WWW usage. In R. Agarwal (Ed.), *Computer personal research* (pp. 195-204). New York: ACM Press.

Liang, Y. (2000). An approach to assessing and comparing OOA methods. *Journal of Object-Oriented Programming, 13*(3), 27-33.

Liang, Y. (2003). From use cases to classes: A way of building object model with UML. *Information and Software Technology, 45*(2), 83-93.

Liang, Y. (2004). Dialogue act-oriented conceptual modelling for Web-based information systems. In J. Rault (Ed.), *Software & systems engineering and their applications* (pp. 41-46). Paris.

Liang, Y. (2005a). Dialogue act modelling for analysis and specification of Web-based information systems. In C. S. Chen, J. Filipe, I. Seruca, & J. Crodeiro (Eds.), *Enterprise information systems* (pp. 89-96). Miami.

Liang, Y. (2005b). Dialogue act modelling: An approach to requirements analysis and specification of Web-based information systems. In N. Callaos, W. Lesso, & E. Hansen (Eds.), *Systemics, cybernetics and informatics* (pp. 369-374). Orlando.

Lynch, J., & Horton, S. (1999). *Web style guide: Basic design principles for creating Web sites*. New Haven, CT: Yale University Press.

Murugesan S., Deshpande Y., Hansen S., & Ginige A. (2001). Web engineering: A new discipline for development of Web-based systems. In S. Murugesan & Y. Deshpande (Eds.), *Web engineering: Managing diversity and complexity of Web application development* (pp. 3-13). Berlin: Springer-Verlag.

Negron, J. A. (1995). The impact of computer anxiety and computer resistance on the use of computer technology by nurses. *Journal of Nursing Staff Development, 11*(3), 172-175.

Nielsen, J. (2000). *Designing Web usability*. Indiana: New Rider.

Norman, D., & Draper, S. (1986). *User-centered system design.* Hillsdale, NJ: Lawrance Erlbaum Associates.

Preece, J., Rogers, Y., Sharp, H., Benyon, D., Holland, S., & Carey, T. (1994). *Human-computer interaction.* Workingham, UK: Addison-Wesley.

Ratner, J. (2003). *Human factors and Web development.* Hillsdale, NJ: Lawrence Erlbaum Associates.

Rumbaugh, J., Blaha, M., Premerlani, W., Eddy, F., & Lorensen, W. (1991). *Object-oriented modelling and design.* Englewood Cliffs, NJ: Prentice Hall.

Searle J. R. (1969). *Speech acts: An essay in the philosophy of language.* Cambridge, UK: Cambridge University Press.

Sommerville, I. (2004). *Software engineering* (7th ed.). Harlow, UK: Addison-Wesley.

Suzuki, K. (1997). *Prologue to cross-cultural psychology* (in Japanese). Tokyo: Brian.

Szewczak, E., & Snodgrass, C. (2002). *Human factors in information systems.* Hershey, PA: IRM Press.

Szyperski, C. (2002). *Component software beyond object-oriented programming* (2nd ed.). London: Addison-Wesley.

Winograd, T. (1988). A language/action perspective on the design of cooperative work. *Human-Computer Interaction, 3,* 3-30.

Winograd T., & Flores F. (1987). *Understanding computers and cognition: A new foundation for design.* Harlow, UK: Addison-Wesley.

Zelnic N. (1998). Nifty technology and noncomformance: The Web in crisis. *Computer, 31*(10), 115-116, 119.

Chapter IX

Challenges in Building a Culture-Centric Web Site

Tom S. Chan
Southern New Hampshire University, USA

Abstract

This chapter discusses the challenges in constructing a culture-centric Web site. The Internet has expanded business opportunities into global marketplaces that were virtually unreachable in the past. With business Web sites reaching international audiences, cultural differences are an important issue in interface design. Global Web sites must be culture-centric, taking into account the attitude, technology, language, communication, sensibility, symbolism, and interface usability of targeted communities. Site design and development also should follow the Unicode standard for multilingual support with implementation done on UTF-8-enabled operating systems and applications. Globalization has led many people to become more sensitive to cultural diversity. The author hopes that understanding and awareness of international user needs, limitations, and expectations will lead to global Web sites with improved usability and sensitivity.

Copyright © 2007, Idea Group Inc. Copying or distributing in print or electronic forms without written permission of Idea Group Inc. is prohibited.

Introduction

The Internet has revolutionized international business and global marketing. Roughly speaking, the Internet is a network of computers interconnected throughout the world and operating on a standard protocol that allows data to be transmitted. Until the early 1990s, the Internet was primarily the domain of the military and academia. The development of new software and technologies turned the Internet (Net) into a commercial medium that has transformed businesses worldwide. There is a strong international market, and businesses are taking note. From 106.4 million online buyers worldwide in 2000, the number is expected to hit 464.1 million by 2006 (Campanelli, 2004). Along with incorporating user-centric design, a business Web site must be culture-centric. A U.S. Web site designer must understand that international users' needs and expectations may be different than U.S. users. Differences in cultural attitudes, technological limitations, linguistics, communication, aesthetic sensibility, symbolism, and interface usability all must be well thought out. Furthermore, the computer platforms also must have multilingual supports.

Background

Constructing successful global Web sites involves three knowledge domains: business operation, technical standard, and interface design. Business issues facing global e-commerce operations are abundant. Chun, Honda, and Schwane (2005) show that these issues include logistics and distribution, financial and technological infrastructures, legal frameworks, and strategic business alliances. The technical standard domain deals with issues such as HTML, Unicode, character set, operating system, application, and browser supports; these are prerequisites for proper multilingual content implementation. This article focuses on both technical standard and interface design issues for global Web sites.

Research in user interface design has been centered on layout, navigation, and performance issues (Lynch & Horton, 1999, Spool, Scanlon, Schroeder, Snyder, & DeAngelo, 1999). While the research provides good guidance for page design, it does not address global site design issues. Scheiderman (1998) proposed a universal accessibility concept that addresses user diversity. Marcus and Gould (2000) studied interface design in terms of cultural perception of information content, images, icons, and symbolism. Huang and Tilley (2001) examined content and structure challenges associated with multilingual Web sites. These research efforts provided a glimpse at both the complexity and opportunity associated with Web usability design in a global economy. The research is still in its early stages, and more investigations are needed on usability, cultural, and linguistic issues.

The Digital and Cultural Divides

The term *digital divide* traditionally has described inequalities in access to technology among social or cultural groups. Although there has been a huge increase in Net speed in the industrialized world, the trend is not global. A recent survey revealed about 36 million broadband lines in the U.S. compared to 295,000 in India (Point Topic, 2005), and India is a high-tech powerhouse with a population 30 times that of the US. Many international users also use older technologies (including older computers and browsers) or lack items such as video and sound cards. Net access is only one aspect of the digital divide. The quality of connections and auxiliary services, the processing speed and the capabilities of the computer used, along with many other technical factors are also important issues (Davison & Cotton, 2003). Designers who ignore users' technological capabilities by creating sites that must be viewed with the latest technologies or with the fast connections are dooming their projects from the outset.

Culture is the collective programming of the mind that distinguishes members of one group from another group. It is also the set of values and attributes of a given group (Hofstede, 1997). The Internet is a new medium, and in a survey as recent as five years ago (Walsh, McQuivey, & Wakeman, 1999), it was reported as the least trusted medium. The situation is improving, and 25% of consumers now express a great degree of trust in online purchases (*Business Journal*, 2003). Yet, cultural differences exist in the degree of trust. For example, U.S. users trust information from the Net more than their UK cousins. However, both groups are unlikely to use Web advertisements in making purchasing decisions (Rettie, Robinson, Mojsa, & Parissi, 2003). Ethnocentricity is the tendency to look at the world primarily from the perspective of one's own culture. While the Net both facilitates and limits interaction by its virtual representation, ethnocentrism survives in cyberspace (Halavais, 2000). It is crucial to understand this cultural divide when designing Web sites for users in another region.

North American users are more likely than the rest of the world to access the Net from home. In many developing countries, access is limited by resource constraints. Cybercafés are popular in these countries. Not surprisingly, public terminal users use the Net for recreational or social purposes more than for instrumental or business purposes. Users from developing countries also have the strongest sense of online community, followed by users from other industrialized countries, with North America last (Boase, Chen, Wellman, & Prijatelj, 2003). People visit cybercafés to be with their friends, to learn languages, and to socialize with others from different cultures. For these users, the Net is more a socializing point than a venue for commerce and transactions. Even within industrialized nations, cultural divides exist: Americans are more home-oriented than Europeans; Europeans and Japanese have a more mobile life; and Japanese like gaming more than Americans.

A Web site is not the business itself. The business runs the Web site. It is important to understand that a site supports a business; and in order for a business to prosper, it must serve its users well. The first task in site design is defining its objectives. One sometimes assumes that a technically elegant system is a successful system, but this is far from the truth. Many technically sound systems fail because the people side of the system is ignored. Ultimately, the system is only a success if it meets the users' requirements, fits comfortably within their capabilities and sensibilities, and above all, makes the users happy with it.

Content Translation, Human vs. Machine

Although many non-English users can read English, they do not always appreciate it. They are more apt to enjoy the site if it is in their native language. A survey indicated that 80% of Net users have a preference for their native language and are four times more likely to purchase from sites that communicate in their own language (McClure, 2001). To seize this opportunity, a business must localize its documents, Web content, and, in some cases, product interfaces. For an English-only site, this means translation. The easiest approach to translating a site is to hire a professional translating service, but this option is expensive. Computerized translation is available. With a large number of Web pages today generated on the fly using scripting languages from contents stored in a database, a middle-tier translation engine would appear to easily solve the multilingual problem for global Web sites. Yet, computerized translation is not always accurate and should be used with great caution.

Research in machine translation began in the late 1950s. Early systems used pattern matching, semantic and syntactic approaches. Programmers wrote lists of rules that described all possible relationships between verbs, nouns, and prepositions for each language (Arnold, Balkan, Humphreys, Meijer, & Sadler, 1994). These systems were little more than word-for-word translations, not addressing the problems of word sense disambiguation and syntactic rearrangement. The results were often unpredictable and unintelligible. For example, in the sentence, "Fruit flies like a baseball," not all fruit, when thrown, would fly through the air like a baseball, except, perhaps, an apple, orange, or peach. But if one substitutes "peach" for "baseball," there is a new meaning. Then fruit flies are pesky insects crawling on a peach having a feast (Melby, 1995).

With the increase in processing speed and the decrease in storage price of computers, modern example-based systems are devoting more time to source analysis and translation synthesis. The system selects the most likely translation using statistical and mathematical models. The system can learn continuously from past translations (Rupley, 2004). Such systems will have enough intelligence to resolve earlier dilemmas. Another current approach is a computer-aided translation system in which human translators work with a computer, allowing for rapid, efficient, and high-quality translations. As translators type, the system provides them with suggestions to complete a sentence. The suggestions are created based on statistical models of machine translation engines that predict words that will come next (Esteban, 2005). Remember, translation can get complicated with regional and cultural differences within specific languages. For example, Spanish is not quite the same in Mexico as it is in Spain or Argentina, nor is Quebecois the same as Parisian French.

Nonverbal and Contextual Communication

Content is only half of the challenge in translation. Perception is not a passive, objective, and neutral process. It is a subjective experience, deeply rooted in one's personality structure and cultural environment (Maletzke, 1996). A number of nonverbal cues, such as body movements, space organization, eye movement, and touching behavior, plays a crucial role in communication (Dahl, 2004). As the Net incorporates more multimedia, nonverbal cues will play an increasing role in communication.

Translation consists of providing the closest equivalence in terms of both meaning and style. It is always a challenge dealing with a language that has a different feel and nuance embedded more in culture than in literal meaning. Accommodations in translation are often necessary in such instances (Shi, 2004). For example, Asians favor implicature in talking to people as opposed to the direct way favored by Westerners. Adjustments often are needed so that the translation does not irate the readers. Thus, proper copy tone is often critical. In the U.S., it would be fine to say, "Welcome back, Tom." In Japan, public language is much more formal, and a proper translation would be, "We are honored by your return visit, Mr. Chan." Apart from cultural attitudes, another topic deserving no less attention is politics. One must be careful with statements that can be construed to be offensive to local governments. In some regions, such neglect could carry costly consequences.

In low-context societies, communication relies more heavily on the literal meaning of words. Meanings of communication are more explicit. To people from high-context cultures, the bluntness of low-context communication styles seems insulting or aggressive. In high-context cultures, much more of the surrounding context is involved in conveying a message. Meaning in high-context communication has to be interpreted within the context of the social relationship between the individuals (DeVito, 2004). Special care must be taken when translating between these two different types of languages to ensure the preservation of the original text's intended meaning.

With the advent of globalization, the language barrier is a major obstacle to the creation of global communities. Efficient and effective translation can create linguistic transparency within these communities. Remember to translate the site fully, including all messages, labels, icons, prompts, supports, search keywords, and results. Furthermore, different cultures react to layouts differently, such as left-to-right or right-to-left. Site design must take into account text flow and object layout. Apart from semantic and contextual correctness, the text must not be culturally offensive, archaic, or nonsensical. To achieve this objective, the translated site should be validated by a focus group from the targeted local community. A site is not usable if it is merely translated into another language without any thought given to user experience.

Designing Color and Hidden Meanings

The Net has come a long way from being a collection of black and white text pages to a portal of color, image, sound, and animation. Color and images are vital tools that a Web designer must master, as it is impossible to create a site without them. However, they sometimes can have hidden meanings. When designing for international users, be aware that color and images could have symbolic meanings in different cultural settings.

Color helps a site to catch and hold attention. As the primary aesthetic tool, it helps to sustain, reinforce, and enhance a positive experience (Morton, 2001). Colors are also an element of design that people react to viscerally, though often subconsciously. Colors can solicit a universal physical reaction; for example, red has been shown to raise blood pressure. Reaction to color also can be culture-bound; white is for weddings in the U.S., but it is the color for funerals in China (Morton, 2004). Many colors, especially black, white, red, and blue, have strong and varying symbolic or religious meanings in many cultures.

Black is the absence of color. It makes objects appear to shrink in size and makes other colors appear to be brighter. The ancient Egyptians and Romans used black for mourning, as do most Europeans and Americans today. Black often stands for evil or the unknown. Among young people, black is seen as a color of rebellion and as racy and trendy. Of course, in the Wild West, good guys wore white, while bad guys wore black. The color black conveys a sense of elegance, sophistication, and mystery.

White is the presence of all colors. White is a brilliant color that can be blinding. In the West, white is the color for brides. In the East, it is the color for funerals. White is often associated with hospitals, doctors, and nurses. White signifies goodness and purity, and in many cultures, it is the color of royalty or deity (Fact Monster,

2005). The color white conveys a sense of wholesomeness, cleanliness, and softness. Colors such as beige, ivory, and cream have the same attributes but are more subdued and less blinding than pure white.

Red is a stimulant, the hottest of all colors. Studies show that red can increase the rate of respiration and blood pressure. For most cultures, red signifies power; hence, the red carpet for VIPs. But red also signals danger. In some cultures, it denotes celebration and prosperity. The Bolsheviks used a red flag when overthrowing the Tsar; thus, red became the color of communism. The color red grabs attentions, making an object stand out. A little bit of red goes a long way, and smaller dosages are more effective.

Blue is a calming color but too much blue can be depressing. In many cultures, blue is significant in religious beliefs, bringing peace and keeping bad spirits away. Blue conveys importance and confidence without the somber and sinister feeling of the color black; hence, a blue suit for the corporate world. Dark blue is associated with intelligence, stability, and truthfulness, but sometimes it is seen as stale and old fashioned. Blue is often considered the safest global color.

Colors obtain symbolism through cultural references in the society. Colors can have very different meanings depending on the culture. Thus, some colors can be risky because of their cultural significance. Yet, like all things in design, colors go in and out of fashion. Apart from cultural factors, age, gender, and even social class also can affect color preferences. By understanding the audience makeup, one can create a very effective color scheme. Remember that basic design principles still hold. Too much red or blue overwhelms a site, and black fonts on red or blue backgrounds are hardly readable.

Image, Movement, and Symbolism

A symbol is an image that stands for something, and it does not need to be explained or accompanied by words. Symbols often invoke strong feelings. While not necessarily engrained in one's subconscious, some symbols nevertheless will bring on emotional responses. The responses could be from one's upbringing, religious beliefs, or even misconceptions (Liungman, 1994). The same symbols can have different cultural interpretations. For example, in Chinese culture, dogs represent faithfulness; in Islamic culture, they represent impurity.

The Nazi swastika, for obvious reasons, evokes very powerful negative feelings. Yet, the swastika is just one of many versions of a symbol that goes back thousands of years. Apart from the history and meaning given to it by the Nazis, it is also a religious symbol that signifies health, life, and good luck in cultures from Tibet, India, Japan, and China (Badlani, 1997). To avoid problems, one should pick logos and icons from internationally accepted symbols when possible (Smith & Siringo, 1997). One also should look to current research whenever in doubt (Symbol.com, 2005).

Hofstede (1997) distinguishes between cultures in terms of their masculinity. For example, men are supposed to be assertive, tough, and focused on material success, whereas women are supposed to be modest and tender. The masculine-feminine cultural orientation, as described by Hofstede, is also applicable in cyberspace (Dormann & Chisalita, 2002). Masculine societies favor clear and distinct social gender roles, but these roles tend to overlap in feminine societies. Masculine cultures like Middle Eastern favor traditional family and gender distinctions. Thus, male-only or father-and-son images are deemed more appropriate in Web sites. Feminine cultures such as the European show less such distinction, and mixing gender images are totally acceptable.

Human sexuality is always a sensitive subject. Regions vary greatly on the degree of body exposure acceptable on the Net. Some cultures expect human images to have much of their body covered, whereas exposing most of the body is acceptable in others. For example, the same nude image may be perfectly okay in France, deemed inappropriate in North America and Asian Pacific, and downright illegal in the Middle East. Cultural sensitivity associated with female and male images must be taken into account when developing a culture-centric site. Designers are well advised to emphasize the preferred social values when designing sites in different cultural communities.

Research shows that up to 90% of the meaning in a message is transmitted nonverbally (Fromkin & Rodman, 1983). This makes communicating across cultures very difficult. For example, body movements convey specific meanings, and the interpretations are culture-bound. Many gestures have several meanings. Some gestures extend across linguistic boundaries, while others are truly national (Desmond, Collett, Marsh, & O'Shaughnessy, 1979). Many people in the U.S. use the OK gesture, symbolized by the letter O, without even thinking. In South America, the gesture carries sexual connotations, symbolizing female genitalia, and is understood to be obscene (Axtell, 1991). Imagine a Web site that has a photo of happy engineers at a major project celebration giving a classic OK gesture with their hands. Such an image would be highly offensive to Brazilian viewers. Avoid using finger gestures or feet in images because they are high-risk elements (Nieilsen, 1999).

A poor choice of colors and images can send the wrong message to users. Bear in mind that cultural constructs are dynamic and that there always will be exceptions. No two people behave in precisely the same way, nor do people from the same culture all interpret symbols uniformly. Often, some clearly understandable images in one culture are nonsense in another. Sometimes an ordinary gesture or image in one culture can be highly offensive in another.

Page Design and Global Usability

A good Web site must be visually appealing, clear, easy to navigate, and forgiving. For international users, a simple layout is the best, as it makes scanning and translating text into other languages easier. Navigation text should be concise and brief. Confusion irritates users who already may be having difficulty with the language. For multinationals, sites in different countries should be linked. This typically is achieved by using a text-based navigation bar at the bottom of the home page. A site map showing hierarchically structured areas and subareas can facilitate navigation. It can be the most helpful design element for international users.

A standardized template can greatly facilitate communication, reduce efforts, and maintain consistency across sites. It also reinforces the unique corporate branding in a global environment. The template defines all Web elements such as logo, color, navigation, icons, font size, and style, and their positioning on the page layout. It also must be in agreement with corporate branding and be in accordance with site construction best practices.

A font is a complete set of characters of one size, in one style, and of one typeface. A typeface is a set of letters, numbers, and symbols with a common weight, width, and design. A type style is a variation of the typeface, such as regular, bold, or italic. Common fonts such as Arial and Times Roman can be used for Western European languages, but they do not contain characters for languages such as Japanese, Chinese, or Russian. Since each language has its own sets of scripts with various font types, the selection process for global sites is far from trivial. As a general guideline, the selected font should go well with other design elements and with the fonts selected for other languages, and most importantly, it should be readily available in the local browsers.

In order to maintain consistency and to simplify the implementation process, logos and icons should be identical image files for all multilingual sites. Thus, text should not be embedded in a graphic object. Instead, the image should be in the background, and the text should be positioned on top of it. If text elements are part of a Web graphic, save them in separate layers, and save the graphic file in its native format before exporting to GIF or JPEG formats. Most graphic tools support layers allowing the localizable text to go into a separate layer, leaving the image layer unchanged.

The greatest challenge in multilingual site design is screen space that either expands or contracts, depending on the language used. When an interface design allows space for a particular set of words and the text length expands in translation, truncation occurs. Designers often want to get the most information into a limited space. It is common to see translated words truncated in the interface. When designing the interface, one should be mindful that the text could expand or contract. The variance could be as little as 30% for large sections of text or as much as 300% for a single word. Take precautions in form, and label designs, as they certainly will run into truncation problems. Finally, when a browser views a page with multiple languages, the default font size may cause one script to be readable and others to be illegibly small or anesthetically large. Thus, one also should set explicit font size attributes for the text for each language.

Format and Input Validation

Notations and formats are another challenge for multilingual site designers. For example, the metric system is the norm for global measurements, but the US does not use it. The following is a short list of areas to which designers should pay attention:

- **Currency:** The currency symbol and its abbreviation.
- **Text sorting sequence:** Is it alphabetical or by stroke?
- **Number format:** For example, Chinese count in units of ten thousands instead of thousands.
- **Salutation:** Mister, Herr, Señor, or Monsieur?
- **Person's name:** First name first or last name first, and what about middle name?
- **Postal address:** Street first or state first, zip code or no zip code?
- **Date format:** Month/date/year or year/month/date?
- **Time format:** 24-hour or 12-hour clock, how many time zones, daylight savings or not?

The HTML form is perhaps the most important tool in e-commerce. Since no one enjoys filling out forms, it is very important to make the usability of the form easy. Over-zealous and culturally insensitive input splitting and validation are major annoyances to international users. As already noted, formats and notations differ in many countries (Starling, 2001). The number-one complaint in form design is address verification. For example, an Australian living in South Australia has a postal code of 5000. This user should not receive a prompt stating that "SA is not a valid state" or be forced to enter "05000" as a postal code. Further, some countries such as Singapore do not have state notation, and Canada's postal codes are alphanumeric and six characters long (Cruz, 2005). Other validation insensitivities include limiting phone numbers to numeric; not accepting e-mail addresses other than .com, .net, or .org; and month/date vs. date/month formatting.

Over-splitting input fields also can be annoying to international users. For example, breaking an address into apartment, flat, street number, and street name requires four separate inputs instead of one. Likewise, using state or province to label state input expands its understanding across national boundaries. Consider using a single-name field, and let users enter what they feel comfortable with, whether that is a full name, first name first, last name first, or even a nickname.

Determine if the priority is to correctly fill up the database or to provide a positive use experience. The gospel according to Customer Relationship Management would say both! Over-restricted requirements often lead to user frustration, and often, users needing to pass through form validation will just enter junk. As with all design, simplicity works best for international users. If possible, encourage users to input the correct data rather than implementing overly strict validation.

Unicode and Multilingual Standards

As the Net grows ever more internationalized with an increasing multicultural audience, the standards and protocols supporting the Web are showing limitations in terms of multilingual supports. The original Web was designed around the ISO/IEC 8859-1, or Latin-1 character set that supports only Western European languages. This system of character encoding, ANSI, used a 256-characters set. The first 128 characters will be identical to the ANSI standard, but the second set of 128 will be taken by a different character set, which is lingual-dependent. The same number can represent a different character in different alphabet systems. Unlike reading hard copy, reading electronic documents from another country, particularly one with a different alphabet, poses a very serious challenge for Web site design and supporting technology.

An early approach to handling multilingual Web pages was to use the <FONT FACE> attribute of the HTML tag to denote non-Latin scripts. This approach has several serious drawbacks. First, an HTML document is transferred on the Net as a sequence of coded characters with each value corresponding to a standard character that the application can interpret and display. By using <FONT FACE> to specify a different text, the browser is being lied to about the identity of the characters that supposedly are identified by the standard codes in the client's computer, perhaps causing logic errors. Second, as a matter of good style and practice, just as one should use <Hn> instead of the <FONT SIZE> tag to denote hierarchy levels of the text, language is a logical and not a physical markup. For example, Chinese is logically a different type of language script and not a Latin script within a different physical layout. However, the most serious consequence is that the document now totally depends upon the availability of the particular font. Users will have to download the particular font if it is not on their machine. Asking a user to download a unique font prior to visiting the Web site is like asking a guest to bring his or her own plate and utensils for dinner. This additional viewing requirement is likely to be ignored along with the site.

The Unicode Consortium was incorporated in January 1991 to promote the Unicode Standard as an international encoding system for information interchange, to aid in its implementation, and to maintain quality control over future revisions. Unicode assigns a unique number to each character in each of the major languages of the world. It is intended to be used with a large set of special characters in all computer systems, not just Windows, and all languages, not just English. Unicode is designed to allow a single document to contain text from multilingual scripts and characters and to allow those documents to remain intelligible in electronic form, regardless of the operating system. It is an ideal language for the World Wide Web. By assigning a unique identifier to each character from the value of 0 to 65535 or x'FFFF' in hexadecimal notation, different language scripts can coexist in the same document without conflicts. The current version (4.1) of the Unicode Standard defines 97,720 characters covering the scripts of the world's principal written languages and many mathematical and other symbols (Unicode, 2005).

There are different techniques to represent a unique Unicode character in binary using either UTF-8, 16, or 32 encoding schema. To meet the requirements of byte-oriented and ASCII-based systems, UTF-8 is the most popular standard in the Net today. The variable-width encoding schema minimizes the number of bytes required to store Unicode characters and allow for efficient string parsing commonly used to encode Web contents. Each character is represented in UTF-8 as a sequence of up to 4 bytes, where the first byte indicates the number of bytes to follow in a multibyte sequence. Since plain ASCII files are already Unicode compliance, these legacy files do not need to be changed when viewing under the UTF-8 standard. The UTF-16 standard uses a 16-bit representation. The first 63,486 characters are represented in 16 bits, while the remaining 2,048 combine with a second 16-bit value to represent another 1,048,544 characters as a pair of 16-bit values. Finally, the UTF-32 standard stores Unicode characters as 32-bit integers, allowing a simple one-to-one correspondence of Unicode character to an integer.

The large number of characters in Unicode naturally poses a severe problem for font vendors and on storage requirements for systems that use the standard. Yet, Web standards are built to ensure interoperability. While standards are still evolving, the design of an internationalized Web site should respect existing standards interpreted within a multi-linguistic framework. Proper implementation of Unicode is a very important task. Even when building a U.S.-only site, the advice is to look at these issues up front in order to make the globalization process at a later date much easier.

HTML Support for Unicode

The HTML 4.0 uses Unicode as its base character set. It adopted the Universal Multiple-Octet Coded Character Set (UCS), equivalent to Unicode standard 3.0 (W3C, 1999). It also has provision for language direction, such as Arabic and Hebrew that are written right-to-left, for appropriate punctuation, and for combining of letters and diacritics (RFC 2070 - Internationalization of the HyperText Markup Language). In 2002, W3C reformulated HTML 4 as XHTML 1.0, an XML application. XHTML 1 is backward compatible with full UTF supports (W3C, 2002). Naturally, the interest here is not in teaching readers to use HTML. The focus will be on the tags and attributes relevant to the creation of internationalized multilingual HTML documents instead.

It is not necessary to have to write all Web documents in Unicode. However, the browser must be able to recognize and interpret the encoding standard used. Therefore, the encoding standard should be declared at the beginning of a document. For an HTML or XHTML file that contains Unicode data, the character encoding is specified in the *char set* parameter of a meta tag in the <head> of an HTML document; for example:

<meta http-equiv="content-type" content="text-html; char set=utf-8">

The HTTP specification mandates the use of ISO-8859-1 standard as the default character code over the Internet. The HTML specification also is formulated in terms of ISO-8859-1, and an HTML document that is transmitted using the HTTP protocol is by default ISO-8859-1 prior to HTML 4.0. UTF-8 is the current character-encoding standard for any HTML file. It allows any of the characters in the document character set to be included, while others such as ISO-8859-1, Latin-1 Western European standard, only allow for subsets.

While *char set* defines the base language of a document, HTML 4 also allows the use of the *lang* attribute to specify language and the *dir* attribute to specify direction of text flow. These attributes can be used in any HTML tag. Naturally, whether it is relevant for a given attribute depends on the syntax and semantics of the attribute and the operation involved. The following example consists of text in three languages: Greek, Chinese, and Japanese.

```
<HTML lang="el">
<HEAD></HEAD>
<BODY>
...Interpreted as Greek...
<P lang="zh">...Interpreted as Chinese... </P>
<P>...Interpreted as Greek again... </P>
<P>...Greek text interrupted by <EM lang="ja">some
   Japanese</EM> Greek begins here again... </P>
</BODY>
</HTML>
```

Language tags can be used to indicate the language of text in HTML and XML documents. An element uses the language attribute based on what is set for the element itself, or the closest parent element's language setting. Information is inherited along the document hierarchy when inner attributes overwrite outer attributes.

A language tag is composed of a primary tag, followed by zero or more optional tags, separated by hyphens. The primary tag represents a language, and any following tags serve to qualify the dialect of the language. In the example, the primary language of the document is Greek ("el"). One paragraph is declared to be in Chinese ("zh"), after which the primary language returns to Greek. The following paragraph includes an embedded Japanese ("ja") phrase, after which the primary language returns to Greek. The list of valid language codes for the primary tag is controlled and maintained by the ISO 639-2 Registration Authority (Library of Congress, 2005).

Optional tags can be added to indicate geographic, dialectal, script, or other refinements to language defined in the primary tag. The 2-letter optional tag must be valid ISO 3166 country codes (ISO, 2005). Any number of optional tags can follow the primary tag, although it is unusual to see more than one. Currently, there are no rules for any third and subsequent optional tags that are being used. For example, one can use an optional tag to specify English "en" on the web page as either the U.S. version of English "en-US" or the Great Britain version of English "en-GB."

```
<P lang="en-US">...In United States, my car needs gasoline to run ... </P>
<P lang="en-GB">...In Britain, I need to find some petrol instead ... </P>
```

It is also possible to register both primary and optional language tags with IANA, and the IANA tags can have three- to eight-letter optional tags (IANA, 2005). Using IANA-registered tags can be beneficial in some instances. For example, under the ISO standards, Simplified Chinese is defined using "zh-CN" for Mainland China as compare to Traditional Chinese using "zh-TW" for Taiwan. Apart from the possibility of mislabeling or misleading, one could not guarantee that others would recognize these conventions or even follow them. For example, some people would use "zh-HK" (Hong Kong) to represent Traditional Chinese. IANA defines the "zh-Hans" tag for Simplified and the "zh-Hant" tag for Traditional Chinese, which definitely would eliminate confusion and increase interoperability. On the other hand, IANA tags may be deprecated as new codes are added to the ISO standard. For this reason, there may be some risk to long-term interoperability when using IANA-registered tags.

Future Trends

With continuing deployment and improvements in technology, more people than ever are interconnected. As the global village shrinks and cultures collide, we must become more sensitive to the myriad cultures surrounding us. Constructing an effective culture-centric Web site requires both technical and design understandings. The technical aspect is becoming more transparent as the multilingual frameworks are well-standardized; and XML/UTF-enabled platforms and applications are widely available. The design aspect remains a challenging and learning process that requires designers keeping current with researches on the subject. In recent years, however, Western and contemporary ideas have become more globally popular. These ideas either have influenced or even replaced some traditional values. Understanding human behavior is getting even trickier nowadays.

Conclusion

Web sites are usually an international user's first glimpse of a company. A good first impression is important to a company, as business is about capturing attention. Globalization has led many people to become more sensitive to cultural diversity. Most know that different things have different meanings to people in different places. This is especially important to businesses that market products and services across geographical boundaries. A Web site designer must be aware of international user needs, limitations, and expectations, and must be culture-centric. Globally, Net usage is growing enormously. Because of the competitive nature of business, cultural usability and sensitivity are critical components for any global business site. The success or failure of the site, and consequentially the business, ultimately is determined by evaluations of real users from the targeted communities.

References

Arnold, D., Balkan, L., Humphreys, L., Meijer, S., & Sadler, L. (1994). *Machine translation: An introductory guide*. London: Blackwell Publishing.

Axtell, R. (1991). *Gestures: The do's and taboos of body language around the world*. New York: John Wiley & Sons.

Badlani, C. (1997). Nazi swastika or ancient symbol? Time to learn the difference. *An End to Intolerance, 5*. Retrieved August 17, 2005, from http://www.iearn.org/hgp/aeti/aeti-1997/swastika.html

Boase, J., Chen, W., Wellman, B., & Prijatelj, M. (2003). Is there a place in cyberspace: The uses and users of public Internet terminals? *Proceedings of the Association of Internet Researchers Conference*, Toronto, Canada.

The Business Journal. (2003, January 7). Survey: Consumer trust of Internet grows. *The Business Journal*. Retrieved August 4, 2005, from http://southforida.bizjournals.com/southflorida/stories/2003/01/06/daily54.html

Campanelli, M. (2004). A world of goods. *The Digital M, 3*(2). Retrieved August 3, 2005, from http://www.digitalm.biz/Issues/April04/WorldOfGoods.html.

Chun, M., Honda, G., & Schwane, C. (2005). An uphill battle. *Graziadio Business Report, 8*(2). Retrieved August 22, 2005, from http://gbr.pepperdine.edu/053/china.html

da Cruz, F. (2005). Frank's compulsive guide to postal address. *The Kermit Project*. New York: Columbia University. Retrieved August 21, 2005, from http://www.columbia.edu/kermit/postal.html

Dahl, S. (2004). Intercultural research: The current state of knowledge. *Middlesex University Discussion Paper No. 26*. Retrieved August 16, 2005, from http://ssrn.com/abstract=658202

Davison, E., & Cotton, S. (2003). Connection discrepancies: Unmasking further layers of the digital divide. *First Monday, 8*(3). Retrieved August 5, 2005, from http://firstmonday.org/issues/issue8_3/davison/index.html

DeVito, J. (2004). *The interpersonal communication book*. Boston: Allyn & Bacon.

Dormann, C., & Chisalita, C. (2002). Cultural values in Web site design. *Proceedings of the ECC11 & SAFECOMP 2002 Usability Forum Conference*, Catania, Italy.

Dsmond, M., Collett, P., Marsh, P., & O'Shaughnessy, M. (1979). *Gestures: Their origins and distribution*. New York: Stein and Day.

Esteban, J. (2005). Computer aid ensures speedy, high-quality translations. *Information Society Technologies*. Retrieved August 15, 2005, from http://istresults.cordis.lu/index.cfm/section/news/tpl/article/BrowsingType/Features/ID/73666.

Fact Monster. (2005). What colors mean. *Fact Monster from Information Please.* Retrieved August 17, 2005, from http://www.factmonster.com/ipka/A0769383.html

Fromkin, V., & Rodman, J. (1983). *An introduction to language.* New York: CBS College Publishing.

Halavais, A. (2000). National borders on the World Wide Web. *New Media and Society, 2*(1), 7-28.

Hofstede, G. (1997). *Cultures and organizations: Software of the mind.* New York: McGraw Hill.

Huang, S., & Tilley, S. (2001). Issues of content and structure for multilingual site. *Proceedings of the Annual ACM Conferences on Systems Documentation,* Santa Fe, New Mexico.

IANA. (2005). *Tags for the identification of languages. Internet assigned numbers authority.* Retrieved September 20, 2005, from http://www.iana.org/assignments/language-tags.

ISO. (2005). English country names and code elements. *International Organization for Standardization.* Retrieved September 20, 2005, from http://www.iso.org/iso/en/prods-services/iso3166ma/02iso-3166-code-lists/list-en1.html

Library of Congress. (2005). Codes for the representation of names of languages. *Library of Congress.* Retrieved September 20, 2005, from http://www.loc.gov/standards/iso639-2/langcodes.html

Liungman, C. (1994). *Thought signs: The semiotics of symbols—Western non-pictorial ideograms.* London: IOS Press.

Lynch, P. J., & Horton, S. (1999). *Web style guide: Basic design principles for creating Web sites.* Boston: Yale University Press.

Maletzke, G. (1996). *Interkulturelle kommunikation.* Opladen: Westdeutscher Verlag.

Marcus, A., & Gould, E. (2000). Cultural dimensions and global user interface design. *Proceedings of the 6th Conference on Human Factors and the Web,* Austin, Texas.

McClure, S. (2001). Language matters. *IDC Viewpoint.* Retrieved August, 5, 2005, from http://www.idc.com/getdoc.jsp?containerId=VWP000061

Melby, A. (1995). *Why can't a computer translate more like a person?* Retrieved August 15, 2005, from http://www.ttt.org/theory/barker.html

Morton, J. (2001). *Color voodoo for e-commerce.* Color Voodoo Publications. Retrieved September 20, 2005, from http://www.colorvoodoo.com/cvoodoo7.html

Morton, J. (2004). *Global color: Clues & taboos.* Color Voodoo Publications. Retrieved September 20, 2005, from http://www.colorvoodoo.com/cvoodoo2_globalcolor.html

Neilsen, J. (1999). *Designing Web usability.* Indianapolis: New Riders Publishing.

Point Topic. (2005). World broadband statistic. *Point Topic.* Retrieved August 5, 2005, from http://www.point-topic.com.

Rettie, R., Robinson, H., Mojsa, M., & Parissi, E. (2003). Attitudes to Internet advertising: A cross cultural comparison. *Proceedings of the Academy of Marketing Conference,* Aston, UK.

Rupley, S. (2004, July). Scaling the language barrier. *PC Magazine.* Retrieved August 15, 2005, from http://www.pcmag.com/article2/0,1759,1612204,00.asp

Scheiderman, B. (1998). *Designing the user interface: Strategies for effective human-computer interaction.* Boston: Addison Wesley.

Shi, A. (2004). Accommodation in translation. *Translation Journal, 8*(3). Retrieved August 15, 2005, from http://www.accurapid.com/journal/29accom.htm

Smith, A., & Siringo, M. P. (1997). The acceptability of icons across countries. *Proceedings of the 7th International Conference on Human Computer Interaction,* San Francisco, California.

Spool, J., Scanlon, T., Schroeder, W., Synder, C., & DeAngelo, T. (1999). *Web site usability: A designer guide.* San Francisco: Morgan Kaufman.

Starling, A. (2001). Usability and HTML forms. *Web Developer's Virtual Library.* Retrieved August 21, 2005, from http://wdvl.internet.com/Authoring/Design/Basics/form1.html

Symbol.com. (2005). *Online encyclopedia of graphic symbols.* Stockholm, Sweden: HME Media. Retrieved August 21, 2005, from http://www.symbols.com/index.html

Unicode. (2005). *Unicode 4.1.0.* Retrieved September 20, 2005, from http://www.unicode.org/versions/Unicode4.1.0

W3C. (1999). The HTML 4.01 Specification. *W3C Recommendation.* Retrieved September 20, 2005, from http://www.w3.org/TR/html401

W3C. (2002). XHTML 1.0 The Extensible HyperText Markup Language (2nd ed.). *W3C Recommendation.* Retrieved December 20, 2005, from http://www.w3.org/TR/xhtml1

Walsh, E., Mcquivey, J., & Wakeman, M. (1999, November). Consumers barely trust Net advertising. *Forrester Research.*

Section IV

Knowledge Management Challenges

Chapter X

Information and Knowledge Management for Innovation of Complex Technologies

Ning Li
University of Guam, Guam

Don E. Kash
George Mason University, USA

Abstract

This chapter investigates the role of information and knowledge management in innovation of complex technologies. A conceptual framework for three patterns of technological innovation (normal, transitional, and transformational) is presented, and the process of information and knowledge management in accessing and using knowledge is analyzed. Particularly, emphasis is put on the cultural impact on the information and knowledge management processes. Five case studies of evolving technologies carried out in the United States, Japan, Germany, India, and China are used to elaborate the conceptual framework and key points presented in this chapter. Lessons for managers and public policymakers concerned with facilitating the innovation of technologies are discussed.

Copyright © 2007, Idea Group Inc. Copying or distributing in print or electronic forms without written permission of Idea Group Inc. is prohibited.

Introduction

Complex technologies are the most valuable technologies in the global market. In the 21st century, the capability of repeated innovation of complex technologies is crucial in order for a country to become or to remain advanced. Networks are essential to the innovation of complex technologies, as networks involve linkages and relationships that make it possible to access, create, synthesize, and diffuse the diverse tacit and codified knowledge.

The capacity of effective management of information and knowledge needed for innovation is a basic requirement for firms that want to be competitive in both domestic and international markets. Information and knowledge management is defined as the collection of processes of the creation, dissemination, and utilization of information and knowledge. Accordingly, information and knowledge management for innovations refers to creating, diffusing, learning, and using information and knowledge to promote technological innovation.

This chapter investigates the role of information and knowledge management in innovation of complex technologies. A discussion of three patterns of technological innovation is presented, and the role of information and knowledge management in accessing and using knowledge is analyzed. Particularly, emphasis is put on the cultural impact on the information and knowledge management processes. This chapter also suggests lessons for managers and public policymakers concerned with facilitating the innovation of technologies.

The research method employed in the chapter is case study. Given the problems in measurement and the availability of empirical data regarding technological innovation and cultural dimensions, a quantitative analysis is hard to achieve. On the other hand, however, the case study method, despite its disadvantage in generalization, provides more insights that help to understand the dynamics of causes and effects. Five case studies of evolving technologies carried out in the United States, Japan, Germany, India, and China are used to elaborate the conceptual framework presented in this chapter. The previous five countries represent five different cultures.

Innovation Patterns of Complex Technologies

Complex Technology

The significance of the capacity to carry out technological innovation in long-term economic growth long has been recognized. As Freeman and Soete (1997) point out, in the classic works of Smith, Richardo, Marx, and Malthus, invention and innovation are regarded as the most dynamic contributors to expanding markets. Modern growth theories, such as the neoclassical growth model, the new growth theory, and evolutionary economics, also heavily address the crucial role of innovation through introduction, diffusion, and continuous improvements of products and processes in economic growth (Nelson & Winter, 1982; Romer, 1986; Solow, 1956).

This chapter postulates that a broad-based advanced economy requires the capacity to carry out repeated innovations of complex technologies. Complex technologies are those that cannot be understood adequately by an individual so that the technology can be communicated across time and distance for precise reproduction. In contrast, simple technologies can be so understood (Rycroft & Kash, 1999).

Complex technologies are the largest and a growing component of the high value and high value-added technologies traded in the world. Simple goods produced with simple manufacturing processes are mostly commodity products (sold on the basis of price), and competitive advantage usually is associated with low labor costs. Alternatively, the source of competitive advantage for most complex technologies comes from the capacity to use synthesis to carry out repeated innovations ahead of or in parallel with competitors (Osborn & Hagedoorn, 1997; Senker & Sharp, 1997). In 2001, among the top 30 most valuable manufactured traded globally, complex manufactured goods represented 86% (Kash, Auger, & Li, 2004).

Note that complex technology goods are different than high-tech goods. High-tech goods are those that have significant R&D intensity, usually measured by the ratio of R&D expenditure to output value (Hatzichronoglou, 1997). The auto industry, the most valuable manufactured goods export sector in 2001, is not high-tech but rather is complex. Typical examples of complex manufactured products are aircrafts, autos, transistors, pharmaceuticals, and telecommunication equipment. In contrast, shoes, clothing, and toys are of simple products.

Networks

The role of networks in technological advances is addressed by scholars in the study of innovation patterns and processes. Various systematic and synthetic innovation models, such as the system model by Soete and Arundel (1993), the system integration and networking (SIN) model by Rothwell (1994), and the complex model by Rycroft and Kash (1999), state that synthetic innovation has become the most dynamic component of our production system and has caused fundamental changes in our physical and organizational reality. These models view technological innovation as a process of co-evolution between the technologies and the organizational networks.

Networks are composed of multiple organizations that, at a minimum, range from the holders of core capabilities to the suppliers of complimentary assets to user organiza-

Figure 1. Organizational network

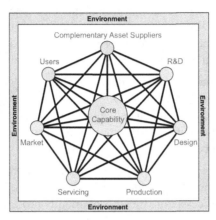

tions (see Figure 1). Networks involve linkages and relationships that make it possible to access, create, synthesize, and diffuse the diverse tacit and codified knowledge necessary for the innovation of complex technologies. No organization acting alone has the capability to carry out the innovation of complex technologies.

Networks are essential to the innovation of complex technologies because they provide the capacity to do what neither conceptual systems (theories) nor individuals can do. Networks offer a way to access and use diverse knowledge that is both codified and tacit in order to accomplish what has not been done before without an understanding of how to do it (Kline, 1995). Networks thus offer two critical capabilities. First, they provide a way to access, create, and use (synthesize) diverse knowledge that is located in multiple organizations and individuals. Second, they provide the structure needed for the range of people and organizations that hold the diverse knowledge to participate in decision making (Sing, 1997).

Innovation Patterns

Although it is not possible to understand the cause-and-effect relationships essential to the innovation of complex technologies, it is possible to associate technologies and innovation processes with one of three patterns (see Figure 2). The patterns are distinguished by different kinds of innovations and by the ways in which the networks access, create, and utilize knowledge and information and make decisions in carrying out the innovations.

The first pattern—transformational—is distinguished by the innovation of first-of-a-kind technologies. The transformational innovation pattern is represented in Figure 2 by the square at the beginning of the trajectories. When the first-of-a-kind

Figure 2. Generic innovation trajectories

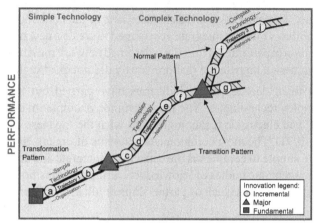

technology is complex, a network of organizations that range from complementary asset suppliers to end users must be formed in order to incorporate the diverse knowledge needed for the innovation. Thus, within the transformational innovation pattern, both a new network and a new technology must be designed and built, in part, through a trial-and-error learning process. If the first-of-a-kind technology is simple, then a network of organizations, if not required, and the sources of knowledge and information for innovation are less diverse.

The second pattern—normal—is distinguished by innovations that deliver enhancements of established technology designs; for example, the ever more powerful microprocessors emerging from incremental innovations. The normal innovation pattern is represented in Figure 2 by the circles on the trajectories. Incremental innovations are predominately the product of cumulative learning within the network that was established to innovate the first-of-a-kind technology. Normal innovations also require diverse knowledge and participation by organizations that range from complementary asset suppliers to end users. However, during the normal innovation pattern, the linkages among the participants in the network already exist, and the network is able to generate internally most of the knowledge needed to carry out the incremental innovations. Thus, during the normal pattern, new innovator competitors can work within an established design (e.g., VHS design for VCR systems), and they can work in an environment where there is general agreement on what will be the next increment in performance (Kash & Rycroft, 2002). Under normal innovation pattern conditions, the innovation challenge is to figure out how to achieve the next step in performance within the established design.

The third pattern—transitional—is distinguished by innovations that deliver major redesigns; for example, the synthesis of electronics and mechanics. The transitional innovation is represented in Figure 2 by the triangles that connect the trajectories. Like the technologies, the networks that carry out transitional innovations must undergo major redesigns. The most distinctive characteristic of the network redesign is that new core capabilities must be integrated or merged to create a new network with the needed innovation capabilities. Thus, the redesigned network must be able to access, create, and synthesize knowledge that previously did not exist in the network.

Over the last three decades many industries have carried out transitions from simple to complex technologies. The most common example involved the fusion of mechanical and electronic technologies into what the Japanese call *mechatronics* (Kodama, 1991). The most common characteristic of organizations making the transition from simple to complex is the capacity to integrate and achieve synthesis from previously separate bodies of knowledge. During the transition, the formation of organizational networks, such as partnerships or alliances, is required.

Information and Knowledge Management for Innovation of Complex Technologies

Effective and efficient management of information and knowledge is essential in facilitating synthesis processes in technological innovation. It is very important to build up the capacity of organizational learning through information and knowledge management (Dutrenit, 2000). It is also very important to realize that, as technologies develop and become more complex, the information and knowledge requirements for innovation increase dramatically.

One of the main goals of information and knowledge management is to ensure the optimal use of information and knowledge. For a company, the ideal situation would be that any information or knowledge generated internally and externally should optimally flow freely to all parts of the company where it can be of use. However, information and knowledge never flow completely and freely within and between organizations for optimal use. Well-organized and efficient information and knowledge management is needed to promote the flows of information and the exchanges of knowledge within innovation networks and to reduce the costs associated with the processes of flows and exchanges.

Information management in firms evolves with the advancement of information and communication technology and the changing environment of business. The main propensity is toward more diversified systems and more involvement by people. Keen (1995) identifies four stages of the evolution of information management: data processing (DP), management information systems (MIS), information innovation and support (IIS), and business integration and restructuring (BIR). The DP stage focuses on data processing with expensive mainframe computers and application software programs that are individually written. The MIS stage shifts the focus to building reporting systems for management with packaged software, which helps decision making and end-user computing based on database management. At the IIS stage, PCs and software packages are widely available and applied to such processes as data resources sharing, computer-integrated manufacturing, just-in-time inventory management, and customer service. Rather than control, coordination becomes the emphasis of information management. For the BIR stage, the driving force is not merely advance in information technology. Instead, information management is organized according to the need of business process and relies more on interactions that involve people and machines. The purpose of information management is to promote survival and prosperity of enterprise.

The evolution of information management also co-evolves with technological innovation. Simple technology often is associated with organizational hierarchy and specialization of labor. Information management for innovations of simple technologies does not require free information flows to every branch of the organization. In contrast, complex technology is associated with networks and organizational learning. Information management for innovations of complex technologies requires not only flows of information but also exchange of codified and tacit knowledge within the network.

With regard to sharing of idea and organizational learning, knowledge management has become a hot topic in recent years. In the core of knowledge management is organizational learning. Learning is the process of creating knowledge and acquiring technological capacities by innovative firms. Organizational learning and network learning are critical to innovation of complex technologies, as complex technologies cannot be understood fully by any individual.

The process of organizational learning includes the creation of tacit and codified knowledge and the promotion of conversion between the two dimensions of knowledge. Codified knowledge can be articulated in formal language, while tacit knowledge is more personal and embedded in individual experience and, thus, cannot be transferred easily among people. Knowledge management is to facilitate the creation of and conversion between the two kinds of knowledge.

Nonaka and Takeuchi (1995) identified four modes of knowledge conversion and their implications for knowledge creation of each of the modes. Conversion from tacit to tacit knowledge is done through the process of socialization by sharing experiences and creating tacit knowledge. Conversion from tacit to codified knowledge is the process of externalization by articulating tacit knowledge into codified forms. Conversion from codified to codified knowledge is the process of combination, which involves diverse bodies of codified knowledge. This is the main task of traditional information management. Conversion from codified to tacit knowledge is the process of internalization by which an individual embodies codified knowledge into his or her personal knowledge base.

Note that innovation of complex technologies has different patterns and the requirement for information and knowledge management for these patterns also differs to a large extent. This is because of different learning styles required in different innovation patterns. Organizational learning in the normal pattern is local or internal. Learning is largely dependent upon routines and heuristics that have been done in the past. Local learning is a dynamic process that involves integrating new knowledge from a variety of projects occurring along the same trajectory.

Organizational learning during transitional innovation often is associated with the transfer of a knowledge base from a narrow but deep one to a broader one. Learning is regional, which means that external information and knowledge sources are involved, although the primary technological area remains the same. The shift of learning from local to regional is particularly difficult and risky. Trial and error is the basic strategy for learning and an important source of feedback.

Learning during transformational innovation can be characterized as cosmopolitan. Technological area shifts to a new one, and learning occurs in areas that are unrelated to past learning experiences. It often is associated with redefining organizational strategy and structure as well as existing routines and decision-making procedures. Information and knowledge sources are diverse, and searching for useful sources becomes critical.

Cultural Impact on Information and Knowledge Management for Technological Innovation

Technological innovation is a sociotechnical process and is not free from impacts of society. Culture—whether national or company—can facilitate or impede the ability of networks to use information and knowledge and, thus, the ability of networks to synthesize and successfully carry out the innovation of complex technologies. Culture here is defined as learned and shared values and attitudes and the way of thinking and acting in a society or organization, or among a group of people.

Culture has many different dimensions and can be studied through the ways people solve problems of such cases as relationships with others, attitudes toward time, and attitudes toward environment. For example, Hall (1976) identifies differences in the ways people use languages while communicating with others. Low-context cultures are explicit in using the spoken and written word, while high-context cultures use work to convey only part of the spoken or written message, leaving the rest to be inferred or interpreted from the context. The inference or interpretation normally is done through body language, physical setting, and past relationships. In this regard, many Asian languages are considered high-context, whereas most Western cultures are low-context.

Scholars of culture studies have done some systematic research on cultural differences. A five-dimension framework developed by Hofstede (2001) provides one approach for understanding how value differences can influence human behavior. The five dimensions are as follows: (1) power distance as the willingness of a culture to accept social status and power differences; (2) uncertainty avoidance as the tendency toward risk taking and discomfort with ambiguity; (3) individualism-collectivism as the tendency of a culture to emphasize either individual or group interests; (4) masculinity-femininity as an indicator to value masculine or feminine traits; and (5) long-term/short-term orientation as a tendency to emphasize values in the future.

In the field of technological innovation, although the point that culture plays a significant role in promoting or impeding innovations is widely accepted, detailed studies of how culture influences innovation processes are surprisingly rare. Kline (1990, 1991) pioneered the study by comparing differences in innovation styles between Japan and the United States but failed to develop a conceptual framework. The Hofstede framework aims at general human behavior and does not work very well for the study of the role of information and knowledge management in technological innovation. In order to determine the role that national and company cultures play in facilitating or impeding the use of information and knowledge in promoting innovation of complex technologies, this chapter develops a new framework to accommodate issues related to technological innovation. The new framework is based on Hofstede's work but with certain revision and expansion. It includes the following cultural characteristics: trust, comfort with tacit vs. codified knowledge, equality vs. inequality, individualism vs. collectivism, and decision-making style.

Trust is a great contributor to efficiency, as high trust means low transaction cost. Organizations that can rapidly access, create, and integrate the needed codified and tacit knowledge have great competitive advantages. In high trust systems, knowledge is moved, created, and used rapidly, because those who give the knowledge are confident it will be used to their advantage or, at a minimum, not to their disadvantage. Further, those who give knowledge assume that they will receive knowledge in return at some future time.

Cultures vary in the degree to which they are comfortable with tacit knowledge. For example, American culture is comfortable with codified knowledge but uneasy with both tacit and synthetic knowledge, the kind produced by systems integration. Both Japan and China appear to be more comfortable with tacit knowledge; certainly, they are societies that find synthetic knowledge comfortable.

Inequality appears in many forms (e.g., family, organizational position, race or religion, age, education, experience, etc.). A culture that emphasizes inequality among people is likely to be a barrier to rapid knowledge transfer and use. High speed of transfer occurs when those who have the knowledge and those who need it communicate directly and on the base of mutual respect.

Successful innovations require that members of organizational networks exchange knowledge freely and rapidly. A society that emphasizes individualism has barriers to giving knowledge away. In an individualistic society, being recognized as the creator or source of valuable knowledge is how one gets recognition and status. At a minimum, individualistic societies slow down the movement of knowledge.

Decisions may be made by individuals, vote, or consensus. Systems integration succeeds on the basis of a decision-making process that selects and tries solutions to problems that emerge from individual and group interactions. Detailed decisions by individuals in positions of organizational authority often are costly, because individuals always have inadequate knowledge. Individual decision making reduces the exchange of knowledge and information.

Where national culture tendencies appear to facilitate innovation, companies can create and maintain cultures that are consistent with the national culture. Where national cultural tendencies impede innovation, companies can create and maintain cultures that compensate for the broader norms. Similarly, public policies can be adopted that take into consideration the necessary organizational requirements and cultural requirements. These policies can take advantage of cultural traits conducive to innovation and attempt to mitigate the effects of cultural traits that impede innovation.

The times that require policy changes or company strategy changes are the transformation and transition periods. If a country or company has a culture that supports the use of tacit knowledge and diffuses decision making and can act quickly on what are effectively consensus decisions, then it will have an advantage similar to the one that Japan traditionally has had with respect to the incremental innovation of complex technologies. For example, Intel has created a company culture that has many of the characteristics of Japanese culture. Intel's company culture emphasizes the importance of giving information and knowledge away quickly, diffusing decision making, and teamwork. It has fostered a cultural environment of "disagree and commit": once decisions are made, everyone actively supports the decision. In Japan, these kinds of explicit company rules are not required, as the traits of the national culture already support these organizational requirements.

Five Cases, Five Cultures

Only detailed case studies provide insight into how companies move along technological trajectories while managing information and knowledge. The five case studies summarized in the following seek to trace the development of networks and their capabilities of synthesis over the history of the five technologies. Figures 3-7 show the development trajectories and different innovation patterns of five technologies

Figure 3. Bosch innovation trajectories

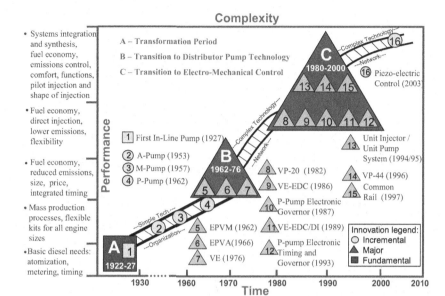

produced by organizational systems centered in five countries: the Bosch (Germany) diesel fuel injection technologies; the Sony (Japan) audio compact disc (CD); the Hewlett-Packard (HP) (U.S.) cardio-imaging technology; the TATA Consultancy Services (TCS) (India) software services and products; and the Haier (China) line of home appliances (Kash, 2001, 2002a, 2002b; Kash, Lieberman, & Schaller, 2000; Kash & Schaller, 2000).

Some complex technologies have evolved from simple technologies, as is the case of the Bosch diesel fuel injection technology (Figure 3) and the TCS software (Figure 6). Some first-of-a-kind technologies arrive as complex systems, as is the case of the Sony CD (Figure 4) and the HP cardio-imaging technology (Figure 5). Some technologies have not crossed the threshold from a simple to a complex technology, as is the case of the Haier home appliances (Figure 7).

Simple technologies do not require organizational networks for their innovation, while complex technologies always do. One of the most difficult challenges faced by companies and countries occurs when it is necessary to make the organizational adaptation from the innovation of simple to the innovation of complex technologies. The simple to complex adaptation requires replacing a hierarchical organization that manages a narrow range of knowledge with a network that synthesizes a diverse range of knowledge. Bosch carried out this adaptation, for example, when it evolved from the innovation of all mechanical technologies to electro-mechani-

Figure 4. Sony CD systems innovation trajectories

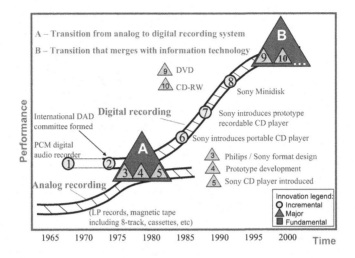

Figure 5. HP cardio-imaging innovation trajectories

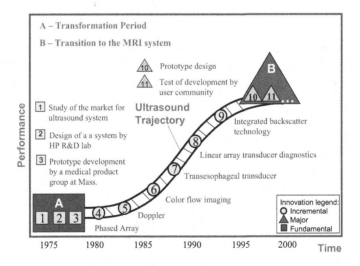

cal technologies. This was a period in which a network had to be built that could synthesize electrical, electroni,c and mechanical knowledge.

For those networks already successfully carrying out the innovation of complex technologies, the major adaptation challenges occur when it is necessary to move from one innovation pattern to another. These adaptations commonly require either

Figure 6. TCS software technology trajectories

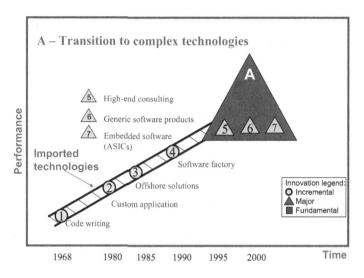

Figure 7. Haier appliances technology trajectories

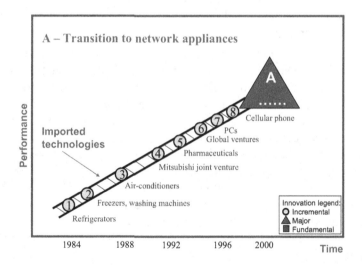

the integration of new organizations with new core capabilities into an existing network or the merger of existing networks for the purpose of producing a major technology redesign. This is evident in all of the case studies except that of Haier.

Following is a detailed illustration of the five case studies. Elaboration of innovation in the five cases is organized in line with different innovation patterns (transformational, transitional, and normal). As one of the major indicators of changing patterns in knowledge management of complex technologies, increasing diversity of knowledge is discussed through the cases. Influences of cultures on knowledge management and innovation are of particular interest in this chapter, and the case studies demonstrate the significance of cultural impact.

Patterns of Innovation

Transformational Pattern

The Bosch diesel fuel injection technology case and the HP cardio-imaging technology case include transformational innovations. In the Bosch case, the transformational innovation pattern encompassed the period in which a simple mechanical technology was produced. In the HP case, the innovation pattern encompassed the period in which a complex electro-optical-mechanical technology was produced.

The Bosch innovation resulted from the identification of a market opportunity through personal interactions between Robert Bosch and the designers of a new engine—the diesel. An initial injection technology design was purchased from a Swiss company, and skilled mechanical craftsmen within the Bosch workshop produced a simple mechanical design for a diesel fuel injection system that remained the dominant design for nearly four decades.

The HP cardio-imaging innovation resulted from a search for a new market for the company's medical electronics core capabilities. A market consultant hired by HP recommended development of the cardio-imaging technology, even though most of the potential users—medical doctors—envisioned little need for it. The first-of-a-kind design required developing a number of first-of-a-kind subsystems (e.g., phased array transducer). A network was created that linked the HP R&D laboratory to the manufacturing facility and those two organizations to lead-user medical doctors and subsystem suppliers. Knowledge that ranged from microwave radar to digital electronics and from signal processing to cardiovascular diagnosis had to be synthesized. Once a workable technology had been designed, it was necessary for the network to link with medical training facilities, medical schools, and hospitals in order to train the doctors and technicians who would be the end users of the technology.

Normal Pattern

The normal innovation pattern, in which established networks carry out incremental innovations that enhance existing designs, is the most characteristic pattern evident in all five case studies. Innovations that occur within the normal pattern produce the major economic payoffs for complex technologies, as innovation-to-obsolescence cycles structure market success for complex technologies (Rycroft & Kash, 1999). Each incremental improvement creates new demands and reduces or kills the demand for existing technologies. The key to success, then, is to carry out the incremental innovations ahead of or in parallel with competitor networks.

Within the normal pattern, the lack of understanding that networks must deal with concerns how to accomplish the next innovation. Alternatively, there is seldom lack of understanding concerning what the performance characteristics of the next innovation will be. Perhaps the most widely recognized example of this is Moore's Law in the semiconductor industry. The networks that carry out normal pattern innovations have high levels of confidence that they will be able to accomplish the next innovation objectives. This confidence is brought about from the networks' repeated experiences in carrying out incremental innovations. The networks have confidence because they have developed network routines and heuristics that guide their problem-solving, trial-and-error activities (Nelson & Winter, 1982).

All of the case studies include multiple incremental innovations. In the Haier case, the incremental innovations involved a simple technology, the associated decision-making was centralized (restricted to company managers), and the knowledge needed to carry out the innovations was narrow. In two of the cases, Bosch and TCS, some of the innovations were concerned with simple technologies and some with complex technologies. Like the Haier case, the simple technology innovations involved centralized decision making and the use of a narrow knowledge base. For the incremental innovations of complex technologies, the decision making required broad participation of those holding the diverse expertise needed to carry out the synthesis required in producing the innovation. This same decision-making and knowledge pattern applied to all of the incremental innovations associated with the HP and Sony cases.

Transitional Pattern

The cases of Bosch, Sony, HP, and TCS include transitional innovations. In the Bosch case, there were two transitions: one from a mechanical in-line pump to a mechanical distributor pump, and one from a mechanical injection system to an electromechanical system. Between these two transitional innovations, the technology crossed a threshold from simple to complex, and the organizational structure evolved from a single hierarchical organization to an organizational network.

In the Sony case, the transition was from a complex electromechanical system (i.e., analog magnetic tape audio system) to an even more complex electro-optical mechanical system (i.e., digital audio CD). The transition was the result of a consensus in the audio community that a major redesign was necessary. Two developments brought about the consensus: a saturation of the market for tape systems and the rapid development of digital electronics. Faced with these developments, the industry, encouraged by the Japanese Ministry for International Trade and Industry, evolved a consensus that the choice of the redesign should be made by an industry-created committee composed of the major manufacturers and not by a market competition. In need of a transitional innovation, Sony entered into a partnership with Philips (The Netherlands) to design an audio CD. This innovation required the merging of two networks in which Philips had the leadership in optics and Sony in audio systems. The industry committee selected the Sony-Philips design (from three competing designs) as the industry standard. At that point, Sony and Philips terminated their partnership, and each created its own network to carry the design to the market. The Sony network linked with suppliers of subsystems, such as lasers, as well as with recording industry companies.

In the HP case, the transition was from a complex ultrasound-based system to an even more complex magnetic resonance imaging- (MRI-) based system. The transition resulted from market saturation of the ultrasound system and the capacity of the MRI system to provide diagnostic data that previously only could be obtained from physically invasive techniques (e.g., inserting a fiber optic tube in the vascular system). This capacity would create a new market for the major redesign of the technology.

The decision to pursue the MRI-based redesign began within the HP ultrasound network. A search was undertaken to find a partner with MRI capabilities. In this sense, MRI capabilities required an already existing network that could put together the complex subsystems necessary to make the overall system work. HP recognized from the beginning that making an MRI system work required substantial tacit knowledge. While there was confidence that the existing HP network could gain the tacit knowledge through trial and error, it would take too much time in an environment in which competitor networks were working on the same design concept. Consequently, the management of HP's medical group signed an agreement that merged its ultrasound network with Philips' (The Netherlands) established MRI network. The design of the new system was carried out by a team composed of people from the two networks. This merger differed from the one associated with the CD (discussed previously) in that it called for the merged network to continue throughout the period of incremental innovations that would occur over the life of the redesign.

In the TCS case, the transition involved moving from a simple technology and single organization primarily concerned with applying software expertise to the problems of specific customers to the development of networks that exist for varying periods of time and that merge substantive industry knowledge with information technology expertise in order to provide high-level consulting services, as well as to design-embedded software—software used on application-specific integrated circuits (ASICs). In its early period, TCS supplied people who worked as subcontractors to write software code as one component of larger-scale, problem-solving teams.

TCS now is assembling teams that link computer makers and software product providers such as Oracle with industry-specific expertise to provide companies with high-level consulting.

Growing Diversity of Knowledge

The case studies provide illustrations of how knowledge processes occur. Especially, they offer insight into the information and knowledge required for each of the three innovation patterns. A review of the knowledge needed to carry out the innovations of Bosch's diesel fuel injection technology over the last 80 years indicates a trend toward knowledge diversity that has been accelerating at an ever more rapid rate since the 1960s. The growing diversity of knowledge has two dimensions: it includes both greater knowledge depth (e.g., a capacity for executing ever finer machining) and breadth (e.g., the inclusion of electronics into a mechanical technology). The Bosch organization's reservoir of both codified and tacit knowledge has increased over its lifetime, and that increase has been accelerating in the last two decades. The knowledge is located in a growing number of organizations, both in other Bosch units and in organizations linked to Bosch through the diesel fuel injection network. The organizational network has developed a body of routines and heuristics that are essential to the successful interactions of the different organizations. The routines and heuristics exist as both codified and tacit knowledge embedded in the Bosch-centered network.

From 1922 to 1960, the knowledge needed in Bosch to carry out the innovation of the in-line pumps was predominately mechanical. During the period of the transitional innovation that produced the first mechanical distributor pump (1960-1976), a more diverse knowledge base was needed. The first distributor pump design required a more sophisticated body of technical knowledge. This began the period when injection technology began a movement across a threshold from a simple to a complex technology. The transition pattern ended with the commercial success of the distributor pump initially used on a large scale in the Volkswagen Golf. The mechanical distributor pump, however, represented a more sophisticated and complex technology than the previously dominant in-line pump.

The next period was after the introduction of the distributor pump, in which the incremental innovations were improving both it and the in-line mechanical pump. The improvements during this period would be of major importance to the next transitional innovation: the electro-mechanical design. This period was characterized by an expansion of knowledge, especially an increase in depth; it involved the kind of learning commonly associated with innovations that fit the normal pattern.

The period of 1980-2000 was when the transitional innovations were being carried out that produced the UIS/UPS, VP-44, and Common Rail systems. The period had two phases that were distinguished, among other things, by different electronic controls: solenoid actuators and solenoid valves. The period saw a very great increase in the diversity of knowledge needed. Specifically, it required integrating electronic and electrical knowledge into what quickly became an organizational network of substantial complexity. During this period, the innovation of the electro-mechanical systems required both diverse knowledge and diffuse decision making. The technology that was innovated during this transitional pattern required the full participation of holders of diverse expertise. The complexity of the technology made the existence of a single designer impossible.

The present period of innovation (2000 to the present) is characterized by the normal pattern, by repeated incremental innovations that produce specialized configurations of the multiple injection system designs that are in current production. Incremental innovations are being carried out on four different designs. The diversity of knowledge held by the Bosch-centered network makes it possible to a rapid design that fits the specific needs or desires of diverse customers.

The other four cases also demonstrate the growing diversity of knowledge in line with the growing complexity of technologies and expanding of networks. In the case of the Sony CD, the predecessor to the CD—tape systems—represented a fairly diverse body of knowledge. As the decision to partner with Philips for the design of the CD was made by top Sony management, it began the period of the transitional innovation pattern. As the transition progressed, there was movement to more diversity of knowledge. With joint Sony-Philips prototype development, the diversity of knowledge required (and, thus, accessed and synthesized) increased significantly.

The HP cardio-imaging technology began the transformational innovation pattern with a diverse body of knowledge. As the period of transformation advanced, the diversity of needed knowledge increased, which was due to the fact that the HP medical products division linked with other HP organizations (e.g., the R&D labs) as well as with members of the end-user community, for prototype design and development. This path of increasing diversity of knowledge continued in these directions with the subsequent incremental innovations. During the transitional period of innovation when HP decided to move to an MRI-based system, the creation of the HP-Philips partnership significantly enlarged the knowledge base and the innovation network. The subsequent joint prototype development and incremental innovations were carried out by a self-organizing network using diffuse decision-making processes to synthesize a very diverse range of tacit and codified knowledge.

In the case of TCS and its innovation in software services and products, the technology is just beginning a transitional innovation pattern after a long period of incremental (normal) innovation. When the technology is simple, the diversity of knowledge is significantly less than that seen with complex technologies. During its period of normal innovation pattern, TCS has increased its diversity of knowledge. After the transition to a complex technology, however, a trend of increased diversity of knowledge will be required.

Copyright © 2007, Idea Group Inc. Copying or distributing in print or electronic forms without written permission of Idea Group Inc. is prohibited.

Haier's innovation of household appliances is concerned solely with simple technology. It represents the least diverse body of knowledge of the technologies examined in our case studies. As it has expanded its line of household appliance products, Haier has increased its diversity of knowledge incrementally.

The Influence of Culture

In the Bosch case, the process of transformational innovation—the initial design of a diesel fuel injection system—is consistent with the hierarchical, codified culture of Germany at that time. The technology required craftsmen with narrow specialized technical expertise—both highly regarded in German culture. Once in the normal pattern of innovation, the dominant culture of the company and its organizational network is one of established routines and heuristics that have either engineered around or taken advantage of cultural traits. The normal pattern of innovation of diesel fuel injection systems appears to fit nicely with the German preference for ever greater depth of specialized technical knowledge—both tacit and codified. Significant stress on the organizational system occurred at the second transition point when the technology transitioned into an electro-mechanical system. It is at this point that Bosch implemented organizational changes in an effort to synthesize new knowledge and to overcome the hierarchy and highly specialized narrow focus of its personnel. During this period of transition, the company was no longer able to organize itself as a symphony orchestra (Gannon, 2001)—assembling a number of highly specialized experts led by a master designer. Rather, the transition required a merging and synthesis of expertise in which the interacting linkages and boundaries between disciplines were no longer clear. Thus, the company implemented a new organizational structure that established multidisciplinary teams.

The Sony innovation trajectory begins with what can be characterized as a self-organizing network; it has a style of operation that utilizes tacit knowledge to synthesize diverse types of knowledge. This is highly consistent with Japanese culture more generally. When the transition point comes, there is a need to link with Philips for purposes of designing a CD system that integrates digital electronics and photonics. The literature states that Japanese culture generally emphasizes the importance of relationships being mutually beneficial. This trait is obviously advantageous in the innovation of complex technologies, as an organization or network cannot operate in a tacit knowledge environment over a period of time without this attitude. Thus, this cultural characteristic is quite beneficial to the Sony-Philips partnership over the design phase.

Having a design in place, the Japanese focus was on getting the product to market quickly. Sony believed in its tacit knowledge and experience, so it approached future problems that would arise with incremental innovations as something to be dealt with later after the product was on the market. Philips, however, preferred codified knowledge and understanding in the scientific sense. It wanted to ensure that the prototype included all necessary specifications to enable all follow-up incremental innovations.

In the HP cardio-imaging case, the company, in the early period, can be characterized as having a not-invented-here syndrome—a rejection of knowledge and expertise from outside organizations. This was particularly the case when it needed radar knowledge for the transducer. The inability of HP personnel to recognize the importance of tacit knowledge that was obtainable from other organizations reflected the traditional American cultural preferences for codified knowledge. This substantially slowed down the innovation process, but HP did learn from this failure.

When the MRI transition point approached, HP management decided to join forces with Philips in order to save time. This approach worked with the Philips' culture, which was organized into distinct technology areas. Thus, the merging of the networks was facilitated easily. It is important to note that HP was a relative risk taker; it was willing to develop a technology that a consulting company said would sell but for which few cardiologists saw a need. One of the reasons for this risk-favorable organizational culture were the norms instilled by the founders, Hewlett and Packard, who emphasized the importance of developing state-of-the-art technology rather than pursuing short-term profits. By the time that the MRI transition took place, the company had learned the value of tacit knowledge, but it was also operating in an economic policy environment that emphasized short-term profits. Thus, the company culture reflected the broader culture of the American marketplace. Within the normal pattern, the HP-centered network was basically self-organizing and used internal learning and rather diffuse decision making to innovate incrementally.

In the case of TCS, the Indian cultural tradition that emphasizes mathematics, philosophy, and hierarchy works well when the software services and products involve a simple technology. However, as the company transitions to a complex technology and a complex organizational network, it will have to create a company culture that is different than traditional Indian culture. It will need a culture that allows people to act without centralized decision making and rigid hierarchical organizational structures and one that promotes greater equality. Indian culture is advantageous in the innovation of complex technologies in the sense that it favors a synthesized view of knowledge rather than distinct specialization. TCS instilled a company culture that necessitates the codification of much tacit knowledge when it recognized the need for both tacit and codified knowledge in the innovation of complex technologies.

In Haier's innovation of household appliances, the company has had to create a company culture not entirely consistent with Chinese culture. It has created a very rigid hierarchical management structure with rigid codified personnel and technical performance requirements that are strictly enforced. In order to mitigate the characteristically Chinese culture of creating networks based on familial and friendship ties, Haier has substituted a hierarchical merit-based authoritarian system in order to ensure better efficiency. While it is still concerned with the innovation of simple technologies, Haier can continue to utilize codified knowledge and centralized decision making. However, in order to transition to complex technology innovation, it will need to utilize consensus decision making that can handle both codified and tacit knowledge. While Chinese culture generally is comfortable with tacit knowledge, Haier will need to institute strategies and actions that allow linkages with organizations that have the needed knowledge rather than with family or friendship ties.

Conclusion and Lessons

Public policies and organizational strategies designed to manage information and knowledge in facilitating commercial technology innovation benefit from recognizing that very different organizational structures and processes are associated with the success of simple vs. complex technologies. The significance of the capacity to carry out technological innovation in long-term economic growth long has been recognized by researchers. For a broad-based advanced economy, the capacity to carry out repeated innovations of complex technologies is a necessary condition, as complex technologies are the largest and a growing component of the high value and high value-added technologies traded in the world.

Technologies and innovation processes are associated with one of three patterns: transformational, normal, and transitional. The three patterns are distinguished by different kinds of innovations and by the ways in which the networks access, create, and utilize knowledge and information and make decisions in carrying out the innovations. Transformational innovations are distinguished by first-of-a-kind technologies. Normal innovations are incremental technical advances that deliver enhancements of established technology designs. Innovations that are transitional are radical improvements distinguished by major redesigns.

Effective and efficient management of information and knowledge management is essential in facilitating synthesis processes in technological innovation. Well-organized and efficient information and knowledge management is needed to promote the flows of information and the exchanges of knowledge within innovation networks and to reduce the costs associated with the processes of flows and exchanges. It is important to note that requirements for management of information and knowledge for technological innovation vary in terms of simple vs. complex technologies and in terms of different patterns of innovation.

The innovation of simple technologies requires a relatively narrow knowledge base composed of both tacit and codified knowledge, and centralized decision-making processes are used to guide their innovation. A few key individuals or units within a single organization (or occasionally a very limited number of organizations) possess the total requisite knowledge and capabilities and make the necessary decisions in carrying out the innovation of simple technologies. Simple technology often is associated with organizational hierarchy and specialization of labor. Information management for innovations of simple technologies does not require free information flows to every branch of the organization.

The innovation of complex technologies, however, requires the synthesis of a diverse body of tacit and codified knowledge that resides in many different individuals and types of organizations that link to form a network. Information management for innovations of complex technologies requires not only flows of information but also exchange of codified and tacit knowledge within the network. Thus, public policies and organizational strategies that facilitate and encourage cooperation among individuals and organizations are critical to the innovation of complex technologies.

Furthermore, public policies and organizational strategies designed to facilitate information and knowledge management for innovation of complex technologies need to reflect the different knowledge acquisition and use processes associated with transformational, normal, and transitional innovation patterns. The transformational pattern is distinguished by organizational networks with the ability to access, create, and synthesize knowledge from sources and with organizational arrangements that are new. Learning how to synthesize knowledge that is new and that is from new sources requires repeated trial and error and is thus heavily dependent on learning from failure. The distinctive characteristic of transitional pattern networks is the ability to integrate previously separate organizational networks and bodies of knowledge into a new, more complex organizational network and body of knowledge held by this new network. New network routines and heuristics are developed by trial and error, and cultural differences pose challenges in the integration of both networks and knowledge.

Thus, public policies and organizational strategies for information and knowledge management that facilitate the rapid development of trust among the participants in networks are particularly valuable during the transformational and transitional innovation patterns. These patterns represent time of change and uncertainty and require the active participation of company managers and public policymakers to guide and facilitate the innovations. For technologies that gain economic benefits from having a single universal design, public policies and organizational strategies that facilitate nonmarket choices of redesign appear to be beneficial.

The normal pattern of innovation is distinguished by networks with the ability to self-organize in order to facilitate internal learning necessary to carry out incremental innovation of existing technology designs. Research and development are rarely the most important and often are not significant contributors to the knowledge necessary for normal pattern innovation. Normal pattern innovation is guided by internal, often tacit, knowledge and consensus decision-making concerning the direction of innovation. Normal pattern innovation requires less management and public policy attention than the transformational and transitional patterns.

The capacity to carry out the synthesis of diverse kinds of knowledge can be developed only by organizational networks. Synthesis is a key core capability in the innovation of all complex technologies and must be viewed by public policymakers and company managers as a factor of equal importance to capital, R&D, labor, or intellectual property protection. At the heart of the capacity of organizational networks to carry out rapid and repeated syntheses is the development of network routines and heuristics. As the process of synthesis is something that networks learn how to do but never understand, important routines and heuristics always will be network-held tacit knowledge.

Innovations of complex technologies are sociotechnical processes. Both national and company cultures have significant impacts on the way in which people manage information and knowledge. Public policies and business strategies need to be designed to take advantage of cultural traits conducive to effective management of information and knowledge and to mitigate the effects of cultural traits that impede the exchange and use of tacit and codified knowledge. In general, a culture that emphasizes a high level of trust, collectiveness, equality, and consensus decision-making, and that is comfortable with both codified and tacit knowledge is highly desired for organizational learning.

Finally, this chapter benefits a lot from five carefully conducted case studies that represent five different cultures. The cases are the Bosch (Germany) diesel fuel injection technologies, the Sony (Japan) audio compact disc (CD), the Hewlett-Packard (HP) (U.S.) cardio-imaging technology, the TATA Consultancy Services (TCS) (India) software services and products, and the Haier (China) line of home appliances. This chapter has proved that conducting case study is an effective research method, as it can provide more insights, especially when empirical data are not readily available.

References

Dutrenit, G. (2000). *Leaning and knowledge management in the firm*. Northampton, MA: Edward Elgar.

Freeman, C., & Soete, L. (1997). *The economics of industrial innovation* (3rd ed.). Cambridge, MA: MIT Press.

Gannon, M. J. (2001). *Understanding global cultures: Metaphorical journeys through 23 nations*. Thousand Oaks, CA: Sage Publications.

Hall, E. T. (1976). *Beyond culture*. New York: Doubleday.

Hatzichronoglou, T. (1997). Revision of the high-technology sector and product classification. *OECD STI Working Papers*, No. 59918.

Hofstede, G. (2001). *Culture's consequences: International differences in work-related values* (2nd ed.). Beverly Hills, CA: Sage Publications.

Kash, D. E. (2001). *Haier company innovation in household appliances* (School of Public Policy Working Paper). Fairfax, VA: George Mason University.

Kash, D. E. (2002a). *Bosch fuel injection systems for diesel engines* (School of Public Policy Working Paper). Fairfax, VA: George Mason University.

Kash, D. E. (2002b). *TATA consultancy services: A case study* (School of Public Policy Working Paper). Fairfax, VA: George Mason University.

Kash, D. E., Auger, R., & Li, N. (2004). An exceptional development pattern. *Technological Forecasting and Social Change, 71*(8), 777-797.

Kash, D. E., Lieberman, S. L., & Schaller, R R. (2000). *Innovation of cardio-imaging technology at Hewlett-Packard and HP/Philips* (The Institute of Public Policy Working Paper). Fairfax, VA: George Mason University.

Kash, D. E., & Rycroft, R. W. (2002). Emerging patterns of complex technological innovation. *Technological Forecasting and Social Change, 69*(6), 581-606.

Kash, D. E., & Schaller, R. R. (2000). *Innovation of the audio compact disc player* (The Institute of Public Policy Working Paper). Fairfax, VA: George Mason University.

Keen, P. G. (1995). *Every manager's guide to information technology*. Boston: Harvard Business School Press.

Kline, S. J. (1990). *Innovation styles in Japan and in the United States: Cultural bases; implications for competitiveness* (The 1989 Thurston Lecture, Report INN-3). Stanford, CA: Stanford University, Dept. of Mechanical Engineering.

Kline, S. J. (1991). Styles of innovation and their cultural basis. *Chemtech, 21*(6), 472-480.

Kline, S. J. (1995). *Conceptual foundations for multidisciplinary thinking*. Stanford, CA: Stanford University Press.

Kodama, F. (1991). *Analyzing Japanese high technologies: The techno-paradigm shift*. London: Pinter.

Nelson, R., & Winter, S. G. (1982). *An evolutionary theory of economic change*. Cambridge, MA: Belknap Press.

Nonaka, I., & Takeuchi, H. (1995). *The knowledge-creating company*. New York: Oxford University Press.

Osborn, R. N., & Hagedoorn, J. (1997). The institutionalization and evolutionary dynamics of interorganizational alliances and networks. *Academy of Management Journal, 40*(2), 261-278.

Romer, P. M. (1986). Increasing returns and long-run growth. *The Journal of Political Economy, 94*(5), 1002-1037.

Rothwell, R. (1994). Industrial innovation: Success, strategy, trends. In M. Dodgson & R. Rothwell (Eds.), *The handbook of industrial innovation* (pp. 33-53). Cheltenham, UK: Edward Elgar.

Rycroft, R. W., & Kash, D. E. (1999). *The complexity challenge: Technological innovation for the 21st century*. London: Pinter.

Senker, J., & Sharp, M. (1997). Organizational learning in cooperative alliances: Some case studies in biotechnology. *Technology Analysis and Strategic Management, 9*(1), 35-51

Sing, K. (1997). The impact of technological complexity and interfirm cooperation on business survival. *Academy of Management Journal, 40*(2), 339-367.

Soete, L., & Arundel, A. (1993). *An integrated approach to European innovation and technology diffusion policy: A maastricht memorandum*. Luxembourg: Commission of the European Communities, SPRINT Program.

Solow, R. M. (1956). A contribution to the theory of economic growth. *Quarterly Journal of Economics, 70*, 65-94.

Chapter XI

Knowledge Workers as an Integral Component in Global Information System Design

Michel Grundstein
MG Conseil & Paris Dauphine University, France

Abstract

Analyzing an information system through knowledge's point of view causes a change that inevitably engenders the awareness of the importance of the human and socio-cultural dimensions to improve its performances. This chapter, based on a model for global knowledge management within the enterprise (MGKME) helps to stress the need to integrate the knowledge worker, both at once, as a component and a user of the information systems within extended companies. Actually, this challenge, put out by a knowledge management's vision that is centered on the core business processes and the people, ought to be included in the strategies for information resources management.

Introduction

Under the influence of globalization and the impact of information and communication technologies (ICT), the Enterprise turned into Extended Company. That implies a cross-cultural impact of its members, who are the components and users of information systems (IS) disseminated all over the world. As knowledge workers, the users are in the heart of the IS that is implemented. Therefore, their knowledge, which is rooted in their individuals' culture, are beginning to be a key factor of success for the design, development, implementation, and utilization of information systems. More and more, employees' knowledge must be considered as an asset as it was as early as 1990 when Carnegie Group, Inc., Digital Equipment Corporation, Ford Motor Company, Texas Instruments, Inc., and US WEST Advanced Technologies, Inc. launched the Initiative for Managing Knowledge Assets. They defined for the first time the notion of knowledge assets as follows: "Knowledge assets are those assets that are primary in the minds of company's employees. They include design experience, engineering skills, financial analysis skills, and competitive knowledge" (IMKA, 1990).

As of that date, numerous research works were carried out, enterprise applications were deployed, and an abundant literature enriched the domain of knowledge management (KM). Nevertheless, KM could not exist without information technologies, which have drastically changed our relationship with space and time. In turn, from the KM point of view, IT cannot be efficient without the knowledge of users. Gradually, the concept of KM has highlighted a broad range of topics and has become a fuzzy concept taking as many senses as people speaking about it. For instance, in his editorial preface titled "What is Knowledge Management?" M. E. Jennex (2005a) gathered some authors' definitions that show that there is no common evidence about what KM is. In our research group, we consider that knowledge cannot be processed as an object independently of the person who has to act. Thus, we think that KM must address activities that utilize and create knowledge more than knowledge by itself. With regard to this question, since 2001, our group of research has adopted the following definition of KM:

KM is the management of the activities and the processes that amplify the utilization and the creation of knowledge within an organization, according to two additional goals closely interlinked, and their underlying economic and strategic dimensions, organizational dimensions, sociocultural dimensions, and technological dimensions: (1) a patrimonial goal; (2) a sustainable innovation goal.

As we can see, this definition, directly addresses some elements of corporate governance and one of its subsets, IT governance, except that there is not a unifying pattern of reference upon which a KM governance can be established. Therefore, the focus of our ongoing research is to establish such a pattern that fits with our definition of KM. It is the purpose of the model for global knowledge management within the enterprise (MGKME).

In this chapter, after having introduced a brief description of the Extended Company, we open a discussion based on the concepts that have been used to establish the MGKME. Doing so, we stress the links between information resources management and KM. Thus, we successively introduce the following topics: distinguishing the concept of knowledge from the concepts of data and information; analyzing the impact of the cultural factors on the knowledge management systems (KMS); and introducing the concept of KM governance. Then we describe the MGKME. This model suggests a systemic approach of KM and considers the knowledge workers both at once as components and users of the KMS. Finally, we put in perspective how this model could help to establish people-focused KM governance principles linked with corporate and IT governance.

The Extended Company

A Brief Description

The enterprise increasingly develops its activities in a planetary space with three dimensions:

1. **A global space:** This space covers the set of the organization that is the geographic places of implantation.
2. **A local space:** This space corresponds to the subset of the organization situated in a given geographic zone.
3. **A space of influence:** This space is the field of interactions of the company with other organizations.

The hierarchical company locked up on its local borders is transformed into an Extended Company without borders, opened and adaptable. Furthermore, this Extended Company is placed under the ascendancy of the unforeseeable environment that leads toward uncertainty and doubt. One can remark that, depending on the geographic place of implantation of the local space (e.g., U.S., South America, Europe, and Asia), the vision of the global space is dependent on the local culture of members.

Figure 1. The formal and informal digital information networks within the extended company

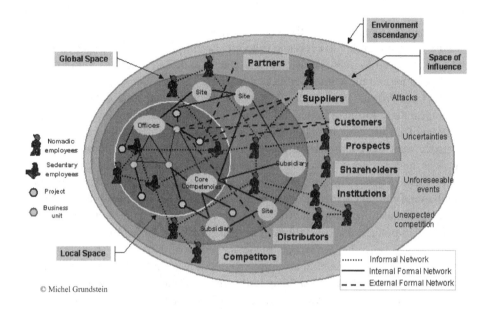

Behaviors may be different, and understanding of things can cause a hierarchical antinomy in the importance of the decisions and choices that are made.

The Extended Company meets fundamental problems of information exchange and knowledge sharing among its formal entities distributed in the world (offices, core competencies, business units, projects) on the one hand and the company's employees (nomadic or sedentary), bearers of diversified values and cultures according to the places of their implantation, on the other hand.

Two kinds of information networks overlap (see Figure 1):

1. A *formal information network* among the internal or external entities in which circulate data and codified knowledge. This network is implemented under intranet and extranet technologies.
2. An *informal information network* among members, nomadic or sedentary employees, in which information exchanges and tacit knowledge sharing take place. This network is implemented through communication technologies.

The Evolution of the Employees' Role within the Extended Company

In the Extended Company, initiatives and responsibilities are increasing, whatever the individuals' hierarchical levels and roles are. Employees are placed in situations in which they need to make decisions. They become decision makers who utilize and create more and more knowledge as a basis for their efficiency. On this basis, Davenport and Prusak (1998) state, "What makes knowledge valuable to organizations is ultimately to make better the decisions and actions taken on the basis of knowledge" (p. 170). Therefore, knowledge is the crucial factor enabling employees to improve their decision-making processes and to enhance their competencies. To answer their missions, these individuals, commonly pointed out as knowledge workers, have to access information, knowledge, and skills widely distributed in the global and influence spaces of their organization. They must rely on the formal and informal information networks of the company through their sedentary or mobile computerized workstations. The computerized workstation becomes a window opened onto the company's planetary space of activities. Thus, beyond the technical infrastructures that are implemented, the essential role of the digital information system is to provide relevant information to each employee at all levels of the hierarchy so that he or she can control his or her situation, make decisions, and undertake actions. Nevertheless, we have to distinguish between "the knowledge of knowers and the codification of that knowledge" (Haeckel, 2000, p. 295). Accordingly, information systems must supply means to share personal knowledge with distant colleagues and to enable access to essential codified knowledge in order to solve problems out of routine.

Background Theory and Assumptions

Data, Information, and Knowledge

As highlighted by Snowden (2000), "The developing practice of KM has seen two different approaches to definition of knowledge. One arises from information management and sees knowledge as some higher-level order of information, often expressed as a triangle progressing from data, through information and knowledge, to the apex of wisdom. Knowledge here is seen as a thing or entity that can be managed and distributed through advanced use of technology. ... The second approach sees the problem from sociological basis. These definitions see knowledge as a human capability to act" (pp. 241-242).

Table 1. Data, information, and knowledge

Data	Fundamental and unbiased qualitative or quantitative element upon which reasoning or realization of treatments are based
Information	Data set structured and organized to give shape to a message resulting from a given context and so perfectly subjective
Knowledge	New information acquired by an intelligent process, study, or practice

The Information Management Approach of Knowledge

The information management approach of knowledge is the most widespread. This approach focuses on a technological and applications perspective. It induces to consider knowledge as an object and so to disregard the importance of people. Envisaged under the angle of the system of information, knowledge implicitly is treated as an object independent of the knower that creates it and utilizes it. More often than not, although the authors take care of proposing a definition intended to distinguish the concepts of data, information, and knowledge, these three concepts quickly are expressed in terms of data processing, knowledge being only a type of information. As an example (see Table 1), let's see the definitions extracted from the report of the Information Management Club for the largest French Enterprises CIGREF (2000, pp. 15-17).

These definitions show an underlying positivist paradigm that leads to characterize and to organize the concept of data, information, and knowledge into a hierarchical vision of objects. Thus, the authors who join this perspective mainly are interested in the contents of the knowledge of the organization. They focus on building and managing stocks of codified knowledge that can be stored, retrieved, and transferred by information systems.

The Sociological Approach of Knowledge

In the information management approach of knowledge, one places too little emphasis on knowledge-creating activities that, as mentioned by Davenport and Prusak (1998), "take place within and between humans" (p. 6). This approach does not underline that the knowledge resides primarily in the heads of individuals and in the social interactions of these individuals, as pointed out by Cohen and Prusak (2001) under the social capital concept. They define the social capital concept as follows: "Social capital consists of the stock of active connections among people: the trust, mutual understanding, and shared values and behaviors that bind the members of human networks and communities and make cooperative action possible" (p. 4).

In the same way, we suggest an approach that is built upon the assumption emphasized by Tsuchiya (1993) concerning knowledge creation ability. He states, "Although terms 'datum', 'information', and 'knowledge' are often used interchangeably, there exists a clear distinction among them. When datum is sense-given through interpretative framework, it becomes information, and when information is sense-read through interpretative framework, it becomes knowledge" (p. 88). He emphasizes how organizational knowledge is created through dialogue and highlights how commensurability of the interpretative frameworks of the organization's members is indispensable in order for an organization to create organizational knowledge for decision and action. Here, commensurability must be understood as the common space of the interpretative frameworks (e.g., cognitive models or mental models) of each member. Tsuchiya states, "It is important to clearly distinguish between sharing information and sharing knowledge. Information becomes knowledge only when it is sense-read through the interpretative framework of the receiver. Any information inconsistent with his interpretative framework is not perceived in most cases. Therefore, commensurability of interpretative frameworks of members is indispensable for individual knowledge to be shared" (p. 89). In other words, we can say that tacit knowledge that resides in our brains results from the sense given through our interpretative frameworks to data that we perceive among the information that is transmitted to us (see Figure 2).

In another way, Wiig (2004), who highlights a discontinuity between information and knowledge, describes this process clearly: "The process by which we develop new knowledge uses prior knowledge to make sense of the new information and, once accepted for inclusion, internalizes the new insights by linking with prior knowledge. Hence, the new knowledge is as much a function of prior knowledge as it is of received inputs. A discontinuity is thus created between the received information inputs and the resulting new knowledge" (p. 73).

Figure 2. Creation of individual's tacit knowledge

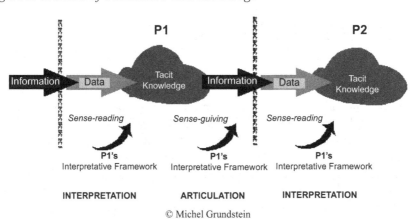

Figure 3. Commensurability of interpretative frameworks (I.F.) and individual sense making

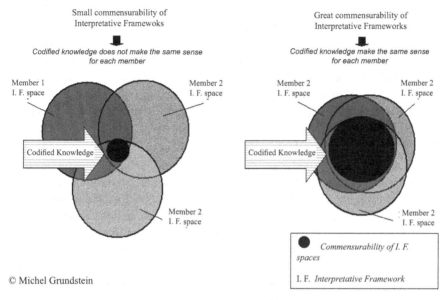

To conclude, we can say that codified knowledge is not more than information. It can be considered as knowledge when members having a large commensurability of their interpretative frameworks commonly understand it. For example, such is the case for members having the same technical or scientific education, or members having the same business culture. In these cases, codified knowledge makes the same sense for each member (see Figure 3).

Impact of Cultural Factors on Knowledge Management Systems

In this section, we introduce the idea of considering the knowledge workers as integral components of knowledge management systems (KMS). Then we analyze the impact of an individual's culture on KMS. Finally, we suggest key indicators to assess the individual's culture.

Considering Knowledge Workers as an Integral Component in Global Information System Design

KM becomes a reality in the implementation of systems that are, paraphrasing Joel de Rosnay (1975, p. 93), "a set of components in dynamic interaction organized according to a purpose." These systems often are called knowledge management systems (KMS). Although this term "does not seem to have a consensus definition" (Jennex, 2005b, p. i), it usually answers to the technological approach of KM.

From our point of view, this approach considers that knowledge can be handled as if it were an object, without any relationship to an individual's perspective. Although this idea is greatly shared, we argue that the design of a KMS should integrate the role of the knowledge workers and consider them as integral components of the system. Thus, in their study on collaborative knowledge management system (CKMS) design, Chua and Brennan (2004) emphasize that "one of the most important components of CKMS is the knowledge workers, who are also the users of the system, and the workspaces they are associated with" (p. 172).

Like Chua and Brennan (2004), we insist on the importance of integrating the knowledge worker as a component and user of the information system. In fact, relying on Tsuchiya's works, we argue that knowledge is dependent on the individual's interpretative frameworks, the context of his or her actions, and the situation that activated them. Interpretative frameworks are key processors that transform data into information and information into knowledge, which, in turn, changes the prior interpretative framework. These frameworks have been elaborated all along the life of each employee from his or her birth to the date of the context and situation that led him or her to make decisions based on his or her sense-reading and sense-giving capabilities to process information (see Figure 2).

Impact of Individual's Culture on KMS

Individuals' cultures are the bases on which employees' interpretative frameworks are deeply rooted. Consequently, we must stress the role of cultural factors whenever social interactions and sharing of information and knowledge are essential to enable efficiency in the global economy. In the same way, Chua and Brennan seem to observe the same phenomenon: "It is generally assumed that CKMS activities such as the forwarding of information, collaboration among workers in different departments, storing of information, the monitoring and tracking of purposes are supported, and facilitate learning and development of knowledge workers within this collaborative environment. Unfortunately, while CKMS generally support these activities, the influence of worker's knowledge and behavioral patterns to the successful working of the CKMS is often greatly underestimated. It is therefore important to design CKMS with the knowledge worker's constraints in terms of their qualifications or previous knowledge but also in term of other factors such as their cultural background, in mind. If we do not consider these factors accordingly, systems will not be suitable for some of the CKMS tasks that involve and rely on the knowledge workers" (p. 175).

To conclude, as employees' interpretative frameworks are transducers that give sense to information and codified knowledge, we argue that individual's culture is a key factor to enable employees to make sense of information that they access from information systems and so internalize it and transform it into action. Thus, in order to make the information and the codified knowledge efficient, one must consider the knowledge workers's culture.

Key Indicators to Assess an Individual's Culture

"Numerous IS researchers have been interested to explain differences in adoption and/or usage of IT/IS by considering the impact of culture—primary national culture on human behavior and the way humans use the IT" (Corbitt, Peszinski, Inthanond, & Hill, 2004, p. 65). For the purpose of our research, in order to highlight the link between national cultural dimensions and the individual's culture, we have chosen to report on the research undertaken from 2003 to 2004 by the Knowledge Management Forum organized by Henley Management College in the UK (Darby, Herbolzheimer, & van Winkelen, 2004). This research was launched to find perceptions of the impact of cross-cultural factors and issues relating to KM. Research included six case study companies that employ more than 100,000 employees worldwide in the telecommunications, pharmaceutical, and business consulting sectors. In order to provide a geographical spread, the dataset includes five U.S., three South American (Brazil and Uruguay), seven European (France and UK), and four Asian (China) sites. In their research, Darby, Herbolzheimer, and van Winkelen (2004) state that "one of the main difficulties in the analysis of culture and its impact on KM initiatives is to separate the business from the national culture" (p. 233). They concentrate on five national cultural dimensions: power distance, tolerance of ambiguity, individualism-collectivism, time orientation, and doing-thinking (pp. 241-242). We have converted their descriptions under an analysis framework that lists key indicators to assess the individual's culture (see Table 2).

Table 2. Key indicators to assess an individual's culture

National Cultural Dimensions	Description	Key Indicators to Assess an Individual's Culture
Power Distance	Willingness of a society to accept hierarchical power systems. Focuses on relationships between people of different statutes **High:** Decision making is top-down with strong hierarchical command systems. **Low:** Decentralizes management systems with encompassing and emerging systems to attract goals and means for the firm.	- Degree of dependence of subordinates - Preference for consultation - Emotional distance - Role between center and affiliates network

Table 2. Continued

Tolerance of Ambiguity	Measure of the society's willingness to take risks. Is risk taking encouraged or discouraged? **High:** Planning and sharing of information on goals and means can be more informal, and there is less need to cater to anxiety among workers about risk taking. **Low:** Management systems should reduce anxiety about risks by detailed planning and sharing goals and means.	- Tolerance in interaction with strangers - Formality of interactions - Communication regulated by rules and social norms - Tolerance of deviant behavior
Individualism/ Collectivism	Refers to the relationship between the individual and the collectivity that prevails in society (high vs. low context differences). It is reflected in the way we live and work together. **Individualism:** A social pattern that consists of loosely linked individuals who view themselves as independent of collectives and who are motivated by their own preferences. Individuals see information as independent of its context. **Collectivism:** A social pattern that consists of closely linked individuals who see themselves belonging to one or more collectives and who are motivated by norms that are imposed by the collectives. Priority to the goals of the collectives over their own personal goals. Duties and obligations; emphasize historical and contextual information.	- Tacit vs. explicit knowledge - Systemic vs. independent - Social environment - Degree of rationality
Time Orientation	Attitudes toward time, which shape the way in which people structure their actions; represents trade-offs made between long long-term value and short-term profitability. **LTO (future—long-term orientation):** Stands for the fostering of virtues oriented toward future rewards. Deferred gratification of needs, long-term virtues—frugality, perseverance, building a strong market position, structured problem solving, saving, and investing in particular. **STO (current—short-term orientation):** Stands for the fostering of virtues related to the past and present; in particular, respect for tradition, preservation of "face," and fulfilling social obligations, but also considers immediate gratification, short-term focusing, spending, bottom line, and analytical thinking.	- Immediate vs. deferred gratification - Spending vs. saving - Quick results vs. perseverance - Sense of shame and losing face
Doing/Thinking	Propensity for action vs. analysis and reflection **High: Propensity for action** and driven by immediate results **Low: Space for reflection**, experimentation, and evaluation	- Action orientation - Reflection and analysis, philosophizing - Resource-driven vs. process-driven

The key indicators listed in Table 2 provide empirical evidence that they have an impact on the individual's cognitive abilities and so develop different interpretative frameworks. Accordingly, when considering the knowledge worker as an information system component, we suggest regarding these indicators as criteria to develop a cultural analysis study in order to conceive, realize, and implement collaborative information systems.

Knowledge Management Governance

After having considered the corporate governance and the information technology governance concepts, we attempt to tackle with a knowledge management governance perspective, drawing a link to the corporate and IT governance principles.

The OECD Corporate Governance

Organization for Economic Co-operation and Development (OECD) corporate governance principles originally were issued in 1999. They since have become the international benchmark for corporate governance. OECD governments in April 2004 agreed with the new Principles and defined corporate governance as shown in Figure 4 (OECD, 2004, p. 11).

The COBIT® IT Governance

Control Objectives for Information and Related Technology (COBIT®, 2000, 2002, 2005) initially was published by the Information Systems Audit and Control Foundation, Inc. in 1996. Guldentops (2004) states that "COBIT® presents an international and generally accepted IT control framework enabling organizations to implement an IT Governance structure throughout the enterprise" (p. 277). A fourth edition was edited in 2005. In the Executive Summary, IT Governance is defined, as shown in Figure 4 (COBIT®, 2005, p. 6).

IT governance provides the structure that links IT process, IT resources, and information to enterprise strategies and objectives. To achieve success, corporate governance and IT governance no longer can be considered separate and distinct disciplines. The COBIT® Management Guidelines help to support these needs. They have identified specific critical success factors, key goal indicators, key performance indicators, and an associated maturity model for IT governance.

KM Governance Perspectives

Corporate governance and IT governance do not explicitly mention considering intellectual capital as a resource in enterprise strategies. Even so, as pointed out by Edvinsson and Malone (1997), "The core of the so-called *knowledge economy* is

Figure 4. KM governance perspective

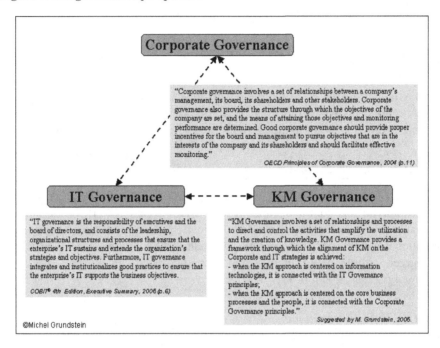

huge investment flows into human capital as well as information technology" (p. 12). However, we think that the knowledge economy will be obliged to take into account Intellectual Capital. Consequently, we need to study the link between KM, corporate governance, and IT governance. To enable such a study, we must refer to a KM pattern of reference to elaborate KM governance principles.

Toward a Unified KM Pattern of Reference

Despite the fact that numerous knowledge management frameworks have been suggested all over the world, there is not a unifying pattern of reference supporting our definition of KM, as described in the introduction. For example, let us consider *The European Guide to Good Practice in Knowledge Management* (CEN-1, 2004). The project team collected, categorized, and analyzed more than 140 KM Frameworks. We notice that this work has produced a high-quality practical outcome that is a reference point to achieve a good understanding of KM. Nevertheless, as contributors to this project, we underline the predominant positivist paradigm and the information management approach of KM that have inspired the project team. Moreover, we have observed that few of them were people-focused, as Wiig (2004) states: "[O]ur emphasis is on people and their behaviors and roles in enterprise operations" (p.

XXV). Furthermore, we have distinguished two main approaches underlying KM: (1) a technological approach that answers a demand of solutions based on the technologies of information and communication (ICT); and (2) a managerial approach that integrates knowledge as resources contributing to the implementation of the strategic vision of the company.

Therefore, we suggest two KM governance perspectives depending on the first or second approach (see Figure 4).

On the one hand, the technological approach leads to reduce knowledge to codified knowledge that is no more than information. In that case, we can manage KM projects in the same way as IS projects. Specific criteria inherent to KMS must connect KM governance and IT governance principles. On the other hand, the managerial approach that integrates knowledge as a resource focuses on the core business processes and the people. Corporate governance principles must integrate the risks linked to the utilization and creation of knowledge.

These aspects involve elaborating management governance guidelines for KM as COBIT® is for IT. The aim of the model for global knowledge management within the enterprise (MGKME), described hereafter, is to contribute to elaborating a guiding framework that serves as a pattern for KM governance guidelines.

MGKME Description

The MGKME supports our full meaning of KM as defined in the introduction of this chapter. It is an empirical model. It should result in the implementation of knowledge management systems that take into account knowledge workers as components of the systems and that allow them to be autonomous and to achieve their potentialities. The MGKME consists of seven key elements shared out in two main categories: the Underlying elements and the Operating elements. Hereafter, we describe each of those elements (see Figure 5).

The MGKME's Underlying Elements

The core knowledge is embodied in people's heads and their abilities to utilize them and generate new knowledge at the same time. The information technologies and the tangible technical resources enhance their competence, while value-added processes and organizational infrastructures are structuring their activities. Nevertheless, their social interactions are essential factors that leverage their potentialities and actually enable them to achieve effective results. Therefore, from our perspective, sociotech-

Figure 5. Model for global knowledge management within the enterprise (MG-KME)

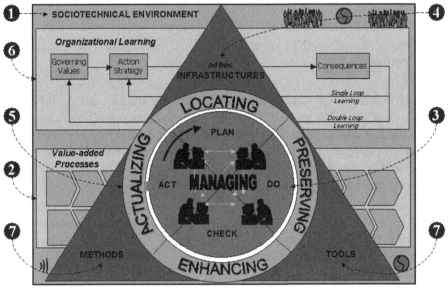

© Michel Grundstein

nical environment *(1)* and value-added processes *(2)* are fundamental components of the Knowledge Management System. As a result, they constitute the underlying elements of the MGKME, upon which the other elements should focus.

Sociotechnical Environment of the Enterprise (1)

The sociotechnical environment constitutes the social fabric in which autonomous individuals supported by ICT and tangible resources interact and are conversing through physical or virtual places (coffee machines, collaborative workspaces, weblogs, wikis, CoPs). Interacting is not enough. Stewart (2001) stated, "Making time to converse at every level of an organization is not an indulgence, not a luxury, it is an imperative" (p. 1). He observes what happens when interacting without conversing: "Stories are not told and associated sense of adventure is lost; knowing is not shared because questioning is not fostered; people become isolated, angry, resentful and do what they do with no real joy; while a business may be profitable it is likely that it is not operating at anywhere near its potential" (p. 17).

Value-Added Processes of the Enterprise (2)

Value-added processes represent the organizational contexts for which knowledge is an essential factor of performance. It is in this context that is implanted a KM initiative. Tonchia and Tramontano (2004) state that "Process Management, with the concepts of internal customers and process ownership, is becoming one of the most important competitive weapons for firms and can determine a strategic change in the way business is carried out." They specify that "Process Management consists in the rationalization of processes, the quest for efficiency/effectiveness, a sort of simplification/clarification brought about by common-sense engineering" (p. 20). As process management engenders structural changes, when doing Business Process Reengineering, we should consider KM activities in order to identify knowledge that is an essential factor to enable value-added processes to achieve goals efficiently.

The Operating Elements

The operating elements of the MGKME constitute the operating components of the KMS. They focus on the underlying elements and consist of managerial guiding principles *(3)*, ad hoc infrastructures *(4)*, generic KM core processes *(5)*, organizational learning processes *(6)*, and methods and supporting tools for KM *(7)*.

Managerial Guiding Principles for KM (3)

The managerial guiding principles for KM bring a vision aligned with the enterprise's strategic orientations and suggest KM governance guidelines by analogy with COBIT®, which concerns IT principles of coordination and control of performances. In particular, we must establish KM indicators. Numerous publications and books relate to that subject (Bontis, Dragonetti, Jacobsen, & Roos, 1999; Moore, 1999; Morey, Maybury, & Thuraisingham, 2000; CEN-4, 2004). In order to monitor a KM initiative, we suggest two main categories of indicators: (1) a category of indicators that focuses on the impacts of the initiative favoring enhancement of intellectual capital and measuring the level of maturity of KM within the enterprise; (2) a category of indicators that insures monitoring and coordination of KM activities, measuring the results and insuring the relevance of the initiative.

In addition, we propose a way to get a good articulation between Deming's cycle and the organizational learning (see Figure 6). First, we refer to the PDCA cycle of activities: plan, do, check, and act (Martin, 1995). This cycle, first advocated by Deming (1992), is well-known as the Deming's cycle by quality management practitioners. The PDCA cycle has inspired the ISO 9004 (2000) Quality Standards in order to get a continuous process improvement of the quality management sys-

Figure 6. Deming's cycle and organizational learning articulation

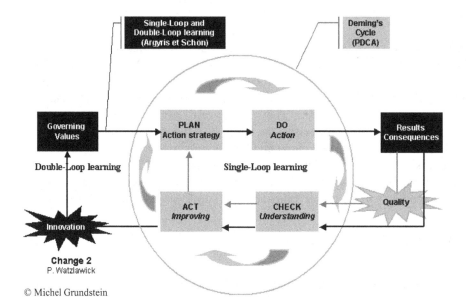

tem. Well-known in quality management, the Deming's PDCA Wheel presents a simplified cycle to achieve design and continual improvement of a product or a process. The cycle consists of four stages: (1) PLAN in order to induce a structured program of actions; (2) DO in order to develop the program and test conformity to protocols; (3) CHECK (or STUDY) in order to verify, analyze, and understand the results; and (4) ACT in order to react, propose, and decide the modifications and improvements.

Second, we refer to the Single-Loop Learning and Double-Loop Learning in the organizational learning theory defined by Argyris and Schön (Argyris & Schön, 1996) as "Learning that results in a change in the values of the theory-in-use, as well as in its strategies and assumptions" (p. 21). Thus, we point out the key contribution of knowledge management to Change 2 defined by Watzlawick, Weakland, and Fisch (1975).

Ad Hoc Infrastructures for KM Implementation (4)

The ad hoc infrastructures are adapted sets of devices and means for action. Beyond a network that favors cooperative work, it is important to implement the conditions that will allow sharing and creating knowledge. Relevant infrastructures must be set up according to the specific situation of each company and the context of the

Figure 7. Semi-opened infrastructure for the implementation of organizational learning

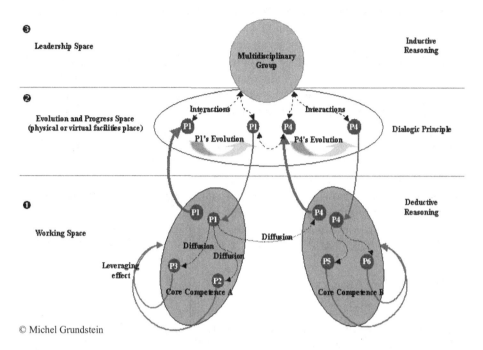

© Michel Grundstein

envisaged KM initiative. These infrastructures could be inspired by the Japanese concept of *Ba* that "can be thought as a shared space for emerging relationships" (Nonaka & Konno, 1998, p. 40). As an example, we present the semi-opened infrastructure of work that has been implemented in order to deploy knowledge-based systems (KBS) within a large French nuclear power plant company (Grundstein, de Bonnières, & Para, 1988).

The aim of the semi-opened infrastructure of work was to encourage the individual and collective apprenticeship to favor knowledge acquisition, to leverage emergence of new products, and to implement computer applications using artificial intelligence technologies. To expand, the semi-opened infrastructure of work requires a multidisciplinary group and the existence of an evolution and progress space. We describe briefly the functioning principle below (see Figure 7).

The leadership space *(3)* was constituted of engineers, organizers, and sociologists accustomed to doing inductive reasoning. This Multidisciplinary Group was in charge to deploy KBS over the whole company.

The working space *(1)* represents two operational units (Core Competence A and B), whatever is their geographical localization, where P1 and P4 are employees whose roles are to communicate on KBS and to implement applications in their own unit. P1 and P4 are using deductive reasoning.

The evolution and progress space *(2)* represents a place, which was a physical room situated at headquarters where P1 and P4 had to work and learn in interaction with the multidisciplinary group. In the evolution and progress space, P1 and P4 have had to practice they own deductive reasoning and to learn to work with the multidisciplinary group practicing an inductive reasoning. So people were interacting following two ways of reasoning. It is what Edgar Morin called dialogic principle that "combines two principles or notions that must be mutually exclusive, but that are integral parts of the same reality" (Morin & Le Moigne, 1999, p. 264). In the evolution and progress space, learning was especially effective, and interpretative frameworks of P1 and P4 were evolving. Arrows show (1) how P1 and P4 evolved in the evolution and progress space; and (2) how P1 and P4 disseminated their new knowledge in their own unit and how organizational learning was deployed. The evolution and progress space has proved to be a place of contacts, a field of multiple cultures in which the potentialities of each knowledge owner have been capitalized.

Generic KM Core Processes (5)

The generic KM core processes answer the problem of capitalizing on company's knowledge defined in the following way: "Capitalizing on company's knowledge means considering certain knowledge used and produced by the company as a storehouse of riches and drawing from these riches interest that contributes to increasing the company's capital" (Grundstein, 2000, p. 263).

Several problems co-exist. They are recurring problems with which the company was always confronted. These problems constitute a general problematic that has been organized in five categories (Grundstein, 2000). Each of these categories contains subprocesses that are aimed to contribute a solution to the set of overall problems (see Figure 8). Thus, beyond the category named "MANAGE", we have identified four generic KM core processes corresponding to the resolution of the other categories of problems. We describe these processes next.

The locating process deals with the location of crucial knowledge; that is, know-how (explicit knowledge) and skills (tacit knowledge) that are necessary for decision-making processes and for the progress of the essential processes that constitute the heart of the activities of the company. It is necessary to identify it, to locate it, to characterize it, to make cartographies of it, to estimate its economic value, and to classify it. One can mentioned an approach named GAMETH® (Grundstein, 2000; Grundstein & Rosenthal-Sabroux, 2004) specifically aimed to support this process.

Figure 8. Generic KM core processes and capitalizing on company's knowledge assets problems

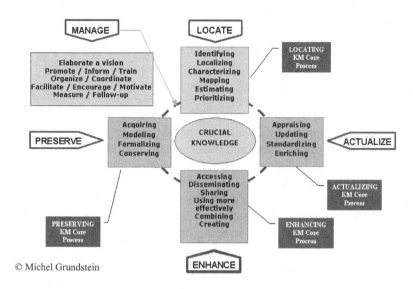

© Michel Grundstein

The preserving process deals with the preservation of knowledge and skills. When knowledge can be articulated into words, it is necessary to acquire it with the bearers of knowledge, to represent it, to formalize it, and to conserve it. This leads to knowledge engineering activities notably described in (Schreiber et al., 2000). When knowledge cannot be articulated, then interactions through communities of practice or other types of networks must be encouraged.

The enhancing process deals with the benefit of knowledge and skills. It is necessary to make them accessible according to certain rules of confidentiality and safety, to disseminate them, to share them, to use them more effectively, to combine them, and to create new knowledge. Here is the link to innovation processes.

The actualizing process deals with the actualization of knowledge and skills. It is necessary to appraise them, to update them, to standardize them, and to enrich them according to the returns of experiments, the creation of new knowledge, and the contribution of external knowledge. Here is the link to business intelligence processes.

Organizational Learning Processes (6)

Organizational learning processes underlay the whole generic KM core processes. The aim of the organizational learning process is to increase individual knowledge, to reinforce competencies, and to convert them into a collective knowledge through interactions, dialogue, discussions, exchange of experience, and observation. The main objective consists of fighting against the defensive routines that make barriers to training and change. Therefore, it is a question of helping the members of the organization to change their way of thinking by facilitating an apprenticeship of a constructive way of reasoning instead of a defensive one.

Methods and Supporting Tools for KM (7)

The methods and supporting tools relevant for KM can be determined only when considering the enterprise context and the envisaged KM initiative. One can find the descriptions and the characteristics of technologies, methods, and supporting tools relevant for KM in many publications such as, for example, Baek, Liebowitz, Prasad, and Granger (1999), Becker (1999), Huntington (1999), and Wensley and Verwijk-O'Sullivan (2000). Among all these tools, the information and applications portal that supplies a global access to the information can meet the needs of KM. In that case, the functional software and the tools answering the ends of KM are integrated into the digital information system.

From our viewpoint, the digital information system, centered on the knowledge worker, requires a human-centric design approach in order to place the knowledge worker into the heart of the design process (Rosenthal-Sabroux, 1996). The design must not disassociate the knowledge worker, stakeholder of different functional and organizational groups and lines of business or projects, from the professional processes in which he or she is engaged, the actions he or she performs, the decisions he or she makes, and the relations he or she has with his or her company environment (persons and artifacts). As such, the design must integrate individuals as components of the system.

Consequently, the conception of the digital information system has to take into account the nature of the information that the individual, as a decision maker, must be able to access. We distinguish three natures of information: mainstream data, source-of-knowledge data, and shared data (Grundstein & Rosenthal-Sabroux, 2003).

Mainstream data make up the flow of information that informs us on the state of a company's business process. Shared data constitute the information handled with the new ICT devices. These devices cause a break with older technologies, a rupture linked to the relationship of humans to space and time. They give us the capacity to be ubiquitous and take us from the real to a virtual world, from the manipulation of concrete to abstract objects. Source-of-knowledge data are the result of a knowledge engineering approach that offers techniques and tools for acquiring and representing knowledge. This knowledge is codified knowledge. It can be transferred easily through information systems. It is immediately intelligible by knowledge workers who share the same interpretative frameworks with the knower who had been asked to have his or her knowledge acquired, represented, and then codified.

Conclusion and Future Trends

The model for global knowledge management within the enterprise (MGKME) supports our full meaning of KM, as defined in the introduction of this chapter. It is a guiding framework for KM that supplies a people-focused KM as proposed by Wiig. It establishes a complementary approach based both on our experience within the industry and on our research works. It presents an attempt to articulate the Deming's cycle with the single-loop learning and double-looplearning defined in the Argyris and Schön's organizational learning theory.

In this chapter, based on the MGKME, we have emphasized the need to incorporate the employees' knowledge within the information and knowledge management systems in such a way that the knowledge of the knower should be distinguished from the flows of codified knowledge that are processed within the systems. To do so, it appears that users of the information systems should be considered as knowledge workers and be thought of as components of the systems.

However, the individual's culture may engender the incommensurability of the employees' interpretative frameworks (e.g., cognitive models or mental models) and so have a key influence on the design, development, implementation, and utilization of the systems. To avoid such risks, we must envisage establishing KM governance principles that would incorporate the role of knowledge workers as components and users of the information systems. On the one hand, these KM governance principles should be connected with IT Governance principles by means of specific criteria inherent to knowledge management systems. On the other hand, they should be connected with corporate governance principles by taking into account the risks linked to the utilization and creation of knowledge. That could be one of the IRM challenges for the coming years.

We expect that the MGKME will serve as a pattern of reference for establishing KM governance principles. Moreover, we hope to succeed to consolidate it in order to become an open framework that enables each enterprise to assess the maturity level of its knowledge management system and to adapt its KM program. In the future, we should complete and validate the MGKME by developing our researches in that sense. That will result in developing appropriate methods and constructing a set of qualitative indicators and specific tools.

References

Argyris, C., & Schön, D. A. (1996). *Organizational learning II. Theory, method, and practice*. Reading, MA: Addison-Wesley.

Baek, S., Liebowitz, J., Prasad, S. Y., & Granger, M. (1999). Intelligent agents for knowledge management, toward intelligent Web-based collaboration within virtual teams. In J. Liebowitz (Ed.), *Knowledge management handbook* (ch. 11, pp. 1-23). Boca Raton, FL: CRC Press.

Becker, G. (1999). Knowledge discovery. In J. Liebowitz (Ed.), *Knowledge management handbook* (ch. 13, pp. 1-27). Boca Raton, FL: CRC Press.

Bontis, N., Dragonnetti, N., Jacobsen, K., & Roos, G. (1999). The knowledge toolbox: A review of the tools available to measure and manage intangible resources. *European Management Journal, 17*(4), 391-401.

CEN-1. (2004). Knowledge management framework. In *European guide to good practice in knowledge management* (part 1). Brussels: CEN, CWA 14924-1:2004 (E). Retrieved June 19, 2004, from ftp://cenftp1.cenorm.be/PUBLIC/CWAs/e-Europe/KM/CWA14924-01-2004-Mar.pdf

CEN-4. (2004). Measuring KM. In *European guide to good practice in knowledge management* (part 4). Brussels: CEN, CWA 14924-4:2004 (E). Retrieved June 19, 2004, from ftp://cenftp1.cenorm.be/PUBLIC/CWAs/e-Europe/KM/CWA14924-04-2004-Mar.pdf

Chua, B. B., & Brennan, J. (2004). Enhancing collaborative knowledge management systems design. In D. Remenyi (Ed.), *Proceedings of the 5th European Conference on Knowledge Management* (pp. 171-178). Dublin: Trinity College.

CIGREF. (2000). *Gérer les connaissances. Défi, enjeux et conduite de projet* (Report No. ATTJ8KE4, pp. 15-17). Paris: CIGREF, Club informatique des grandes entreprises françaises. Retrieved May 06, 2004, from http://www.cigref.fr

COBIT®. (2000, 2002, 2005). *Gouvernance, contrôle et audit de l'information et des technologies associées* (Trans. of *Control objectives for information and related technology. information systems audit and control*, (3rd ed.). (Trans. into French by AFAI, the French Chapter of the Information Systems Audit and Control Association). Rolling Meadows, IL: IT Governance Institute.

Cohen, D., & Prusak, L. (2001). *In good company: How social capital makes organizations work*. Boston: Harvard Business School Publishing.

Corbitt, B. J., Peszinski, K. J., Inthanond, S., & Hill, B. (2004, July-September). Cultural differences, information and code systems. *Journal of Global Information Management, 12*(3), 65-85.

Darby, R., Herbolzheimer, E., & van Winkelen, C. (2004). Cross-cultural context in the implementation of knowledge management: A cross-cultural case study analysis. In D. Remenyi (Ed.), *Proceedings of the 5th European Conference on Knowledge Management* (pp. 231-242). Dublin: Trinity College.

Davenport, T. H., & Prusak, L. (1998). *Working knowledge. How organizations manage what they know.* Boston: Harvard Business School Press.

Deming, W. E. (1992). *Out of the crisis.* Cambridge, MA: MIT Press.

de Rosnay, J. (1975). *Le macroscope. Vers une vision globale.* Paris: Éditions du Seuil.

Edvinsson, L., & Malone, M. S. (1997). *Intellectual capital. Realizing your company's true value by finding its hidden brainpower.* New York: HarperCollins.

Grundstein, M. (2000). From capitalizing on company's knowledge to knowledge management. In D. Morey, M. Maybury, & B. Thuraisingham (Eds.), *Knowledge management, classic and contemporary works* (pp. 261-287). Cambridge, MA: MIT Press.

Grundstein, M., de Bonnières, P., & Para, S. (1988). *Les systèmes à base de connaissances systèmes experts pour l'entreprise.* Paris: Afnor Gestion.

Grundstein, M., & Rosenthal-Sabroux, C. (2003). Three types of data for extended company's employees: A knowledge management viewpoint. In M. Khosrow-Pour (Ed.), *Information technology and organizations: Trends, issues, challenges and solutions, 2003 IRMA proceedings* (pp. 979-983). Hershey, PA: Idea Group Publishing.

Grundstein, M., & Rosenthal-Sabroux, C. (2004). GAMETH®, A decision support approach to identify and locate potential crucial knowledge. In D. Remenyi (Ed.), *Proceedings of the 5th European Conference on Knowledge Management* (pp. 391-402). Dublin: Trinity College.

Guldentops, E. (2004). Governing information technology through COBIT®. In W. V. Grembergen (Ed.), *Strategies for information technology governance* (pp. 269-309). Hershey, PA: Idea Group Publishing.

Haeckel, S. H. (2000). Managing knowledge in adaptive enterprises. In C. Despres & D. Chauvel (Eds.), *Knowledge horizons* (pp. 287-305). Woburn, MA: Butterworth-Heinemann.

Huntington, D. (1999). Knowledge-based systems: A look at rule-based systems. In J. Liebowitz (Ed.), *Knowledge management handbook* (ch. 14, pp. 1-16). Boca Raton, FL: CRC Press.

IMKA. (1990). *IMKA technology technical summary.* Carnegie Group, Inc.; Digital Equipment Corporation; Ford Motor Company; Texas Instruments Inc.; US West Advanced Technologies Inc.

ISO 9004. (2000). *Système de management de la qualité; ligne directrices pour l'amélioration des performances (Quality Management Systems, Guidelines for Performance Improvements).* Paris La Défense: AFNOR.

Jennex, M. E. (2005a). What is knowledge management? *International Journal of Knowledge Management, 1*(4), i-v.

Jennex, M. E. (2005b). Knowledge management systems. *International Journal of Knowledge Management, 1*(2), i-iv.

Martin, J. (1995). *The great transition. Using the seven disciplines of enterprise engineering to align people, technology, and strategy.* New York: AMACOM.

Moore, C. R. (1999). Performance measures for knowledge management. In J. Liebowitz (Ed.), *Knowledge management handbook* (ch. 6, pp. 1-29). Boca Raton, FL: CRC Press.

Morey, D., Maybury, M., & Thuraisingham, B. (2000). *Knowledge management, classic and contemporary works.* Cambridge, MA: MIT Press.

Morin, E., & Le Moigne, J.-L. (1999). *L'intelligence de la complexité.* Paris: L'Harmattan.

Nonaka, I., & Konno, N. (1998). The concept of "ba": Building a foundation for knowledge creation. *California Management Review, 40*(3), 40-54.

OECD. (2004). *OECD principles of corporate governance.* Retrieved September 2005, from http://www.oecd.org/document/49/0,2340,en_2649_34813_31530865_1_1_1_1,00.html

Rosenthal-Sabroux, C. (1996). Contribution méthodologique à la conception de systèmes d'information coopératifs: Prise en compte de la coopération homme/machine. *Mémoire HDR.* Paris: Université Paris-Dauphine.

Schreiber, A.Th., Akkermans, J. M., Anjewierden, A. A., de Hoog, R., Shadbolt, N. R., Van de Velde, W., et al. (2000). *Knowledge engineering and management. The commonKADS methodology.* Cambridge, MA: MIT Press.

Snowden, D. (2000): The social ecology of knowledge management. In C. Despres & D. Chauvel (Eds.), *Knowledge horizons* (pp. 237-365). Woburn, MA: Butterworth-Heinemann.

Stewart, A. (2001). *The conversing company, its culture, power and potential.* Retrieved June 2004, from http://www.knowledgeboard.com/download/3343/conversing-company.pdf

Tonchia, S., & Tramontano, A. (2004). *Process management for the extended enterprise.* Berlin-Heidelberg: Springer-Verlag.

Tsuchiya, S. (1993). Improving knowledge creation ability through organizational learning. In J.-P. Barthes (Ed.), *Proceedings of the International Symposium on the Management of Industrial and Corporate Knowledge* (pp. 87-95). Compiègne, France: Université de Technologie de Compiègne.

Watzlawick, P., Weakland, J., & Fisch, R. (1975). *Changements: Paradoxes et psychothérapie.* Paris: Éditions du Seuil.

Wensley, A. K. P., & Verwijk-O'Sullivan, A. (2000). Tools for knowledge management. In C. Despres & D. Chauvel (Eds.), *Knowledge horizon: The present and the promise of knowledge management* (pp. 113-130). Woburn, MA: Butterworth-Heinemann.

Wiig, K. (2004). *People-focused knowledge management. How effective decision making leads to corporate success*. Burlington, MA: Elsevier Butterworth-Heinemann.

Chapter XII

Cultural Impact on Global Knowledge Sharing

Timothy Shea
University of Massachusetts Dartmouth, USA

David Lewis
University of Massachusetts Lowell, USA

Abstract

This chapter introduces how culture impacts global knowledge sharing. Effective knowledge sharing (KS), one of the four interdependent dimensions of knowledge management (KM), is particularly important in today's global environment in which national cultural differences are negotiated all the time. Knowledge sharing is described along six dimensions and national culture along four dimensions. A model is presented, which provides guidelines for effectively sharing different types of knowledge within different cultural environments. Several examples are presented to illustrate the model's effectiveness. Using the model as a guide, the authors believe that decision makers will increase the chances that information and knowledge will be shared successfully.

Copyright © 2007, Idea Group Inc. Copying or distributing in print or electronic forms without written permission of Idea Group Inc. is prohibited.

Introduction

Information and communications technology—in particular, when it is used to gather and utilize knowledge—are key to the growth of today's dynamic and highly competitive world economy (OECD, 1996). Companies are under constant pressure to have the right knowledge at the right time in the hands of the right person in order to help increase productivity, be more innovative, and increase competitiveness. Whether a company is coordinating activities among its various manufacturing and sales operations around the world, managing far-flung outsourcing relationships, or engaging in other value chain activities with geographically dispersed companies, companies of the 21st century operate in environments with insatiable needs for collaborating, sharing, and organizing knowledge. As an indication of the magnitude of the situation, more than one-third of world trade is conducted by multinational companies.

Given how organizations today have an unprecedented ability for information coordination through IT developments in ERP systems, global communications, and the like, effective information resource management (IRM) is paramount. One IRM tenet is that organizations need to be able to share information in ways that improve both efficiency and effectiveness. Today, knowledge management (KM) provides a methodology for defining and measuring knowledge needs and implementing an appropriate technical solution. Within KM's typical processes, effective knowledge sharing is particularly important in order for the various business networks to be effective both within companies and between companies (Moller & Svahn, 2003). A company's competitiveness is influenced more and more by its ability to identify and apply its specialized knowledge resources (Bhagat, Kedia, Harveston, & Triandis, 2002). For example, innovations and new ideas often are born out of having the right people in touch with one another.

So, how does an organization create and manage effective KM systems? Moffett, McAdam, and Parkinson (2003) suggest that there has been an over-emphasis on technology issues. Nemati (2002) reaffirms that technology is just one of five major factors that influence knowledge management from a global management perspective: culture, firm strategy and structure, IT infrastructure, organizational/managerial, and industry specifics.

To date, there has been limited research on how cross-cultural issues relate to knowledge management or to knowledge sharing (Ford & Chan, 2003). Therefore, this chapter will focus on cultural impacts and their role and impact on knowledge sharing; specifically, national culture issues, not organizational culture issues. The chapter will review the evolution of KM and knowledge sharing (KS), provide a background on national culture, and explore the impact of national culture on knowledge sharing. The chapter will end with a discussion of future trends in KM/KS and KM/KS research.

Background: Literature Review

Knowledge management/knowledge sharing and national culture are areas of research in which each has a large body of work. This section begins with an overview of knowledge management and knowledge sharing and follows with a discussion of literature related to ethnic or national culture.

Knowledge Management

Knowledge management (KM) initiatives by companies seek to "achieve knowledge integration and benefit from the collective knowledge of the organization through learning" (Mason, 2003, p. 31). More specifically, KM "facilitates the creation, capturing, organization, accessing, and use of an enterprise's knowledge capital," consisting of human capital (e.g., knowledge and skills), structural capital (e.g., systems, processes, and methods), and relational capital (e.g., relationship with customers, suppliers, and external organizations) (Cloete & Snyman, 2003, p. 237). This section discusses what knowledge is, the components of KM, where knowledge sharing fits in, and the goals of knowledge management systems.

The KM literature is comprised of varying views of what KM entails. Early studies referred to improving the information value chain (Rayport & Sviokla, 1995), organizational memory (Walsh & Ungson, 1991), and organizational learning (Huber, 1991). More recently, Grover and Davenport (2001) discussed the KM processes of creation, codification, transfer, and realization. Others describe knowledge acquisition (KA), knowledge documentation (KD), knowledge sharing (KS), and knowledge application (KP) as the four interdependent basic dimensions of the KM process. Nevertheless, organizations are likely to practice KM differently by adopting various tactical and operational implementations of knowledge-related activities (Bhatt, 2001).

Knowledge takes different forms and types, such as explicit and tacit, objective and experience-based, organizational routines and procedural knowledge, general and domain-specific, individual and organizational knowledge, as well as external and internal knowledge. Perhaps the most familiar distinction in the knowledge management domain has been between explicit and tacit knowledge (Nonaka, 1991). Explicit knowledge refers to knowledge that is easily formalized and documented through different tools such as information technology, rules, and procedures. Tacit knowledge, on the other hand, is personal and remains in the human mind, behavior, and perception. It exists in the form of people skills, competences, experiences, and expertise, know-how, and even organizational and national culture. While tacit knowledge is difficult to describe using words (Karhu, 2002), it is argued that since it is difficult to document or transfer, it may be central to a firm's competitive advantage (Ambrosini & Bowman, 2001).

Knowledge management systems (KMS) are information systems that support knowledge management initiatives within and between companies. Today, especially with the impetus of the Internet (and intranets), there is a wide variety of technology tools that support KM, ranging from low-cost and easy-to-install options to expensive options that require a major system development effort (Fichter, 2005; Moffett, McAdams, & Parkinson, 2004):

- **Collaborative tools:** Groupware, meeting support systems, knowledge directories, blogs, instant messaging, and wikis
- **Content management:** Web portals, company intranets, agents and filters, electronic publishing systems
- **Business intelligence:** Data warehousing, data mining, group decision support systems, decision support systems, executive information systems, expert systems

Knowledge management systems (KMS) usage can be supported or hindered by organizational factors (e.g., the workplace culture, whether it is a knowledge-sharing culture or not), supervisory control (e.g., user expectations and management's need for control), system characteristics (e.g., usefulness and ease of use), and the role of the KMS (Fichter, 2005; King, 2006). These characteristics also impact which tools to use for knowledge sharing and collaboration in an organization.

Today, communities of practice (CoPs) are an increasingly popular and effective means for implementing KM and KS initiatives. Similar in some ways to knowledge networks mentioned previously, CoPs are groups that "share a concern, a set of problems, or a passion about a topic, and ... deepen their knowledge and expertise in this area by interacting on an ongoing basis" (Wenger, McDermott, & Snyder, 2002, p. 4). A community of practice has three main characteristics:

- **Domain:** A CoP is not just a group of friends. Involvement in the community requires some knowledge and some competence in the focus area, or domain. The domain is the definition of the area of shared inquiry and of the key issues.
- **Community:** Members of the community interact and learn together. The community is the relationships among members and the sense of belonging.
- **Practice:** The CoP develops as members interact, especially as they solve problems. The practice is the body of knowledge, methods, stories, cases, tools, and documents.

Technically, CoPs often are implemented as Web portals that provide a single point of access for personalized, easy-to-navigate internal and external information content; numerous means for connectivity among the community members (e.g., e-mail, forums); and increasingly, KM components such as capture-and-store, collaborate, and solve-or-recommend. These portals can serve as a network-based memory for the community (Cloete & Snyman, 2003).

Increasingly, communities of practice (CoPs), specifically virtual CoPs, are becoming the heart of a knowledge management (KM) system (Furlong & Johnson, 2003), especially for multinational corporations (Ardichvili, Page, & Wentling, 2003). A well-designed and managed online CoP can speed up problem solving, encourage innovation, and support creative thinking (Chung, 2004). It also can enhance the trust especially critical to colocated teams—trust that supports participation, development of team identity, and interpersonal confidence (Kimble & Hildreth, 2005).

The implementation of KM strategies and initiatives are expected to lead to major cost reductions and performance increases (Cavaleri, 2004). What the organization comes to know explains its performance (Argote & Ingram, 2000). In order for an organization to remain competitive, it must effectively practice the activities of acquiring, documenting, sharing, and applying knowledge to solve problems and exploit opportunities (Sharkie, 2003). However, like most information systems, KM initiatives by themselves are not the end but rather the means. Thus, KM initiatives provide organizations with opportunities for organizational learning (Stata, 1989), the development of competencies (Alavi & Leidner, 2001), and knowledge integration (Kogut & Zander, 1996).

Knowledge Sharing

Whether the term employed is knowledge sharing, knowledge transfer, or knowledge dissemination, it is becoming a more and more important part of an organization's competitive advantage. In a world where every year the amount of knowledge is doubling, while at the same time, vast quantities of knowledge are becoming obsolete, and only 5% of employee knowledge is accessible across the company, effective knowledge sharing—in particular, as part of organizational learning—is becoming a more important part of an organization's competitive advantage (Drucker, 1997; Senge, 1997; Wells, Sheina, & Harris-Jones, 2000). Alternatively, knowledge sharing is only worthwhile if the knowledge is worth sharing (Schulz, 2001).

Knowledge sharing (KS) refers to sharing both explicit information as well as tacit information such as beliefs and experiences (Davenport & Pruzak, 1998; Nonaka, 1991). However, Kimble and Hildreth (2005) argue that when the tacit/explicit model is discussed, the primary goal of traditional KM and KS is typically an externalized representation of knowledge; that is, explicit knowledge, such as a report. In their study on the use of CoPs and KM, they found that effective knowledge sharing is actually a duality where sharing harder knowledge, such as a planning document, needs to be integrated with softer knowledge that is accomplished through motivated social processes such as meetings. The goal is to manage a balance of hard and soft processes. According to this viewpoint, the primary value of the KM or KS system comes when the hard knowledge is used as a catalyst for soft knowledge processes such as participation. This is quite different from a context where the focus is predominately on increasing hard or explicit knowledge.

How to support and encourage knowledge sharing (e.g., getting employees to contribute to digital knowledge repositories [the supply side] or to post a question [the demand side]) is an open question. It is a complicated activity that involves resistance; intrinsic and extrinsic motivation; and technical, social, cultural, and organizational issues (Ardichvili et al., 2003; Ciborra & Patriota, 1998; Holthouse, 1998; Osterioh & Frey, 2000; Wakefield, 2005). For example, for individual workers, KS includes evaluating the search and transfer costs before deciding whether to knowledge share or seek knowledge (Hansen, Mors, & Lovas, 2005). Kimble and Hildreth (2005) suggest that common interest, task focus, and deadlines help. Dixon (2000) suggests that people are more willing to share information informally (tacitly) than to contribute to a database (explicitly). McLure and Faraj (2000) suggest that when employees perceive work-related knowledge as belonging to the organization and not to the individual, then knowledge sharing is far more prevalent.

Knowledge sharing (KS) environments and systems can support mutual understanding and trust among different groups that support cooperation and sharing, be it between individuals and team members, within a company, or between companies (Janz & Prasarnphanich, 2003; Larsson, Bengtsson, Henriksson, & Sparks, 1998). Knowledge-sharing applications include databases that enhance the corporate memory through customer data, a repository of past projects, or best practices; facilitating communication among the organization's members, such as shared data maps of internal expertise; and knowledge networks connecting pools of expertise (Alavi & Leidner, 2001; Mason, 2003; Ruggles, 1998). However, current technology is limited. For example, current workflow technology solutions, although quite popular as process-oriented coordination tools, are limited and will not adequately support coordination for knowledge-intensive business processes such as new product development (Marjanovic, 2005).

In sum, knowledge sharing within and across organizations, while generally acknowledged to be critical for business success in today's world, is a complicated and not very well-understood process. When one adds cross-cultural differences to the mix, the water gets even muddier. The next section discusses the cultural aspect and its relationship to knowledge sharing in more detail.

Culture

Culture can be described as the way a group of people does things. Culture includes the values, norms, and attitudes expressed by this group. We can analyze groups along a number of dimensions; the most studied are organizational culture and ethnic culture. Our focus for this chapter is on ethnic culture. When evaluating ethnic culture, an individual's country of residence often is used as a surrogate measure since it is easily quantified. There has been a number of researchers that have tried to identify cultural characteristics, specifically which values and norms are universal and which are dependent on an individual's specific background. We begin by describing three studies and their research findings.

The first major study was conducted by the Global Leadership and Organizational Behavior Effectiveness Research Project Team (Javidad & House, 2001). They identified nine dimensions that distinguish one culture from another: assertiveness, future orientation, performance orientation, humane orientation, gender differentiation, uncertainty avoidance, power distance, institutional collectiveness vs. individualism, and in-group collectivism. To have value in practice, one must be able to relate these dimensions to individual countries. For example, from the Globe Study, the United States was the most performance-oriented, and Russia was the least performance-oriented.

Geert Hofstede performed a much older study in the 1960s and 1970s (Hofstede, 1980). His work identified four dimensions similar to four of the dimensions described previously. These were power distance, uncertainty avoidance, individualism/collectivism, and masculinity. The descriptors of these dimensions are similar to those in the Globe study. His study was large enough so that countries could be placed into clusters; those that reacted similarly to those from other countries were in the same cluster. The Globe study identified nine clusters. Hofstede identified seven with an outlier cluster. Overall, the clustering was very similar.

The third researcher who often is reported in the literature is Fons Trompenaars (1993). He identified four value dimensions: obligation, emotional orientation in relationships, involvement in relationships, and legitimization of power and status. All of these authors and others who since have added their own cross-cultural studies using the same instruments (questionnaires in all cases) have identified cultural profiles for the countries studied. There are relatively few but subtle differences among the country profiles. For example, Harris and Moran (2000) developed a profile of Americans at a glance, including that they tend to be informal, competitive, and individualistic.

While these studies have been useful, there has been some discussion, including whether or not Hofstede's dimensions are appropriate. Walsham (2001) believes that while there are endless individual nuances and differences, there are enough similarities in shared symbols, norms, and values to make national culture a useful distinction. Others believe that using the nation as the unit of analysis is, at best, of limited use; for example, there are often many different cultural groups within a nation. In addition, work-related actions have proved hard to link to cultural attitudes (Myers & Tan, 2002).

More specific to KM, Mason (2003) described how boundary-spanning activities—syntactic, semantic, and pragmatic—could be used to support effective coordination among diverse groups as they developed KM systems. Repositories support the syntactic level by enabling communication of facts, tasks, and actions (knowledge transfer and knowledge sharing). Standardized forms and procedures help to create common standards between groups, supporting the semantic level—knowledge translation. Finally, objectives, maps, and models help to make embedded knowledge explicit, supporting the pragmatic level—knowledge transformation and learning. Kimble and Hildreth (2005) add how social interaction (in particular, face-to-face interaction) is essential to knowledge creation, especially the cultural context.

One study recommends proactive management for more effective cross-cultural work when outsourcing between multinationals (Krishna, Sahay, & Walsh, 2004):

- Using systems such as coordination and control systems to harmonize between outsourcer and supplier.
- Understanding differences in norms and values between cultures, for example, differences in hierarchy/power and business practices.
- Encouraging compromise and a negotiated work culture for cross-cultural teams through training and exchange mechanisms.

While such activities can help, they warn that there are limits to how much one can affect deeply ingrained attitudes and values.

Specific to knowledge sharing, Lam (1997) looked closely at work systems in which a British and Japanese firm had engineer's collaborating in high-level technical work. She found that knowledge sharing and cross-border collaborative work was impeded by many differences in culture, including educational background, approaches to coordination of work, and the way knowledge (in particular, tacit knowledge) was organized and disseminated. Ford and Chan (2003) studied knowledge sharing between a company and its international subsidiary. They found that knowledge sharing was impacted negatively by several cultural differences: multiple languages, heterogeneity vs. homogeneity of the national cultures, and culturally acceptable advice-seeking behaviors.

There are numerous cultural variables that managers must consider when working overseas or sharing information or knowledge overseas. Some of these include management style (democratic vs. autocratic), communication (verbal, nonverbal, noise in the communication process), negotiation and decision making (negotiation styles, negotiator characteristics, characteristics of the negotiation), and motivation or leadership (what motivates, which type of leadership style). The next section explicitly addresses how cultural issues can be addressed when facilitating knowledge sharing.

Relating Cultural Variables to Knowledge Sharing

Sharing of information, like most managerial functions, is more difficult when cross-cultural differences exist. To date, there has been little research on the impact of cross-cultural factors on knowledge-sharing activities. However, Ford and Chan (2003) suggest that "knowledge sharing may also be the most susceptible to effects of cross-cultural difference within a company" (p. 12). For example, if we are a manufacturing company that is developing a new product using a global team, including members from culturally diverse countries such as China, Spain, and the United States, the chances of miscommunication, misinterpretation, and using the incorrect methods of motivation can increase substantially. Cross-cultural differences may be more important than technical variables, organizational culture, and others when examining the failure of complex projects.

However, if we know something about the knowledge that managers or individuals are trying to share and something about the cultures between which information sharing takes place, we can develop a set of guidelines to improve the chances that knowledge sharing is successful. Table 1 provides guidelines, given specific cultural characteristics for which knowledge dimensions are shared more easily. Managers must be very careful about how they share information if they don't have control over the characteristics of the knowledge or if there exist major differences between the cultures of the managers involved.

For purposes of this chapter, we will focus on the four dimensions developed by Hofstede, as they are clearly the most widely used by management theorists to differentiate cultures. Based on the knowledge management literature described in an earlier section, we have selected six major characteristics of knowledge to relate to Hofstede's cultural dimensions: tacit vs. explicit, organizational vs. individual, general vs. specific, external vs. internal, sensitivity to the employee level, and media richness. Table 1 provides guidelines on how knowledge should be packaged in order to successfully share that knowledge, depending on how a country falls on Hofstede's dimensions and the type of knowledge being shared. We introduce abbreviations as shown in Diagram 1.

We then will provide two scenarios that illustrate the model's predictions.

Note that the proposed knowledge-sharing strategies are hypothesized by the authors and must be verified in future research.

Diagram 1.

Knowledge Dimensions			
Exp	Explicit	Tac	Tacit
Org	Organizational	Ind	Individual
Gen	General	Spec	Specific
Ext	External	Int	Internal
Same	Employee Level Same	Diff	Employee Level Different
Rich	Media rich	NotR	Medium not rich

Table 1. Knowledge-sharing guidelines

Cultural Dimensions	Tacit/Explicit Information	Organizational/ Individual	General/ Specific	External/ Internal	Employee Level	Media Richness
Individualistic /Collective						
° Individualistic	Tacit	Ind	ND	Ext	ND	NotR
° Collective	Explicit	Org		Int		Rich
Power Distance						
° High	Explicit	Ind	Spec	Int	Diff	Rich
° Low	Tacit	Org	Gen	Ext	Same	NotR
Uncertainty Avoidance						
° High	Explicit	Ind	Spec	Int	Same	Rich
° Low	Tacit	Org	Gen	Ext	Diff	NotR
Masculine / Feminine						
° Masculine	Explicit	Ind	Spec	Int	Diff	ND
° Feminine	Tacit	Org	Gen	Ext	Same	

* *Note: ND = no difference*

It is important that the reader understand the information in the table. The entries within a cell indicate the relationship between the knowledge and cultural variables. The cell combinations describe recommended strategies for successfully sharing knowledge. If there is a disconnect between the cultural and information variables, it is highly likely that information will be lost, that it will not be shared in either an efficient or effective manner.

An example might help. Managers from two individualistic countries (e.g., the United States and Great Britain) would be less likely to share knowledge, since knowledge often is seen as a form of power, part of the competitive landscape, and an aid to success. Their individualistic cultural perspective would tend toward an individual perspective on who owns the data. Meanwhile, managers from collectivist countries (e.g., Spain and Argentina) would be more likely to see data as belonging to the organization and be more inclined to share the data, perhaps by entering new best practices information based on a recent project experience into a company knowledge portal. Lack of cooperation by someone from an individualistic national culture would not be a matter of ill-intent but rather differing perceptions. Implementation of a common knowledge-sharing technique, such as a knowledge portal, clearly would need to take different forms in the two different cultural environments.

What does this all mean in practice? Using the preferred knowledge-sharing technique will increase the success that knowledge is shared correctly. If countries score similarly on one or more of the cultural dimensions and have control over how knowledge is presented by using the results in the table, they can increase the chances that knowledge will be shared successfully. The following scenarios will illustrate.

Scenario 1

An American automobile firm decides to outsource a number of its business functions. Specifically, it subcontracts its call center to India to handle customer support issues, and it subcontracts its engine production to Indonesia. There is expected to be minimal knowledge sharing between the subcontractors. Mostly, the knowledge sharing will be between the home office of the firm in the United States and its individual subcontractors. Let's examine the type of knowledge that must be shared, using the dimensions defined previously, and cultural characteristics as defined by Hofstede to examine how difficult knowledge sharing will be in this situation.

Before looking at the issue of knowledge sharing, let's first look at the level of cross-cultural differences between these countries. Based on Hofstede's four dimensions, our bases for describing culture, the differences among these four countries can be seen in Table 2.

Immediately we can see that there are potential cross-cultural issues, especially between the home office in the United States and its subcontractors in India and Indonesia. The countries are different on all four cultural dimensions, opposite for three of them. Thus, when looking at the results in Table 1, Knowledge-sharing guidelines, we must be especially careful when the knowledge-sharing strategies between countries are different. Taking care requires us to understand the preferences of the representatives of the other cultures and to try to adapt to their needs as

Table 2. National cultural differences of three countries (Hofstede, 1980)

Cultural Characteristics	Country	United States	India	Indonesia
Individualism/Collectivism		Individualistic	Collective	Collective
Power Distance		Low	High	High
Uncertainty Avoidance		Low	High	High
Masculinity/Femininity		Middle	Masculine	Masculine

much as possible. Since, at least on these general dimensions, India and Indonesia are similar, once we have identified a knowledge-sharing strategy for one, we will be able to use that strategy fairly easily for the other. In Table 3, we replicate the knowledge-sharing guidelines table in order to show the preferred knowledge-sharing techniques applicable to this scenario.

Reviewing the chart above, it can be seen that the types of knowledge sharing are likely to be different between the U.S. and Indonesia as well as between the U.S. and India. One interesting example is the high power distance rating for India and Indonesia compared to a low power distance rating for the U.S. A high power distance rating suggests a more hierarchical organizational structure in which decisions are made by superiors without much consultation with subordinates, and subordinates are fearful of disagreeing or contradicting their superiors. Knowledge sharing is more often one direction, top down, more limited, and, as seen in Table 1, more explicit. A low power distance rating often will lead to flatter organizations and a more participatory style of management. Knowledge sharing is more likely to be both directions, and the sharing of ideas and opinions (examples of tacit knowledge sharing) will be more prevalent. However, explicit knowledge sharing up the organization may be easier in the high power distance culture. Its employees will adapt more readily to the extra data input demands of a new CRM or ERP system, even if the extra time mostly supports new management reports. Low power distance employees may be more likely to grumble and procrastinate in response to such requests.

Since the American firm is the home firm, it should do its best to accommodate the style of India and Indonesia in order to help with the successful sharing of information. In this case, information, if possible, should be shared in an explicit and formal way, respecting the chain of command, targeting the individual level; as further indicated in Table 1, it should be very specific in nature, based on internal facts to the organization, aimed at the specific level of the employee involved; and extra time should be taken to make sure that the media is rich, maybe requiring multiple technologies to increase the successful exchange of information.

Table 3. Knowledge-sharing guidelines between the U.S. and India or Indonesia

	Tacit/ Explicit	Organizational/ Individual	General/ Specific	External/ Internal	Employee Level	Media Richness
U.S. ° Individualistic ° High Power Distance ° Low: Uncertainty Avoidance ° Middle: Masculinity/ Femininity	° Tacit ° Tacit ° Tacit ° ND	° Ind ° Org ° Org ° Org	ND	° Ext ° Ext ° Ext ° Ext	° ND ° Same ° Same ° Same	° NotR ° NotR ° NotR ° ND
° India/ Indonesia ° Individualistic ° High Power Distance ° Low: Uncertainty Avoidance ° Middle: Masculinity/ Femininity	° Explicit ° Explicit ° Explicit ° Explicit	° Org ° Ind ° Ind ° Ind	° ND ° SK ° SK ° SK	° Int ° Int ° Int ° Int	° ND ° Diff ° Diff ° Diff	° NotR ° NotR ° NotR ° ND

(Note: India and Indonesia are considered together since they show similar cultural characteristics on Hofstede's dimensions)

Scenario 2

You are a project manager in a software firm whose headquarters is in Japan. Specifically, you work in new product development. You have been given the leeway to search throughout the world for the most highly skilled programmers, and your current team is made up of three Indians, four Russians, six Chinese, two Japanese who are not directly involved in the technical aspects, and an Australian who is one of the original developers of the language you are using for writing the new software. Again, we will use the previously discussed cultural and knowledge dimensions to analyze the difficulty in sharing information in this situation.

The cross-cultural management situation is more complex here, first because there are more nationalities involved (American, Indian, Russian, Chinese, Japanese, and Australian), and second, because there is likely to be communication and knowledge sharing among all of these individuals.

As a starting point, we reproduce the research results for these nationalities on Hertzberg's four cultural dimensions.

In this case, we potentially will have all team members interacting together. The specific ways in which information should be shared thus become much more complex. The first step would to be to try to identify clusters; that is, countries that

Table 4. Country cultural characteristics, scenario 2

Country Cultural Characteristics	U.S.	Russia	China	Japan	India	Australia
Individualism/ Collectivism	Individualistic	Individualistic	Collective	Middle	Collective	Individualistic
Power Distance	Low	No data	No data	Middle	High	Low
Uncertainty Avoidance	Low	No data	No data	High	Low	Middle
Masculine/ Feminine	Middle	No data	High	High	High	Low

seem to share common characteristics. The first attempt indicates that the U.S. and Australia are very similar, and China and India show some similarities. Japan seems not to follow any specific cluster, and we do not have enough data on cultural tendencies for Russia to make any definitive statement. Thus, the cultural characteristics scenario in Table 4 and the guidelines for knowledge sharing in Table 1 would be a good place to start to sort through the complexities and establish knowledge-sharing strategies.

- **Step 1:** When Americans are sharing with Australians, they should provide opportunities to use tacit information; for example, face-to-face meetings or lunches that promote conversations, brainstorming, and sharing ideas. In addition, they should provide access to external information to illustrate their points, be less concerned with mixing employee level, and not be concerned with the level of media richness.
- **Step 2:** When Indians are talking to Chinese, just the opposite would be recommended: use explicit information; for example, establish a formal, weekly, detailed, progress report; talk at the organizational level; use information external to the firm to illustrate points; target the information being shared to the level of the employee; and use rich media to make sure the point is being made.
- **Step 3:** When Americans are talking to all others (excluding Australians), they should keep in mind the preferred strategies of the other cultures and, as much as possible, try to adapt their preferred strategies to the other cultures in order to increase the chance that knowledge will be shared successfully.
- **Step 4:** The organization should conduct awareness training to all group members. This training would sensitize individual participants about cultural differences and preferred knowledge-sharing strategies. Having been sensitized

to differences, participants will be more able to identify where conflicts may exist and to take the time to adjust and understand the preferred knowledge-sharing strategies of the other countries. Understanding and adaptability would be key to understanding.

These two scenarios illustrate that there is no ready-made prescription to effective knowledge sharing across cultures. Based on six dimensions of knowledge and four dimensions of culture, we can hypothesize the best strategy for cultures that exhibit certain cultural characteristics, the best way of presenting knowledge to them. However, in most cases, differences will exist, and where these differences do exist, sharing of knowledge may be compromised. Understanding these differences and being able to adapt our own preferred type of knowledge to that of the other culture will increase the chances of successful communication and sharing of knowledge.

Conclusion and Future Trends

KM initiatives throughout the last 15 years have been typical of many information system innovations—the focus initially was technology-centric while many of the bugs were worked out, vendors identified marketplaces, and companies explored how to effectively incorporate KM into their organizations. Through the experience of hundreds of companies (along with the additional technical capabilities provided over the past 10 years by ERP systems and their analytics, data warehouses, collaboration tools, Web portals, and readily available high-speed communications), KM is maturing to a point where the focus can move to the how, the why, and the value of KM systems. Nontechnical success factors such as culture (both organizational and the diversity of employees' cultural backgrounds) now can be given the attention they deserve.

However, there is still a need for good theory to help ground research in a cross-cultural context studying information systems, knowledge management, and knowledge sharing. This chapter utilized Hofstede's cultural dimensions as a means for exploring effective knowledge sharing. It also discussed some limitations in using Hofstede's cultural dimensions. One area that may contribute is learning theory, specifically culture-based learning. "Culture is a source of differences in cognition" (Mason, 2003, p. 24). Learning studies have shown that the culture of the individual's initial schooling experience makes a difference in how that individual learns, frames problems, solves problems, and utilizes information (e.g., tables, ordering, plans and maps) (Kozulin, 1998; Mason, 2003). Western-style education's recent experiments with problem-based learning (PBL), especially in the sciences and medicine, are showing promise that it can better prepare students to continue to learn (in different environments) after school than traditional education methods (Jones, Higgs, de Angelis, & Prideaux, 2001; Tien, Ven, & Chou, 2003). Since one of KM's primary goals is to support both individual and organizational learning, learning theory might prove beneficial.

One compelling discussion concerns the ultimate result of the collision of cultures that we are now experiencing. Some argue that the importance of culture and the lasting nature of cultural norms and values will keep the need to understand the impact of organizational and national cultural impacts on cross-cultural work on the front burner for the foreseeable future (Appadurai, 1997; Walsham, 2001). Others suggest that there will be standardization. A small example of standardization would be if English tightens its grip as the standard language for global business. A far greater example would be if cultural differences between organizations and societies largely disappear. If that is the trend, then it would be appropriate for KM implementations to be designed as culture-free (Mason, 2003).

Perhaps culture will be both strengthened and weakened. Fulmer (2003) describes one example in which explicit attention was paid to the native language of the users of a KM initiative. The KM forums supported multiple languages through the use of translators. Shortly, separate regional forums developed, including both Spanish and English language forums. However, over time, the forums standardized on English as the common language for forum participants. This supports Mason's (2003) premise that KM initiatives often include a strong push to standardize and strengthen a shared organizational culture, sometimes at the cost of national culture and ethnic differences.

For the foreseeable future, knowledge management and knowledge sharing in cross-cultural work settings will only increase in importance as globalization continues its relentless march. Those organizations that effectively knowledge share are likely to positively impact their productivity, innovativeness, and competitiveness. However, it appears that the immediate challenge will not be making the information and communication technology pieces work together. Much of that work has been done. It may well be that the organization that understands and incorporates its cultural diversity will have the key that makes the difference.

References

Alavi, M., & Leidner, D. E. (2001). Review: Knowledge management and knowledge management systems: Conceptual foundations and research issues. *MIS Quarterly, 25*(1), 107-136.

Ambrosini, V., & Bowman, C. (2001). Tacit knowledge: Some suggestions for operationalization. *Journal of Management Studies, 38*(6), 811-829.

Appadurai, A. (1997). *Modernity at large: Cultural dimensions of globalization.* New Delhi: Oxford University Press.

Ardichvili, A., Page, V., & Wentling, T. (2003). Motivation and barriers to participation in virtual knowledge-sharing communities of practice. *Journal of Knowledge Management, 7*(1), 64-77.

Argote, L., & Ingram, P. (2000). Knowledge transfer: A basis for competitive advantage in firms. *Organizational Behavior and Human Decision Processes, 82*(1), 150-169.

Bhagat, R., Kedia, B., Harveston, P., & Triandis, H. (2002). Cultural variations in the cross-border transfer of organizational knowledge: An integrative framework. *Academy of Management Review, 27*(2), 204.

Bhatt, G. (2001). Knowledge management in organizations: Examining the interaction between technologies, techniques and people. *Journal of Knowledge Management, 5*(1), 68-75.

Cavaleri, S. A. (2004). Leveraging organizational learning for knowledge and performance. *The Learning Organization, 2*(11), 159-176.

Chung, H. (2004). Deciphering six sigma, KM, and CoP. *Logistics Management, 26*(2), 3-5.

Ciborra, C. U., & Patriota, G. (1998). Groupware and teamwork in R&D: Limits to learning and innovation. *R&D Management, 28*(1), 1-10.

Cloete, M., & Snyman, R. (2003). The enterprise portal—Is it knowledge management? *Aslib Proceedings: New Information Perspectives, 55*(4), 234-242.

Davenport, T. H., & Prusak, L. (1998). *Working knowledge: How organizations manage what they know.* Boston: Harvard Business School Press.

Dixon, N. (2000). *Common knowledge: How companies thrive by sharing what they know.* Boston: Harvard Business School Press.

Drucker, P. F. (1997). Looking ahead: Implications for the present. *Harvard Business Review, 75*(5), 18-24.

Fichter, D. (2005). The many forms of e-collaboration: Blogs, wikis, portals, groupware, discussion boards, and instant messaging. *Online, 29*(4), 48-50.

Ford, D., & Chan, Y. (2003). Knowledge sharing in a multi-cultural setting: A case study. *Knowledge Management Research & Practice, 1*, 11-27.

Fulmer, W. E. (2003). *Buckman laboratories (A)* (Report No. 9-800-160). Cambridge, MA: Harvard Business School.

Furlong, G. P., & Johnson, L. (2003). Community of practice and metacapabilities. *Knowledge Management Research & Practice, 1*, 102-112.

Grover, V., & Davenport, T. (2001). General perspectives on knowledge management: Fostering a research agenda. *Journal of Management Information Systems, 18*(1), 5-21.

Hansen, M., Mors, M., & Lovas, B. (2005). Knowledge sharing in organizations: Multiple networks, multiple phases. *Academy of Management Journal, 48*(5), 3-7.

Harris, P., & Moran, R. (2000). *Managing cultural differences* (5th ed.). Houston, TX: Gulf Publishing Company.

Hofstede, G. (1980). *Culture's consequences: International differences in work-related values*. Beverly Hills, CA: Sage Publishing.

Holthouse, D. (1998). Knowledge management research issues. *California Management Review, 40*(3), 277-280.

Huber, G. (1991). Organizational learning: The contributing processes and literatures. *Organizational Science, 2*(1), 88-115.

Janz, B. D., & Prasarnphanich, P. (2003). Understanding the antecedents of effective knowledge management: The importance of a knowledge-centered culture. *Decision Sciences, 34*(2), 351-384.

Javidad, M., & House, R. (2001). Cultural acumen for the global manager: Lessons from project GLOBE. *Organizational Dynamics, 29*(4), 289-305.

Jones, R., Higgs, R., de Angelis, C., & Prideaux, D. (2001). Changing face of medical curricula. *The Lancelot, 357*(9257), 699-703.

Karhu, K. (2002). Expertise cycle—An advanced method for sharing expertise. *Journal of Intellectual Capital, 3*(4), 430-446.

Kimble, C., & Hildreth, P. (2005). Dualities, distributed communities of practice and knowledge management. *Journal of Knowledge Management, 9*(4), 102-113.

King, W. (2006). Maybe a "knowledge culture" isn't always so important after all! *Information Systems Management, 23*(1), 88-89.

Kogut, B., & Zander, U. (1996). What firms do? Coordination, identity, and learning. *Organization Science, 7*(5), 502-518.

Kozulin, A. (1998). *Psychological tools: A sociocultural approach to education*. Cambridge, MA: Harvard University Press.

Krishna, S., Sahay, S., & Walsh, G. (2004). Managing cross-cultural issues in global software outsourcing. *Association for Computing Machinery, 47*(4), 62-66.

Lam, A. (1997). Embedded firms, embedded knowledge: Problems of collaboration and knowledge transfer in global cooperative ventures. *Organization Studies, 18*(6), 973-996.

Larsson, R., Bengtsson, L., Henriksson, K., & Sparks, J. (1998). The interorganizational learning dilemma: Collective knowledge development in strategic alliances. *Organizational Science, 9*(5), 285-305.

Marjanovic, O. (2005). Towards IS supported coordination in emergent business processes. *Business Process Management Journal, 11*(5), 476-487.

Mason, R. (2003). Culture-free or culture-bound? A boundary spanning perspective on learning in knowledge management systems. *Journal of Global Information Management, 11*(4), 20-36.

McLure, M., & Faraj, S. (2000). It is what one does: Why people participate and help others in electronic communities of practice. *The Journal of Strategic Information Systems, 9*(2-3), 55-173.

Moffett, S., McAdam, R., & Parkinson, S. (2003). An empirical analysis of knowledge management applications. *Journal of Knowledge Management, 7*(3), 6-26.

Moffett, S., McAdam, R., & Parkinson, S. (2004). Technological utilization for knowledge management. *Knowledge and Process Management, 11*(3), 175-184.

Moller, K., & Svahn, S. (2003). Crossing east-west boundaries: Knowledge sharing in intercultural business networks. *Industrial Marketing Management, 33*(3), 219.

Myers, M. D., & Tan, F. B. (2002). Beyond models of national culture in information systems research. *Journal of Global Information Management, 10*(1), 24-32.

Nemati, H. (2002). Global knowledge management: Exploring a framework for research. *Journal of Global Information Technology Management, 5*(3), 1-11.

Nonaka, I. (1991). The knowledge creating company. *Harvard Business Review, 69*(6), 96-104.

OECD. (1996). *The knowledge based economy.* Retrieved November 20, 2005, from http://www.oecd.org/dataoecd/51/8/1913021.pdf

Osterioh, M., & Frey, B. S. (2000). Motivation, knowledge transfer, and organizational forms. *Organization Science, 11*(5), 538-550.

Rayport, J. F., & Sviokla, J. J. (1995). Exploiting the virtual value chain. *Harvard Business Review, 73*(6), 75-85.

Ruggles, R. (1998). The state of the notion: Knowledge management in practice. *California Management Review, 40*(3), 80-89.

Schulz, M. (2001). The uncertain relevance of newness: Organizational learning and knowledge flows. *Academy of Management Journal, 44*(4), 661-681.

Senge, P. (1997). Communities of leaders and learners. *Harvard Business Review, 75*(5), 30-32.

Sharkie, R. (2003). Knowledge creation and its place in the development of sustainable competitive advantage. *Journal of Knowledge Management, 7*(1), 20-31.

Stata, R. (1989). Organizational learning: The key to management innovation. *Sloan Management Review, 30*(3), 63-74.

Tien, C-J., Ven, J-H., & Chou, S. (2003). Using the problem-based learning to enhance student's key competencies. *Journal of the American Academy of Business, 2*(2), 456-458.

Trompenaars, F. (1993). *Riding the waves of culture*. London: Nicholas Brealey Press.

Wakefield, R. (2005). Identifying knowledge agents in a KM strategy: The use of the structural influence index. *Information & Management, 42*(7), 34-38.

Walsh, J. P., & Ungson, G. R. (1991). Organizational memory. *Academy of Management Review, 16*(1), 57-91.

Walsham, G. (2001). *Making a world of difference: IT in a global context*. Chichester, UK: Wiley.

Wells, D., Sheina, M., & Harris-Jones, C. (2000). *Enterprise portals: New strategies for information delivery*. Retrieved November 7, 2005, from www.ovum.com.

Wenger, E., McDermott, R., & Snyder, W. (2002). *Cultivating communities of practice: A guide to managing knowledge*. Cambridge, MA: Harvard Business School Press.

Section V

Information Communication Technology Adoption Challenges

Chapter XIII

The Role of Information and Communication Technology in Managing Cultural Diversity in the Modern Workforce:
Challenges and Issues

Indrawati Nataatmadja
University of Technology, Sydney, Australia

Laurel Evelyn Dyson
University of Technology, Sydney, Australia

Abstract

This chapter demonstrates how managers can use information and communication technology (ICT) more effectively in culturally diverse workforces. Basing our analysis on the cultural dimensions of Hofstede and Hall, we compare a range of ICTs and provide a chart summarizing their strengths and weaknesses. In addition, a framework for developing ICT is proposed, and an example of its application to

Copyright © 2007, Idea Group Inc. Copying or distributing in print or electronic forms without written permission of Idea Group Inc. is prohibited.

a global organization is presented. The study shows that none of the existing ICT tools is perfect in all situations and all cultural contexts. Therefore, managers need to provide a variety of ICTs to their employees, and developers should build flexibility into their ICT designs.

Introduction

With cultural diversity in the modern workforce a reality today, there is a challenge for managers to capitalize effectively on this diversity in order to harness the benefits while avoiding potential problems. To optimize the positive outcomes, information and communication technology (ICT) is an essential tool. Using appropriate technologies is shown to break down cultural barriers and to promote understanding and knowledge sharing among employees of different language backgrounds and, hence, lead to successful collaboration.

First, we will discuss culture and its dimensions according to some of the principal cultural theorists of today: Hofstede, the GLOBE group, Trompenaars, and Hall. We also will highlight the importance of language in modern-day management practices. The second section will analyze the relationship among management, cultural diversity, and ICT, and how managers can improve employee effectiveness, employee empowerment, and decision making, and also facilitate knowledge management within the organization. Third, a range of information and communication technologies will be compared and contrasted, including e-mail, discussion forums, chat rooms, intranet, groupware, teleconferencing, videoconferencing, and mobile technologies. These technologies will be evaluated in the context of the cultural dimensions outlined in the first section to provide a guideline for managers to choose the best-fit ICT for their needs. Finally, we will provide a framework to develop new ICT systems for a multicultural workplace.

In summary, this chapter's objective is to raise the awareness of managers, administrators, information systems developers, and other knowledge workers about the role of ICT in managing cultural diversity in the workforce. Valuing diversity by welcoming, recognizing, and cultivating differences among people so that they can develop their unique talents ultimately will assist in the creation of effective and competitive organizations.

Background: Cultural Dimensions

Culture is the set of key values, norms, and beliefs that members of a society or an organization share (Daft, 2000). Hofstede and Hofstede (2005) state that these values can be described in terms of dimensions, a dimension being defined as "an aspect of culture which can be measured relative to other cultures" (p. 23).

Geert Hofstede is one of the most respected cultural theorists. His famous cultural dimensions theory was based on a six-year survey and analysis of hundreds of IBM employees from 53 countries (Marcus & Gould, 2000). These dimensions have been adopted widely and have been applied to many research studies (Hofstede, 2001):

1. **Power distance:** The degree to which a society accepts inequality in power distribution. High-power distance societies accept that there is an unequal distribution of power. In a low-power distance society, equality and opportunities for all are expected. It has flatter hierarchies, and it is easier for people to move upward in society.

2. **Uncertainty avoidance:** The degree to which the society accepts uncertainty and ambiguity. High uncertainty avoidance societies create many rules, regulations, and procedures to avoid uncertainty and ambiguity. Low uncertainty avoidance societies are less concerned about uncertainty and ambiguity. They are more flexible and more tolerant to different opinions and changes.

3. **Individualism vs. collectivism:** The degree to which the society accepts the individual's rights to look after his or her needs first, in contrast to collectivism in which individuals have to adhere to group needs first. High individualism societies will place individual needs and rights above collective needs and rights.

4. **Masculinity vs. femininity:** The degree to which the society believes in masculine ideals of competition and achievement, in contrast to feminism, which emphasizes relationships, compromise, and quality of life.

5. **Long-term vs. short-term orientation:** The degree to which society prefers working in sequence and its tolerance of interruption. Long-term orientation societies value perseverance, hard work, long-term tradition, and commitment. Short-term orientation societies are more willing to embrace changes and desire immediate results.

More recent research on cultural dimensions conducted by the GLOBE (Global Leadership and Organizational Behaviour Effectiveness) project team identified nine dimensions, six of which originated in Hofstede's work (House, Hanges, Javidian, Dorfman, & Gupta, 2004). The approach only appeared in 2004, and hence, very few studies have been carried out to verify the validity of the model or to apply it to evaluating ICT from a cultural perspective.

Another well-known, cross-cultural theorist is Fons Trompenaars (Deresky, 2006; Trompenaars & Hampden-Turner, 1998). His research was spread over 10 years, representing 47 national cultures. Only a limited number of researchers has adopted Trompenaars' framework when evaluating ICT, and often in conjunction with Hofstede (Gould, Zakaria, & Yusof, 2000).

Edward T. Hall is another respected anthropologist and cross-cultural researcher. One of his important contributions is the concept of communication context. Hall (1976) stated that in low-context societies, communication is direct and explicit. In high-context societies, people have to be aware of verbal and nonverbal cues that carry implicit meaning. In addition, they usually prefer to know the history of relationships and the backgrounds of the communicators. For example, more significance is attached to information coming from a person who has high status or who is respected for his or her knowledge or expertise in the field. Hall's theory has received increasing attention in recent years as more people focus on communication issues with ICT.

Beyond Hall's concept of communication context, there are also other aspects of language that need to be considered in multicultural work environments. Languages have many different parameters, including vocabulary, grammar, word order, intonation, and idiom. For example, in English, intonation applies to whole sentences; however, in Chinese, intonation applies to single words (Manning, 2004). Hence, the same Chinese word with a different intonation can lead to different meanings and different interpretations. These differences can be a major source of misunderstandings and a source of poor communication or collaboration.

Some researchers would question these theories of national culture as stereotyping and forcing people into fixed cultural patterns (Myers & Tan, 2002). Nevertheless, Hofstede notes that one can identify statistical trends and tendencies despite the fact that not every individual complies with the overall cultural patterns of his or her country (Marcus & Gould, 2000). Furthermore, Trompenaars argues that even though within each culture there is a spread of values and assumptions, this spread forms a pattern around the average for that cultural group (Trompenaars & Hampden-Turner, 1998). Some researchers also have questioned the relevancy of cultural dimensions in the era of the global use of ICT. However, Geert and Gert Hofstede argued that "the software of the machine may be globalised, but the software of the minds that use them is not" (Hofstede & Hofstede, 2005, p. 330).

In this chapter, in order to simplify our discussion in what is obviously a very complex area, we will focus on only some of the cultural dimensions previously outlined. Since most researchers in culture and ICT have utilized Hofstede's dimensions, these will form the main basis of our analysis rather than the theories of Trompe-

Figure 1. Dimensions of culture

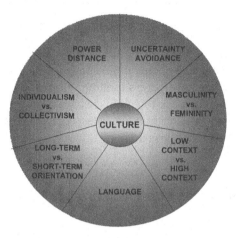

naars or the GLOBE Project. However, because of the fundamental importance of communication in ICT and in managing cultural diversity in the workplace, we also will rely on Hall's theory of communication context in addition to other language issues. These dimensions of culture are summarized in Figure 1.

Management, Cultural Diversity, and ICT

The challenge for management is how to capitalize on cultural diversity. Daft (2000) stated that ICT can provide a better relationship among managers, employees, and their environments. By being aware of the various cultural dimensions and taking into account language differences when implementing and using ICT, managers can achieve the following:

1. Improved employee effectiveness
2. Empowered employees
3. Organizational learning or knowledge management

Improved Employee Effectiveness

Employees can access information they need and also can share information or insights with each other more easily by using ICT. It can provide information about customers, markets, and competitors through databases, news bulletins, and the Internet. Since they have access to more information, managers are able to delegate more challenging work to employees. Consequently, this will lead to better employee effectiveness. For example, salespeople can research pricing over the network and then use this to close deals by offering more competitive prices to customers (Daft, 2000).

Empowered Employees and Decision Making

ICT can contribute to power sharing and break down power distance by providing information to lower-level employees who otherwise will not get access to it. ICT has resulted in the restructuring of many organizations since the introduction of personal computers in the 1980s; there has been a trend toward a flattening of the organization with the elimination of many middle management positions (Oz, 2002). Employees have been given responsibility to make decisions that previously were made by their supervisors. They have been provided with decision-making tools, such as spreadsheets, decision support systems, and business intelligence tools. ICT communications tools such as e-mail also allow every employee to communicate with every other employee, including top managers. This has broadened the decision-making base of the organization and resulted in people from many different cultural backgrounds contributing and having their say. Without the participation of all members, the goals of capturing the best that diversity brings will not be achieved (Joplin & Daus, 1997).

Knowledge Management

Two components have emerged as of prime importance in knowledge management: people as central to the processes of knowledge creation, sharing, and application; and ICT as a vehicle by which the knowledge processes can work (Daft, 2000). Cross-cultural differences will affect how knowledge flows through the organization. A country that is high in power distance tends to have top-down knowledge flows, whereas a country with lower power distance will accept knowledge flows from bottom-up, top-down, and laterally. Furthermore, knowledge flows *between* cross-cultural groups is more likely to happen along formal, business-related issues. However in informal interactions, most knowledge flow exists *within* cultural groups (Ford & Chan, 2003). To encourage better knowledge flows, management should encourage staff to use the intranet and e-mail for formal and informal news by establishing an open communication strategy, either through incentives or policy. For example, in low power distance cultures like the U.S. and the UK, staff was found to be ready to share knowledge when senior management encouraged them to do so (Forstenlechner, 2005).

Further research by Ford and Chan (2003) revealed that knowledge sharing is affected primarily by language barriers. While they found that some knowledge was transferred between the two groups under study, either directly from knowledge holder to recipient or through a translator, a lot of knowledge was lost due to the inability to express the knowledge in the second language or due to translation. ICT can improve this situation by providing asynchronous technologies like discussion forums that allow employees more time to study information before replying by deploying intranets, groupware tools, and collaborative systems that disseminate information efficiently to staff, and by providing translation software.

Cultural Diversity and Workplace Communication Using ICT

Management of cultural diversity involves communication first and foremost. Culture will affect how individuals communicate with others, because culturally learned norms, rules, and beliefs affect the way people communicate as well as predict the effect of their communication behavior on others. Dube and Pare (2001) found that the communication barrier between participants from different cultural and language backgrounds becomes more severe if they communicate using ICT, leading to misunderstandings and loss of information. This occurs because of the lack of contextual clues available in most technologically mediated communication. In addition, the communication style of the dominant culture usually will take over and, consequently, deter people from different cultures participating (Zorn, 2005). However, Massey, Hung, Montoya-Weiss, and Ramesh (2001) proposed that certain ICTs may enhance the ability of some individuals to communicate, depending on the particular communication style inherent in their culture.

Several attributes of ICTs must be considered when assessing their suitability for fostering communication in culturally diverse workplaces. Synchronicity vs. asynchronicity is a simple way of classifying computer-mediated communication. Synchronous ICT tools are those that allow people to talk to each other and to receive quick replies in the same session, similar to a face-to-face conversation. Asynchronous tools usually involve a delay in response, since participants are not necessarily using the tool at the same time and normally do not have any way of knowing when others will be online. Massey et al. (2001) provide a more refined framework of characteristics with which to describe and evaluate ICT tools:

- **Richness:** The capacity to convey nonverbal as well as verbal cues and, thus, to facilitate shared meaning.
- **Social presence:** The degree to which individuals feel close. The richer the ICT, the more social presence it provides.
- **Interactivity:** The rapidity of feedback provided by the medium, thus enhancing interactivity among participants.
- **Rehearsability:** The extent to which messages can be fine tuned before sending or posting.
- **Reprocessability:** The degree to which individuals can re-examine messages sent to them.
- **Flexibility** (what Massey et al. call "symbolic variety"): The number of different modes of communication or choices provided by any communication tool.

Using this framework, Massey et al. (2001) found that three cultural dimensions could assist in understanding cross-cultural communication differences:

- **Individualism-collectivism:** People from a collectivist culture may prefer synchronous technology (chat rooms, teleconferencing, and videoconferencing), because it allows real-time interaction and makes them feel close to each other, leading to a collective feeling and social presence. In an individualist culture, people may be more comfortable doing the work individually, in parallel to others, and can adjust better to asynchronous technology (e-mail, discussion forums, etc.).
- **Uncertainty avoidance:** Massey et al. (2001) believed that synchronous technology is generally a better fit for reducing task ambiguity because of the greater possibilities of interactivity and immediate feedback and, therefore, may be preferred by people from high uncertainty avoidance cultures.
- **High and low context:** The richer the technology, the better it will be for high-context people because it allows for a feeling of social presence and provides nonverbal as well as verbal cues.

Different communication tools thus have different benefits and may not have universal acceptance with all employees. As a result, the organization should be flexible, provide a range of tools, and allow employees to choose whatever mode

Table 1. ICTs and their strengths and weaknesses in communication

ICT	Strengths	Weaknesses	Cultural Dimensions Fit
Email	• Reprocessability provided by record of communicatiton	• Lack of richness and social presence since no non-verbal cues	*Good for:* • Uncertainty avoidance • Individualism *Poor for:* • High context cultures
Discussion Forums	• Reheasability as postings can be edited • Reprocessability provided by record of discussion and sometimes by search facilities	• Low interactivity because of asynchronicity • Lack of richness and social presence since no non-verbal cues	*Good for:* • People with poor oral language skills • Uncertainty avoidance • Long-term orientation • Individualism *Poor for:* • High context cultures
Chat Rooms	• High interactivity provided by synchronicity	• Speed of interaction challenging and no rehearsability	*Good for:* • Short-term orientation • Collectivist cultures *Poor for:* • People with poor written language skills
Intranet	• Quick dissemination of information across cultures		*Good for:* • People with poor oral language skills • Collectivist cultures
Groupware	• Flexibility provided by range of tools • Relationship building across cultures		*Good for:* • Collectivist cultures
Tele-conf.	• Some added richness provided by more verbal cues • High interactivity and feedback	• Lack of rehearsability and reprocessability	*Good for:* • Collectivist cultures *Poor for:* • People with poor oral language skills
Video-conf.	• High level of richness and social presence • High interactivity and feedback	• Lack of rehearsability and reprocessability	*Good for:* • High context cultures • Collectivist cultures
Mobile Tech.	• High interactivity • Flexibility provided by range of functions	• Lack of rehearsability and reprocessability	*Good for:* • Short-term orientation *Poor for:* • High power distance cultures

of communication with which they feel most comfortable. In the following discussion, we examine modern ICTs and outline their potential benefits in the context of a culturally diverse workforce. Table 1 provides a summary of the main tools and their advantages and disadvantages.

E-Mail

Some studies have shown that e-mail as a primary mode of communication can help to avoid culture clashes and can overcome more subtle language barriers because many people from a non-English speaking background can write and read English better than speak it, particularly if they have learned English in a formal classroom setting. Another advantage of using e-mail is that both sides will have a record of the communication, which will help to reduce misunderstandings (Jana, 2000); that is, there is a high degree of reprocessability provided by e-mail message banks.

On the other hand, Pauleen and Yoong (2001) say that in high-context cultures, people will rely heavily on nonverbal gestures and on an understanding of the surrounding context, which may include the backgrounds of the people involved, previous decisions, and the history of the relationship. A degree of personal relationship is important. The nonverbal cues used to judge people's true feelings are not available when communicating through a solely text-based channel such as e-mail. It is not a rich medium. However, people from low-context cultures prefer more objective and fact-based information and so have no problem with e-mail. Therefore, when working across cultures, because of conflicting communication styles, e-mail could be an added barrier.

Discussion Forums

Discussion forums provide another means of written communication that may favor people who have studied English formally and, therefore, are more confident about writing than speaking the language. In one notable respect, forums are superior to e-mail in that people generally have more time to think before posting something, and most forums allow one to edit the posting if mistakes have been made (unlike e-mail in which there is no getting the message back once it has been sent). In other words, discussion forums have a high degree of rehearsability.

In forums, a reply has to be connected to someone else's posting according to a topic (thread), and so, discussions are usually ordered and more formal than many e-mails. Postings can be seen much longer, sometimes even for years, and sometimes can be searched via a search engine, thus providing a useful bank of information (Zorn, 2005). Thus, they have a high degree of reprocessability, allowing individuals to view and consider messages carefully. People from non-English speaking backgrounds may find it easier to follow the more structured discussion, appreciate the extra time that they can spend reading discussion threads, and even feel comfortable with the passive role of reading others' postings without being obliged to post themselves.

The main disadvantages of discussion forums are that there is limited interactivity, because the person must wait for an unspecified period before receiving a reply due to its asynchronous nature, and, as with e-mail, there is limited richness and social presence, which may affect people from high-context cultures.

Chat Rooms

The informal structure of chat, its synchronicity, and ease of use make it a good tool for building relationships among the workforce. Like other synchronous communication tools, it has a high level of interactivity. MSN messenger is one of the free chat software programs that is used widely to share information and to attach documents. However, Zorn (2005) states that in order to participate actively in electronic chat, one has to be able to type quickly and write short, quick sentences. In fact, there is a special chat language (http://abbreviations.virtualsplat.com/ category/chat-abbreviation.asp). In chat rooms, people have to post quickly, because they do not want other chatters posting before they have time to respond. It will make them feel like interrupting, and they might lose the thread of the conversation because it has already moved on to another topic. This puts people who don't know the terms used or who are from other language backgrounds at a distinct disadvantage.

Intranet

Many organizations nowadays have found that a company intranet allows for quick dissemination of information among all employees, such as the posting of minutes, action lists, and news. Jana (2000) reports a successful example of a company, Quadstone, that uses an intranet to link its employees from many nationalities across two continents, which has eliminated the problem of different accents and poor understanding of spoken English. With increasing globalization and more and more companies pursuing a geographically distributed organizational model, intranets are an ideal communication tool.

Groupware

For sharing information to a mass audience, messaging and groupware tools such as Lotus Notes may be an appropriate choice (Topi, 2004). Groupware tools such as Notes enable staff to share documents within the organization and externally via the Web; engage in threaded discussions via discussion forums; send messages over e-mail, receive notification and reminders and, in short, communicate and share resources with fellow workers via its intranet capabilities (Hawryszkiewycz, 2003). They can promote opportunities for informal, spontaneous exchanges, which, in turn, can foster more productive work collaborations than are possible without this personal foundation (Pauleen & Yoong, 2001). This relationship building is particularly important in culturally diverse workplaces. Groupware has the advantage that it is very flexible and provides a range of communication tools from which employees from different cultural backgrounds and with different preferences can select the communication channel that is appropriate for them.

Teleconferencing

Unlike most of the written communication tools already described, teleconferencing provides a high degree of interactivity and allows people to convey emotion relatively easily. However, it is very difficult for participants to actively participate if they are not fluent in the language used in the teleconference. They also will have trouble entering the conversation without seeing the other people and being able to pick up visual cues, and often misinterpretations will occur (Dube & Pare, 2001). It also has the problems of low rehearsability and low reprocessability, both of which give people from other language backgrounds a further disadvantage.

Videoconferencing

Videoconferencing is a good tool for culturally diverse workplaces, because it enables team members to see facial expressions, gestures, and many other visual cues, and so establish a social presence. Pauleen and Yoong (2001) found that "eyeing" people was important in relationship building and that videoconferencing generally enhanced social relationships by allowing people to "put a face to the name" (p. 210). It is better than teleconferencing, which is difficult for people from other language backgrounds. However, Dustdar and Hofstede (1999) found that there were fewer contextual clues in videoconferencing than in face-to-face meetings and proposed that the design of videoconferencing needs to support social protocols and be culturally aware. Futhermore, while teleconferencing and, particularly, videoconferencing are richer communication media than other ICTs, they require a high level of commitment, flexibility, and discipline from the participants (Dube & Pare, 2001); for example, working outside normal business hours to accommodate time zone differences, care in watching for turn-taking cues, and a commitment to including all participants in the conversation.

Mobile Technologies

Mobile technologies, such as cell phones, laptop computers, personal digital assistants (PDAs), two-way pagers, and the blackberry (an e-mail device), have been adopted widely in many organizations in which employees commonly work off-site, roam from one location to another, or need to be on call. They have been responsible for extending a social presence into areas where, before wireless networks, it could not exist before. Generally, they provide a high level of interactivity (either through phone calls or text messages) and an increasing flexibility, as the range of available functions increases every year. Rehearsability and reprocessability, though, are low.

A number of studies have shown that cultural differences affect how at least some of these technologies are used. For example, in the U.S., the number of mobile phone users is high, and there is a preference for graphic mobile technologies such as Palm.Net and Pocket Internet Explorer to send e-mail messages, in preference to text messaging (Urbaczewski, Wells, Sarker, & Koivisto, 2002).

Text messaging has been shown to vary greatly depending on power distance. Sarker and Wells (2003) found that in high-power distance cultures such as Korea, text messaging to supervisors was viewed as highly offensive, whereas in low-power distance cultures such as Norway, text messaging was not offensive, although sometimes it was considered inappropriate in more formal communications due to its reliance on abbreviations and slang.

Cultural variables have a large influence on preferred mobile device interface features. One international study (Choi, Lee, Kim, & Jeon, 2005) showed that Korean and Japanese participants displayed a high degree of uncertainty avoidance in their preference for clear, concrete menu labeling and a large amount of secondary information about the contents available to advise their choice of options. The Koreans also were found to be more collectivist, preferring popularity rankings so that they could choose content that many other people had used. The Finnish participants, on the other hand, were characterized as risk takers who disliked secondary information on menu items, instead being happy to click on options and see for themselves what was available. They liked a low-context, simple design with fewer icons and colors. In order for mobile technology to be effective in fostering communication in the workplace, it needs to be designed to take cultural preferences such as these into account.

Challenges in Developing ICT for the Culturally Diverse Workforce

The design of ICT with a focus on cultural diversity is important if organizations are to optimize the use of ICT in management, knowledge management, and communication. Most system development in the past has been aimed at culturally homogenous groups, but with increasing globalization, this is no longer acceptable.

Figure 2. Framework for developing culturally diverse ICT

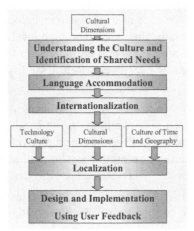

There has been much research into the challenge of developing ICT for culturally diverse user groups, and this work is ongoing. A framework for cross-cultural ICT development is given in Figure 2.

1. **Understanding the culture:** The first step in ICT development is to understand the target cultures (Nakakoji, 1996). The cultural dimensions of Hofstede and Hall provide a systematic basis to begin the description and segmentation of the user group's cultures. If different cultural groups are to use the same system, then this stage also should include the identification of any shared needs, interests, and goals in addition to communication strategies among groups (Bourges-Waldegg, Lafreniere, Tscheligi, Boy, & Prates, 2001).

2. **Language accommodation:** Since language is the most obvious difference between cultural groups, language translation, or at least language accommodation, is the next step. Where there is one dominant language, system designers might decide that the cost of providing systems in multiple languages is not feasible. However, providing features to translate and edit the language and allowing users to check their grammar and spelling are necessary (Dube & Pare, 2001). In addition, ICT designers should try to develop cues such as context-adding smiley faces or emoticons in e-mail and chat rooms in order to avoid misunderstandings and provide social presence (Churchill & Bly, 2000; Zorn, 2005).

   The most important issue for ICT communication tools is in determining the structures of communication; for example, communication procedures, turn taking, methods of interrupting, and other social protocols. Different cultures have different social etiquette and different levels of power distance; transgressions against these can affect the outcome of communication severely (Churchill & Bly, 2000). The developer, therefore, must consider different cultural communication styles and strive to keep the environment tailorable in order to allow people to develop their own social environments. Design should focus on communications and conversations, not on technology (Churchill & Bly, 2000; Zorn, 2005).

3. **Internationalization:** This step ensures that the design is culture-independent, which is particularly important when organizations are doing business globally on the Web and when it is impossible to predict exactly who the users will be. This is also a useful preliminary step in tailoring systems to different known cultural groups. Having identified culturally specific elements of the system, as many of these are eliminated as possible. A good example given by Nakakoji (1996) is the method of dealing efficiently with different languages on screen; rather than embed text into figures, it is stored separately. Figures then can be displayed and the text added in at the time of display in the language appropriate to the user. This avoids the necessity of redrawing figures in each language and results in economies of storage, since the number of figures is reduced to a minimum.

   A variant on internationalization is universal usability. Some designers believe that universal access to ICT is only possible if systems are designed with all possible users in mind (Shneiderman, 2000). They have developed the Universal Design Principles, which include equity, flexibility, simple and intuitive design, legibility of information, error tolerance, low physical effort, and appropriateness to human physical and environmental limitations (Burgstahler, n.d.)

4. **Localization:** Normally, following internationalization, the system is localized for each user group's culture by including elements of local content and style. The design should accommodate different languages, date and number formats, color and graphic representations, appropriate icons, and an appropriate direction of flow (left to right, or right to left), according to the culture (Nakakoji, 1996). Perhaps more important than superficial appearance is the organization and representation of information structures to represent the users' knowledge framework; explicit representation of the information architecture of the system is also a good idea in order to enhance user understanding (Walton, Marsden, & Vukovic', 2002).

   Localization also should accommodate cultural dimensions, since these are also critical for user acceptance. For example, one comparison of Malaysian and U.S. Web sites showed that the Malaysian Web sites reflected the high power distance requirements of their users and a strong emphasis on collective and social goals, whereas the American Web sites for similar products displayed a much lower power distance, a greater stress on individualism, and a more task-oriented approach (Gould et al., 2000). Research into ICT design for Arabic users has proposed that collectivism and high context communication is important in addition to privacy and security issues (Zakaria & Stanton, 2003).

In addition to cultural dimensions, the designer also has to consider issues of technology culture, such as:

- The way of charging for technology access, including Internet and phone, which sometimes prohibits people from being online. In some countries, access is charged per minute, and so charges might be high.
- The technological expertise available. The system should be easy to use, provide a gentle learning curve, and allow for different technological capabilities, particularly where expertise is known to be low.
- The understanding of the use and place of technologies. For example there is no need for passwords in certain societies for individually owned computers (Churchill & Bly, 2000).

Churchill and Bly (2000) also list the culture of time and geography as important considerations:

- Different holidays and working hours.
- Different notions of private and public time. Systems should enable a set of shared expectations about where to leave messages and available times for interaction in order to create a sense of co-presence among participants. They should be capable of providing asynchronous as well as synchronous messages.

5. **Design and implementation:** Refining of the system's design based on user feedback is especially important in multicultural design environments (Nakakoji, 1996). This normally involves reiterations of testing with the user group and sometimes active participation of users on the design team.

Example of Applying the Development Framework

Here we will give a practical scenario to demonstrate how this model might be implemented (Table 2). Let us take the example of a U.S. firm that has subsidiaries in the United Kingdom and Indonesia. It wishes to create a Web site that will provide customers with product and service information and also act as a portal to its intranet and groupware tools for employees in all three countries.

Future Trends

There are many emerging technologies that will impact how cultural diversity is managed in the future. To some extent, ICT will evolve to answer existing challenges and issues in a more effective way. However, new challenges will arise as

Table 2. Example of applying the development framework

Development Phase		USA	UK	Indonesia
Understanding the Culture	Identify Cultural Dimensions (ITIM, 2003; Hall, 1976)	• Low power distance • High individualism • Short-term orientation • Medium to high masculinity • Low context communication		• High power distance • Collectivist • Medium-low masculinity • Medium uncertainty avoidance • Long-term orientation • High context
		• Medium uncertainty avoidance	• Low uncertainty avoidance	
	Identify Shared Needs	Company goals and organizational "culture"		
Language	Language	English		English + Indonesian
	Etiquette	Provide options for turn-taking and other local communication protocols		
Int.	Internalization	Common graphics, layout, tools, functions		
Localization	Local Content and Style	e.g., US spelling; mm/dd/yy date format	e.g., UK spelling; dd/mm/yy date format	e.g., US spelling of English; dd/mm/yy date format; no graphic sexual depiction
	Apply Cultural Dimensions	For example: • Low power distance: focus on customers and their needs • Individualistic: focus on recognition of employee achievements via email or electronic noticeboard		For example: • High power distance: focus on company profile to impress customers • Collectivist: allow anonymous postings on discussion board; emphasize group achievement
	Technology Culture	e.g., Greater complexity in customer interface since higher computer literacy rates		e.g., Less sophisticated, easy-to-use customer interface
	Time/Geog. Culture	e.g., Information and alerts regarding public holidays in USA, UK and Indonesia		
D&I	Design & Implementation	Local user testing and reiterative design approach		

new technologies come on the market and also as the world changes under the increasing impact of globalization.

The use of mobile devices is rapidly increasing, owing to improved wireless technology, cost reduction, and advances in service provision. In addition, people's ways of carrying out their work is changing, with the number of mobile workers rising around the world (Chen & Corritore, 2005). These people rely more and more on mobile devices to keep them connected to the office and connected to clients. For example, laptop PCs are gaining market share. In 1995, there were 31 million, accounting for 13.7% of all PCs in-use, while in 2005, the number was expected to double to 25% of market share (Computer Industry Almanac Inc., 2005 June). However, the design of many mobile devices presents additional and unique challenges that primarily arise from the small size of the interface, limited input and interaction capabilities, slower bandwidth, security risks, and the need to adapt content to the context of where the device is being used (Murugesan & Venkatakrishnan, 2005). The challenge of establishing how one can use these mobile tools effectively in a culturally diverse workplace is magnified by these issues.

New technological developments in other fields also could change the way that modern workforces do business and their preferred ICT. Voice-over IP (VoIP) is already saving costs on international telephone calls for many businesses (Erlanger, 2005-2006). An extension of this technology is video-over IP, which promises to reduce the usage costs by allowing videoconferencing for the price of a local call to the ISP. This is now being employed effectively in some geographically dispersed organizations in which face-to-face meetings are difficult to arrange (Olde, quoted by Housely, 2005). With fewer financial disincentives to using videoconferencing, it well may become the preferred interoffice or interorganizational communication tool, providing a richer medium and a greater social presence than any other ICT. For people from many cultures and, in particular, high-context cultures, this might become the best option after face-to-face communication and meetings. These benefits are likely to be realized as soon as the security challenges of VoIP and video-over-IP are resolved.

Currently, ICT is dominated by the English language. In 2001, the top three languages on the Internet were English (43%), Japanese (8.9%), and Chinese (8.8%) (Gromov, 2002). However, in the future, China may well become the dominant force. By 2005, English showed a marked decline to 31.9%, while China had grown to 12.8% (Miniwatts International, Ltd., 2005). China, which is now in second place in broadband subscriptions, is projected to surpass the U.S. (Computer Industry Almanac Inc., 2005 November), and in mobile telephony, China is already the leader (Computer Industry Almanac Inc., 2005 September). This has enormous implications on the design of Web pages and ICT generally. More than ever, it will be necessary for the developer to be aware of cultural considerations and to follow a rigorous development framework that focuses on cultural issues.

Ultimately, more research is needed into cultural aspects of ICT development and particularly into comparing and contrasting multiple groups from different cultures. Currently, most research is limited to comparisons of ICT between only two national cultures, such as the U.S. and Germany (Junglas & Watson, 2004) or the U.S. and Malaysia (Gould et al., 2000). There are very few case studies involving multiple cultures in international businesses.

Conclusion

With the transformation of the modern workforce into an increasingly culturally diverse one, there is a challenge for managers and administrators in managing this diversity effectively. ICT offers many different tools to assist, each with its own advantages and its own weaknesses. For the manager, there is the challenge of choosing appropriate tools for the workplace to better take into account different communication styles and cultural backgrounds. There is also the challenge for the systems developer in designing new applications that reflect the reality of cultural diversity in the workplace today.

In this chapter, we have discussed the management of cultural diversity using ICT in the context of language and the dimensions proposed by the principal comparative cultural theorists of our day, Hofstede and Hall. Studies have shown that at least some of these dimensions form a valid framework for interpreting and understanding the differences between ICTs and how they can be used most effectively in promoting better work practices, knowledge sharing, and communication in multicultural organizations. These are power distance, uncertainty avoidance, individualism vs. collectivism, short-term vs. long-term orientation, and high-context vs. low-context communication. Only one study discussed Hofstede's dimension of masculinity vs. femininity (Marcus & Gould, 2000) but was not considered by the authors as providing sufficient evidence.

One of the obvious challenges to the manager is the fact that none of the existing ICT tools is perfect in all situations or in all cultural contexts. For example, whereas one tool, such as e-mail, might provide fast, direct, fact-based communication suited to people from low-context cultures, studies show that it is a problem for people from high-context cultures, because it is not a rich medium and gives few nonverbal cues. As a result, the manager should provide a range of communication and other ICT tools and give employees from different cultural backgrounds the opportunity to choose whatever ICT with which they feel comfortable. Likewise, developers will need to build flexibility into their design of new systems, if they are to be culturally appropriate and gain acceptance.

References

Bourges-Waldegg, P., Lafreniere, D., Tscheligi, M., Boy, G., & Prates, R. O. (2001 March-April). Identifying "target cultures"; To what extent is that possible? *CHI '01 Extended Abstracts on HumanFactors in Computing Systems*, 223-224.

Burgstahler, S. (n.d.). *Universal design of instruction*. Retrieved October 14, 2005, from http://www.washington.edu/doit/Brochures/Academics/instruction.html

Chen, L., & Corritore, C. (2005). Nomadic culture and its impact on organizational support for nomadic behaviors and employee job satisfaction. In W. Brookes, E. Lawrence, R. Steele, & E. Chang (Eds.), *Proceedings of the International Conference on Mobile Business* (pp. 9-15). Los Alamitos, CA: IEEE Computer Society.

Choi, B., Lee, I., Kim, J., & Jeon, Y. (2005). A qualitative cross-national study of cultural influences on mobile data service design. *Proceedings of the SIGCHI Conference on Human Factors in Computing Systems* (pp. 661-670). Portland, OR: ACM Press.

Churchill, E. F., & Bly, S. (2000). Culture vultures: Considering culture and communication in virtual enviroment. *ACM SIGGROUP Bulletin, 21*(1), 6-11.

Computer Industry Almanac Inc. (2005, September 26). *China tops cellular subscriber top 15 ranking*. Retrieved January 17, 2006 from http://www.c-i-a.com/pr0905.htm

Computer Industry Almanac Inc. (2005, June 20). *Mobile PCs in-use surpass 200m*. Retrieved January 11, 2006, from http://www.c-i-a.com/pr0605.htm

Computer Industry Almanac Inc. (2005, November 14). *USA leads broadband subscriber top 15 ranking*. Retrieved January 11, 2006, from http://www.c-i-a.com/pr1105.htm

Daft, R. (2000). *Management* (5th ed.). Orlando, FL: Dryden Press.

Deresky, H (2006). *International management: Managing across borders and cultures* (5th ed.). Upper Saddle River, NJ: Pearson Prentice Hall.

Dube, L., & Pare, G. (2001). Global virtual team. *Communication of the ACM, 44*(12), 71-73.

Dustdar, S., & Hofstede, G.J. (1999). Videoconferencing across cultures—A conceptual framework for floor control issues. *Journal of Information Technology, 14,* 161-169.

Erlanger, L. (2005-2006, December-January). VoIP may be vulnerable to threats. *Information Age,* 34-37.

Ford, D. P., & Chan, Y. E. (2003). Knowledge sharing in a multi-cultural setting: A case study. *Knowledge Management Research & Practice, 1*(1), 11-27.

Forstenlechner, I. (2005). The impact of national culture on KM metrics. *KM Review, 8*(3), 10.

Gould, E. W., Zakaria, N., & Yusof, S. A. M. (2000, September). Applying culture to Website design: A comparison of Malaysian and US Websites. *Proceedings of the 2000 Joint IEEE International and 18th Annual Conference on Computer Documentation* (pp. 161-171). Los Alamitos, CA: IEEE Computer Society.

Gromov, G. R. (2002). *History of Internet and WWW: The roads and crossroads of Internet history*. Retrieved January 11, 2006, from http://wwwnetvalley.com/intvalstat.htm

Hall, R. T. (1976). *Beyond culture*. New York: Anchor Press.

Hawryszkiewycz, I. T. (2003). *Developing e-business systems*. Sydney: University of Technology, Sydney.

Hofstede, G. (2001). *Culture's consequences: Comparing values, behaviors, institutions and organizations across nations*. London: Sage Publications.

Hofstede, G., & Hofstede, G. J. (2005). *Cultures and organizations: Software of the mind*. New York: McGraw-Hill.

House, R. J, Hanges, P. J., Javidan, M., Dorfman, P. W., & Gupta, V. (Eds.). (2004). *Culture, leadership, and organizations: The GLOBE study of 62 societies*. London: Sage Publications.

Housely, T. (2005). IP telephony (presentation to the ACS April Branch Forum). Sydney: Australian Computer Society.

ITIM. (2003). *Geert Hofstede: Cultural dimensions*. Retrieved January 23, 2006, from http://www.geert-hofstede.com/hofstede_dimensions.php

Jana, R. (2000). Preventing culture clashes. *InfoWorld, 22*(17), 95-96.

Joplin, J. R. W., & Daus, C. S. (1997). Challenges of leading a diverse workforce. *Academy of Management Executive, 11*(3), 32-47.

Junglas, I. A., & Watson, R. T. (2004). National culture and electronic commerce: A comparative study of U.S. and German Web sites. *E-Service Journal, 3*(2), 3-34.

Manning, A. D. (2004). Universals and variables in cross-cultural communication. *Proceedings of the International Professional Communication Conference* (pp. 36-41). Los Alamitos, CA: IEEE Computer Society.

Marcus, A., & Gould, E. W. (2000). Crosscurrents: Cultural dimensions and global Web user-interface design. *Interaction, 7*(4), 32-46.

Massey, A. P., Hung, Y. T. C, Montoya-Weiss, M., & Ramesh, V. (2001). When culture and style aren't about clothes: Perceptions of task-technology "fit" in global virtual teams. *Proceedings of the 2001 International ACM SIGGROUP Conference on Supporting Group Work* (pp. 207-213). New York: ACM Press.

Miniwatts International, Ltd. (2005). *Internet world stats: Usage population statistics.* Retrieved January 11, 2006, from http://www.internetworldstats.com/stats7.htm

Murugesan, S., & Venkatakrishnan, B. A. (2005). Addressing the challenges of Web applications on mobile handheld devices. In W. Brookes, E. Lawrence, R. Steele, E. Chang (Eds.), *Proceedings of the International Conference on Mobile Business* (pp. 199-205). Los Alamitos, CA: IEEE Computer Society.

Myers, M. D., & Tan, F. B. (2002, January-March). Beyond models of national culture in information systems research. *Journal of Global Information Management*, 24-32.

Nakakoji, K. (1996). Beyond language translation: Crossing the cultural divide. *Software, IEEE, 13*(6), 42-46.

Oz, E. (2002). *Management information systems* (3rd ed.). Boston: Thomson Learning.

Pauleen, D. J., & Yoong, P. (2001). Relationship building and the use of ICT in boundary-crossing virtual teams: A facilitator's perspective. *Journal of Information Technology, 16*(4), 205-220.

Sarker, S., & Wells, J. D. (2003). Understanding mobile handheld device use and adoption. *Communications of the ACM, 46*(12), 35-40.

Shneiderman, B. (2000). Universal usability: Pushing human-computer interaction research to empower every citizen. *Communications of the ACM, 43*(5), 85-91.

Topi, H. (2004). Supporting telework: Obstacles and solutions. *Information Systems Management, 21*(3), 79-85.

Trompenaars, F., & Hampden-Turner, C. (1998). *Riding the waves of culture: Understanding cultural diversity in global business* (2nd ed.). New York: McGraw-Hill.

Urbaczewski, A., Wells, J., Sarker, S., & Koivisto, M. (2002). Exploring cultural differences as a means for understanding the global mobile Internet: A theoretical basis and program of research. *Proceedings of the 35th Annual Hawaii International Conference on System Science* (pp. 654-663).

Walton, M., Marsden, G., & Vukovic', V. (2002). "Visual literacy" as challenge to the internationalization of interfaces: A study of South African student Web users. *Proceedings of the CHI'2002.* New York: ACM Press.

Zakaria, N., & Stanton, J. M. (2003). Designing and implementing culturally-sensitive IT applications: The interaction of cultural values and privacy issues in the Middle East. *Information Technology & People, 16*(1), 49-75.

Zorn, I. (2005). Do culture and technology interact? Overcoming technological barriers to intercultural communication in virtual communities. *ACM SIGGROUP Bulletin, 25*(2), 8-12.

Chapter XIV

From 9 to 5 to 24/7:
How Technology has Redefined the Workday

Linda Duxbury
Carleton University, Canada

Ian Towers
Carleton University, Canada

Christopher Higgins
The University of Western Ontario, Canada

John Ajit Thomas
Carleton University, Canada

Abstract

This chapter explores the use of work-extension technologies such as e-mail, BlackBerry devices, portable computers, and cell phones. After a review of the literature, the chapter presents the usage patterns of these work extension technologies by Canadian knowledge workers and describes how work is being performed in a variety of nonoffice locations outside normal working hours. Our findings with respect to the impact of work extension technology were contradictory. Some technologies were

Copyright © 2007, Idea Group Inc. Copying or distributing in print or electronic forms without written permission of Idea Group Inc. is prohibited.

found to lead to an increase in employee workloads and stress, while others were found to have less of an impact. We also discovered that many respondents reported that technology made them more productive and made their work more interesting. After an analysis of the advantages and disadvantages of these technologies, the chapter concludes with suggestions of ways in which employers and employees can use them more effectively.

Introduction

Technology has changed the workday and business practices of millions of knowledge workers. The hardware and software that revolutionized the ways people worked initially were found only in the workplace and were used only during traditional 9-to-5 working hours because the technology was simply not portable. Such was the situation 15 years ago. Nowadays, however, the act of performing work is not limited to specific hours at a specific location. The latest incarnations of work-related technology for knowledge workers support work outside the confines of the office at almost any time of the day or night.

This chapter focuses on *work extension*, which we define as the act of engaging in work-related activities outside of regular office hours in locations other than the business office. We refer to the technologies that permit work extension as work extending technologies (WET). These include cell phones, BlackBerry devices, portable computers, and PDAs, to name a few. While this term is not currently in use in the academic literature, it has been used by information technology practitioners (IDC Research, 2002).

The following research questions are addressed in this chapter:

- How much do Canadian knowledge workers use WET?
- How do employees use these different types of technology?
- What are the effects on employers and employees of using WET?
- What are the perceived advantages and disadvantages of using WET?
- How can employees and employers manage the use of WET in order to enhance the usefulness and reduce the challenges associated with its use?

While this study was done in Canada, many of the firms that participated in the research were multinationals with operations in the United States, Europe, and Asia. This, plus the literature showing the growth in use of WET internationally, suggests that the findings from this study may be generalizable to employees and employers who compete globally.

The chapter is divided into four sections. It begins with a brief review of the available literature on technology-supported work outside the office. This is followed by a description of the research methodology. In the sections that follow, empirical data are used to provide answers to each of the research questions. The chapter ends with a set of recommendations with respect to the management of WET.

Background

The section gives some background information on WET.

Use of WET

The use and availability of WET are growing rapidly. The CTIA (an American trade association of cellular service providers and phone manufacturers) estimates that there were more than 200 million subscribers to wireless services in the US in February 2006; this compares to 28 million subscribers in June 1995. According to a 2006 report by Gartner Research, global sales of personal digital assistants (PDAs) reached a record 14.9 million in 2005, an increase of more than 14% compared to 2004. The consulting firm Canalsys reported that global shipments of the newest generation of cell phones (smartphones) manufactured by Nokia increased by 142% in the third quarter of 2005 compared to the previous year, to reach more than 7.1 million (Canalys, 2005). In 2005, 225 million North Americans were Internet users, more than double the number in 2000; 50% of the population of Europe was Internet users, compared to 30% in 2000 (IWS, 2006).

Effects of WET Technology

In 1987, Nobel Prize winning economist Robert Solow famously observed that the computer age was evident everywhere except within the productivity statistics. It appeared that the large cost of technology investment has not always been accompanied by a corresponding increase in productivity, a phenomena that has been referred to as the productivity paradox. This paradox refers to the fact that the massive investments in technology made by corporations have not resulted in corresponding increases in productivity. Morgan Stanley's chief economist, Stephen Roach, estimated that U.S. companies spent close to $1 trillion in the 1970s and 1980s on IT without any appreciable gain in productivity (The Edge, June 25, 2001). Researchers have attempted to explain this inconsistency by noting that technology itself is not capable of creating productivity gains but rather acts as an enabling force that makes productivity increases possible only when the investment in technology is aligned with the underlying business processes (King, 2002). McGinn (2002), on the other hand, suggests that the paradox might be due to the fact that we may not have been patient enough and that investments in technology pay off over time. King (2002) feels that the productivity paradox might be due to the fact that training has not kept pace with the rapidity of technological change and concludes that the reason productivity has not kept pace with the adoption of technology may be due to the fact that employees have not learned to use the technology properly.

The productivity paradox also might be due to the fact that the potential gains in productivity offered by WET technology are offset by the human costs associated with the inappropriate or overuse of these technologies. Support for this contention can be found in the work of authors such as Gleick (1999), Bluedorn (2002), and Eriksen (2000), who describe work extension technology as a "thief of time" and an "invader of space." As the title of his book *Faster* implies, Gleick believes that the speed of everything in society is accelerating. The amount of available time is decreasing from the time taken to eat a meal to the amount of time available to complete a task at work. He holds that modern technology is the *sine qua non* of this move toward speed. Gleick claims that the outcome of the ability to complete tasks more quickly is not an increase in the amount of free time available to an individual but rather an increase in the number of tasks to be completed. He argues that the effect of this on the individual is a feeling that there is no time to waste, that he or she should be doing something productive at all times.

Eriksen (2000) explores the concept of slow time and fast time. He defines time taken by an individual for personal gratification as slow time and time spent checking voicemail while waiting for an egg to boil as fast time. Fast time happens when people multitask by using a short period of time when they have nothing to do to carry out another activity rather than simply relaxing. He argues that there is an increasing trend toward fast time and that technology is erasing the distinction between work and leisure. Support for this contention comes from Frase-Blunt (2001), the AMA (2002), and Silver and Crompton (2002). Frase-Blunt (2001) reported that 60% of office workers took a WET device with them on their vacation. More than half of these received a work-related call during their holiday. The AMA observed that 63% of business executives were in touch with their office at least once a week during their vacation, while 26% were in contact with their office on a daily basis (AMA, 2002). Silver and Crompton (2002) noted that 17% of Canadian workers are doing supplemental paid work on Saturday mornings. Unfortunately, little empirical work has been done in this area, and the impact of working while on vacation or on weekends is poorly understood.

Mobile technologies should also lead to an erosion of personal space. Eriksen (2000) supports his point by noting that people check voicemail at a busstop or even in a public bathroom. Brown (2002) observes that "cafés, bars, restaurants all become transformed into sites for work." This marks a further eroding of the work/nonwork boundaries, with third spaces (i.e., cafés) between home and work becoming legitimate places of work (Brown, 2002, p. 13). Even the car now has become a mobile office (Corbett, 1994).

Impact of WET on Organizations and Employees

The following summarizes the current thinking on the advantages and disadvantages of work extension.

Advantages to the Organization

Organizations that supply their employees with WET benefit in the following ways:

- **Increased working hours by employees:** Studies by third-party consultants who were engaged by WET manufacturers (e.g., Ipsos Reid's 2001 study for Blackberry maker RIM; the Gartner Group's 2002 study for Intel) indicate that professional users gain five hours of productivity each week by having a notebook computer for use at home, which results in an annual dollar benefit of $19,200 per employee (Gartner Group, 2001).

- **Improved accessibility of employees:** When employees have laptops, cell phones, and pagers, they always are accessible to their employers. In theory, the employee is able to regulate the pace at which work is performed and balance time for the family with work at home in off-hours. This would be an idyllic situation, if the workload or productivity expectations were held constant. Unfortunately, it appears that employers expect that WET users will deliver an increased amount of work (Gant & Kiesler, 2002; Green, 2001).

- **Additional control mechanisms over employees:** WET provides opportunities for the employer to monitor and control the employee (Fairweather, 1999; Green, 2001). Bassett (2000) contends that it is not necessary that monitoring actually take place in order for it to be an effective source of control; the individual merely needs to feel that he or she may be monitored. He reports on a series of interviews with workers whose cell phones had been supplied by their employer. These employees were reluctant to turn their phones off or to use them for private calls in case they were unavailable when someone from the office called them.

- **Positive image:** Organizations may give the latest technology to their employees for more than purely functional reasons. In one case, employers felt that the mere fact that their employees possessed a BlackBerry would create a more positive image for the organization (Schlosser, 2002).

Advantages to Employees

Suggested benefits of WET use to employees include the following:

- **Practical advantages:** WET may give employees more control over where, how, and when they work. The ability to do work at their own pace in locations of their own choosing can serve as a powerful inducement to adopt these technologies (Ipsos-Ried, 2001).

- **Impression management:** Impression management is "the process by which people attempt to influence the image others have of them" (Bolino, 1999, p. 84). The importance of impression management as a possible benefit for work extenders emerges when we consider their willingness to perform work after hours, on the weekend, and on vacation. The primary drive behind such behavior may be exemplification, in which people seek to be viewed as dedicated.

- **(Self) image management:** Bassett (2000) found that owners of a cell phone perceived that not only did their phone enable them to communicate but it also allowed them "to communicate something to people in the immediate physical locale … that they are people-in-demand, and that they live their lives at a certain speed." (pp. 4-5). Schlosser (2002) and Strom (2002) call mobile devices props that are used to create a specific impression to people in the vicinity.

Disadvantages to the Organization

The potential disadvantages for the employer include the following:

- **Cost:** The cost of WET ownership does not end with its acquisition. Maintenance and operation also must be considered.

- **Potential for loss:** There is a possibility for reduced competitive advantage through the loss of corporate data when technology is misplaced.

Disadvantages to the Employee

Potential disadvantages for the employee include the following:

- **Loss of personal time and space:** The literature indicates that most work extension is done at home, which subsequently becomes an extension of the workplace (Salazar, 2001).

- **Potential for family conflict:** Family members may resent the amount of time that the work extender actually is spending on work and may feel that this is time taken from them. Furthermore, as the work extender works at home, the family members may feel that space is being taken away. Thus, the two dimensions of work extension may contribute to conflict (Bolan, 2001).

Methodology

We used a variety of methodologies to address the research questions as noted next. They were:

1. Examination of data collected as part of our National Study on work, family, and lifestyle (Duxbury & Higgins, 2002) on the use and impact of WET.
2. A survey to collect quantitative data on the use and impact of work extension technologies (i.e., the WET survey).
3. Focus groups with heavy and light users of work extension technologies to examine the pros and cons of using WET.
4. A series of interviews to help us to gain a further understanding of how the different work extension technologies are currently being used, the perceived advantages and disadvantages of these technologies, and how the challenges associated with using them could be minimized and/or the benefits enhanced.

A brief summary of the methodologies is given next.

National Study of Work, Family, and Lifestyle

One source of data comes from our national study of work-life balance (National Study) that was conducted by Duxbury and Higgins in 2000-2001. In total, 31,571 employees who worked in 100 medium- to large-sized (i.e., 500 or more employees) firms completed a 12-page survey instrument. The survey sample is representative with respect to sector of employment, age, region, community size, job type, education, personal income, family income, and family's financial well being, which is reflective of the Canadian workforce. Similarities between Canadian and U.S. workforces suggest that the results could be generalized beyond this study.

The National Study provided data on hours per week spent (1) in work (including overtime work at the office and at home), (2) working at home outside regular hours, and (3) using technology to do job-related work outside regular office hours. We also asked perceptual questions about how technology had affected (1) workload, (2) ability to balance work/personal/family life, (3) stress, (4) productivity, (5) job security, and (6) interest in work. These perceptual questions were measured on a five-point scale on which 1 = decreased, 3 = no change, and 5 = increased. For analysis purposes, the data were collapsed into three groups: (1) decreased (scores of 1 and 2), (2) no change (score of 3), and (3) increased (scores of 4 and 5). A third group of questions (also measured on a five-point scale) asked about job satisfaction (Quinn & Staines, 1979), intent to turnover (Duxbury & Higgins, 2002), job stress (Rizzo, House, & Lirtzman, 1970), perceived stress (Cohen, Kamarck, & Mermelstein, 1983), burnout (Maslach & Jackson, 1986), and life satisfaction (Diener, Emmons, Larsen, & Griffin, 1985).

WET Survey

Data on the use and impact of different types of WET were collected using a Web-based survey conducted in 2004. Virtually all of the scales used in this questionnaire were designed for this study but modeled on psychometrically sound measures that were used in other studies. The questionnaire was divided into six sections: background information, use and impact of laptops, use and impact of cell phones, use and impact of BlackBerry devices, use and impact of PDAs, and use and impact of home PCs. The background section of the survey included basic demographic data as well as information on work demands. To address the issue of the use of WET, we began by asking respondents if they used the various types of WET being considered in this study. For those who indicated yes, we collected data on how much time they spent using the technology within the office, at home, and in locations other than the home or office on both workdays and nonworkdays. We also asked them to indicate how much their use of this technology had changed over the past six months. Data on the impact of various forms of WET also were collected. We asked respondents to consider the impact of each type of work-extending technology on their stress, workloads, productivity, and work-life balance.

Eight hundred and forty-five employees responded to the survey. This sample was well distributed with respect to gender (53% women) and age (32% were under the age of 40, 39% were 40-49, and 27% were 50 or older). Virtually all of the respondents were knowledge workers—employees who perform nonroutine, knowledge-intensive work (Drucker, 1999). At the end of the WET survey, we asked respondents to indicate their willingness to participate in follow-up focus groups or interviews. This generated a list of 117 possible focus group candidates and 136 individuals who were willing to participate in an interview.

The percent of the total workweek spent using WET was used to calculate a variable that we labelled "Dependence on WET". We used this variable to divide the sample into three approximately equal-sized groups:

- LOW dependence on WET (up to 20% of total workweek using WET)
- MODERATE dependence on WET (21% to 60% of total workweek using WET)
- HIGH dependence on WET (more than 60% of total workweek using WET)

Focus Groups

Focus groups were conducted to collect information on the advantages and disadvantages of WET. Two focus groups were conducted: one included nine volunteers with a high dependence on WET, and one included eight volunteers with moderate to low dependence.

Interviews

In-depth interviews were conducted to allow a greater understanding of why people use WET the way they do as well as to explore their feelings and attitudes toward WET. The interview sample was selected so that the demographic characteristics resembled the survey sample. Interviews were done with 30 individuals with moderate to low dependence on WET and 31 individuals with high dependence. The 30-60 minute interviews were conducted by telephone, recorded, transcribed, and content analyzed.

Results

The Use of Work Extension Technology

Just less than half of the 31,571 respondents to our national survey used WET technology to support work outside the office. These respondents spent an average of five hours per week (or 20 hours per month) in work extension activities. Virtually all of these extra hours were unpaid overtime at home.

The WET Survey can be used to give a more detailed estimate of the use of WET. These data (see Table 1) suggest that employees are making significant use of technology to support work extension activities. Half of the sample used WET to perform work at home, while one in three used WET to support work in locations other than work or home.

Another indication of the use of WET in Canada can be obtained by examining employees' dependence on WET, which we defined operationally as the percent

Table 1. Time in technology-supported work (Source: Work Extension Technologies survey, 2004. Sample size = 845 respondents)

	% of sample using WET to work at:		Total hours/week spent in WET supported work at:	
	Alternative Location	Home	Home	Alternative Location
Clerical	23.8 %	34.3 %	1.8	1.1
Technical	35.8 %	43.9 %	2.4	3.1
Professional	33.9 %	49.9 %	3.2	2.1
Managerial	36.3 %	53.3 %	3.8	2.4
Total Sample	33.3%	50.5%	3.0	2.2

Table 2. Total working hours and WET use (Source: Work Extension Technologies survey, 2004. Sample size = 845 respondents)

Dependence on WET	Total hours worked/week at:				Total hours/week spent in WET supported work at:	
	Office	Alternative Location	Home	All Locations	Alternative Location	Home
Low	32.5	5.7	5.1	43.3	1.6	2.1
Moderate	30.8	9.75	11.8	52.4	5.5	9.3
High	31.8	8.3	28.7	68.8	6.7	25.2
Total Sample	32.4	6.0	5.84	44.2	1.87	3.0

of an employee's total workweek spent using WET. Our data indicate that 42% of respondents had a low level of dependence on WET (i.e., spent less than 20% of their work hours using WET), 28% were moderately dependent (i.e., spent 20% to 60% of their work hours using WET), and 30% were highly dependent (i.e., spent more than 60% of their workweek using WET).

Examination of Table 2 shows that employees who have a high dependence on WET spend significantly more hours per week in work-related activities (68.8 hours per week) than employees in the moderate (52.4 hours per week) or low (43.3 hours per week) dependence groups. The difference in workloads cannot be attributed to differences in the amount of time per week spent working at the office, nor can they be credited to time spent working in alternative work locations. Rather, the data are clear; employees with high dependence on WET spend more than twice as many hours per week working from home as those with moderate dependence and more than five times more hours per week working from home as those with low dependence. It is not possible from these data to determine the direction of causal-

Table 3. Use of work extension technologies (Source: Work Extension Technologies survey, 2004. Sample size = 845 respondents)

	Used six months ago	Used now	Change
Cell phone	62.5%	78.9%	+26.2%
BlackBerry	69.5%	88.5%	+27.4%
PDA	63.5%	77.4%	+21.9%
Laptop			
With e-mail	46.8%	45.3%	No change
No e-mail	65.2%	61.2%	(-6.1%)
Home PC			
With e-mail	60.3%	60.3%	No change
No e-mail	77.6%	70.7%	(-8.9%)

ity (i.e., does WET technology increase workloads or do increased workloads lead people to adopt WET as a means of coping with these demands?) The interview data presented later will help us to resolve this issue.

Table 3 shows the use of cell phones, laptops, home computers (with and without access to e-mail), BlackBerry devices, and PDAs. It is clear that the use of cell phones, BlackBerry devices, and PDAs is increasing, while the use of laptops and PCs is stable.

This study also sought to quantify the amount of time per week that employees spent using the different forms of WET (see Table 4). These data provide further support for the conclusion that WET use is associated strongly with job type. For example, BlackBerry use is associated strongly with level—with managers and professionals making significantly more use of these technologies.

Appropriate/Inappropriate WET Use

During the interviews, we asked respondents to describe appropriate and inappropriate uses of WET. Respondents indicated that WET was being used appropriately when employees used these tools to telecommute (30%), to stay in communication with colleagues and clients (22%), to finish up the day's work tasks at home in the evening (22%), to balance work and family (12%), and to connect to the office when traveling for work (12%). Inappropriate use, on the other hand, was identified as using WET to consistently work late in the evening, on weekends, and on holidays (34%); to work during a meeting (22%); to allow friends and family to use the technology (15%); to surf the Web or play games (10%); and to look at

Table 4. Use of WET per week (Source: Work Extension Technologies survey, 2004. Sample size = 845 respondents)

WET	Hours spent using WET - Total Sample			Hours/week spent using WET - Job Type			Hours/week spent using WET: Dependence on WET		
	Workday	Non-Workday	Per Week	Technical	Professional	Mgr	Low	Moderate	High
Cell phone	0.84	0.77	5.74	6.1	5.7	5.3	5.5	6.5	5.3
BlackBerry	1.02	0.59	6.28	3.8	5.9	6.9	5.5	8.2	9.2
PDA	1.03	0.89	6.93	7.6	6.7	6.3	6.7	9.9	5.3
Laptop									
With e-mail	1.37	0.84	8.53	8.8	8.2	9.6	8.4	10.3	7.7
No e-mail	1.60	0.90	9.80	12.4	9.4	9.3	9.4	12.0	14.8
Home PC									
With e-mail	0.89	0.83	6.11	6.2	5.6	6.6	5.9	6.6	N/A*
No e-mail	1.03	0.94	7.03	7.3	6.9	6.8	6.5	10.1	N/A*

(Values for clerical/administrative employees were not calculated due to very small sample size)

* = no observations

pornography (10%). One in four also said that it was inappropriate to keep cell phones and Blackberry devices turned on in meetings. Respondents indicated that one in three of their colleagues and managers made inappropriate use of WET on a daily basis. They blamed inappropriate use on workload pressures (51%), the need to be always available for work (39%), the work culture within their organization (i.e., 25% felt that their company gave promotions to employees who were always accessible and worked long hours, which encouraged inappropriate use), and the fact that technology is fairly new and that people are not aware of what should and should not be done (21%).

How does WET Use Impact Employers and Employees

We used three sets of quantitative data to shed light on the impact of WET use. The first set of data (see Table 5) shows the relationship between dependence on WET and a number of key work and employee outcomes from the National Study. Dependence on WET is associated positively with time at work, stress, burnout, and job stress, and negatively associated with job satisfaction ($p < .05$ in all cases).

The second set of data (see Table 6) also comes from the National Study. The majority of respondents perceived that WET had not affected their job security (80%), their ability to balance work and family (62%), their interest in their work (52%), or their stress levels (51%). The majority, however, did feel that technology had increased their workloads (54%) and productivity (55%). Very few individuals (approximately 10% of the sample) reported that the use of WET had reduced their workloads, their stress levels, and their conflict between work and life. These findings are consistent

Table 5. Impact of dependence on WET and key outcomes (Source: The National Work, Family and Lifestyle Survey, 2002)

	Dependence on WET		
Work Outcomes	Low	Moderate	High
% working > 45 hours per week	12%	29%	58%
% with high job satisfaction	50%	45%	40%
% with high job stress	27%	36%	45%
% thinking of leaving weekly or more	25%	28%	33%
Employee Outcomes			
% with high stress	52%	53%	60%
% with high burnout	28%	31%	37%
% with high life satisfaction	40%	46%	40%

* *Sample size = 31,571*

* *Population benchmarks (Duxbury & Higgins, 2002) were used to identify the percent of each of these three samples that reported high, moderate, and low scores on each of these outcome variables. Only the high percentage scores are given in this table.)*

Table 6. Effect of WET on employees (Source: The National Work, Family and Lifestyle Survey, 2002. Sample size = 31,571)

Impact of Technology on	Manager	Professional	Other	Total
Workload				
° increased	73	52	50	54
° no impact	21	39	40	37
° decreased	6	9	10	9
Balance				
° increased	33	25	25	26
° no impact	47	63	55	62
° decreased	21	11	9	12
Stress				
° increased	57	42	40	43
° no impact	38	52	54	51
° decreased	5	6	7	6
Productivity				
° increased	68	53	53	55
° no impact	21	36	38	45
° decreased	11	11	9	10
Job Security				
° increased	6	7	14	8
° no impact	88	86	75	80
° decreased	6	7	12	12
Interest in Work				
° increased	45	37	43	41
° no impact	48	57	50	52
° decreased	7	6	7	6

with the productivity paradox described earlier and suggest that the productivity gains made possible through the use of WET are being offset by the social impact of this technology. All results are significant at 5%.

The data in Table 6 indicate that the perceived impact of WET is associated strongly with job type, with those in management positions being more likely to report both positive and negative impacts of technology. Managers were the most likely to say that WET had increased their workloads (73%), increased the amount of stress they experienced (57%), and decreased their ability to balance work and family (21%). They also were, however, the group most likely to report that WET had increased their productivity (68%), their ability to balance work and family (33%), and their interest in their work (45%).

While the data from the National Study provide interesting insights into the impact of WET, they do not allow us to determine if different types of technology have different impacts. Follow-up data were collected (see Table 7) to look at impacts. Key observations with respect to how the use of the different types of WET impacted workloads, stress, work-life balance, and productivity are summarized next. Only significant results (p < .05) are presented.

Workload

The use of a BlackBerry device appears to be associated positively with an increased workload for a substantive proportion of the sample (36%). Virtually no one in the sample indicated that access to WET had decreased workloads.

Table 7. Impact of WET on employees: The WET survey (Source: Work Extension Technologies survey, 2004. Sample size = 845 respondents)

	Increased	No Impact	Decreased
Perceived impact of cell phone on:			
>Workload	13%	83%	4%
>Stress	18%	67%	14%
>Productivity	44%	54%	2%
>Work-life Balance	32%	59%	9%
Perceived impact of BlackBerry on:			
>Workload	36%	49%	15%
>Stress	27%	38%	35%
>Productivity	78%	21%	1%
>Work-life Balance	46%	26%	28%
Perceived impact of PDA on:			
>Workload	4%	87%	9%
>Stress	4%	62%	33%
>Productivity	55%	45%	0%
>Work-life Balance	37%	61%	2%
Perceived impact of laptop with e-mail on:			
>Workload	26%	70%	4%
>Stress	22%	67%	11%
>Productivity	34%	60%	6%
>Work-life Balance	18%	65%	17%
Perceived impact of laptop without e-mail on:			
>Workload	26%	68%	7%
>Stress	20%	68%	12%
>Productivity	46%	50%	5%
>Work-life Balance	22%	62%	16%

Table 7. Continued

Perceived impact of home PC with e-mail on:			
>Workload	12%	78%	10%
>Stress	12%	65%	23%
>Productivity	39%	60%	1%
>Work-life Balance	30%	58%	12%
Perceived impact of home PC without e-mail on:			
>Workload	12%	81%	7%
>Stress	12%	61%	27%
>Productivity	42%	55%	3%
>Work-life Balance	33%	53%	14%

Stress

BlackBerry use appears to be linked to increased stress levels for a substantive number of employees (27% reported increased stress). It is also interesting to note that four forms of technology are associated with a reduction of stress: the BlackBerry (35% report reduced stress), the PDA (33% report reduced stress), a home PC with e-mail (23% report reduced stress), and a home PC without e-mail (27% report reduced stress). The varying impact of these technologies on employees is more likely due to how the employee uses the technology rather than the technology itself. This conclusion is supported by the data on the impact of the BlackBerry, where approximately equal proportions of the sample report no association, a positive association, or a negative association between their use of this technology and stress. In other words, it is not the technology per se that contributes to or alleviates stress but how the technology is used by the employee. This conclusion is supported by the findings from the interview study discussed in the next section.

Productivity

Virtually no one perceived that the use of WET had lowered his or her productivity. Of concern, however, is the fact that almost half of the respondents, despite the increased time in work, are not experiencing productivity gains from using WET. In fact, in only one case (BlackBerry) did a clear majority of respondents (78%) indicate that they felt that their use of this type of WET had increased their productivity. These data are consistent with the productivity paradox and suggest again that it is not the technology per se that will increase productivity, but rather how it is used and the culture within which it is used that make a difference. It also suggests that the link between WET use and increased workloads and stress may be reducing possible productivity gains in the workforce.

Work-Life Balance

While relatively few people report that their use of WET has reduced their ability to balance work and life, the majority of respondents feels that the overall impact of WET has been neutral. There is, however, no consensus on the link between BlackBerry use and work-life conflict. While 46% of the sample said that a BlackBerry had increased their ability to manage their work-life balance, 28% said it had the opposite impact. Similar (though not as pronounced) findings can be observed with respect to the laptop.

What are the Perceived Advantages and Disadvantages of Using WET?

The focus groups and interviews give us a number of insights into the perceived advantages and disadvantages of using WET. One focus group consisted of individuals who were highly dependent on WET. The second was made up of employees with moderate to low dependence. Those with a high dependence on WET identified the following three work-related advantages: enhanced teamwork (i.e., enables team members to respond quickly when needed); decreased time spent commuting; and an increased flexibility with respect to when and where work can be done. They also identified two key nonwork-related advantages. They felt that the technology had increased their ability to deal with family and personal emergencies and made it easier to accommodate both work and family demands.

Heavy WET users also were able to identify the dark side of WET use. Work-related disadvantages included increased workloads, increased expectations with respect to availability for work, and privacy issues regarding personal use of the technology (i.e., loss of corporate data). This group was only able to identify two nonwork-related disadvantages: increased expectations with respect to availability for the family and less time for the family.

Participants in the low dependence focus group generated a quite different list with respect to the advantages of WET. They noted the following work-related advantages: time management is easier; increased reach (i.e., WET makes it possible to communicate with large groups of people, hard-to-reach individuals, and people in other locations or time zones); and the work/life boundary is more flexible. This group identified the following as the most important nonwork-related advantages of using WET: the ability to make use of the technology for both work and personal life; an increased ability to manage time; and decreased stress.

The set of the most important work-related disadvantages associated with the use of WET by the low users was very similar to that identified by the high use group. They noted the following disadvantages: increased expectations with respect to speed of response, increased workloads, and increased stress when the WET did not work correctly. The most important nonwork-related disadvantages of using WET were that respondents felt that they were on call all the time; there was no clear boundary between work and home; and real income (dollars per hours worked) had decreased because the use of WET promoted longer working hours.

In-depth interviews with two sets of employees, those with high dependence on WET and those with low dependence on WET generated a second list of advantages and disadvantages of using WET to the employer (Table 8) and the employee (Table 9).

The set of advantages and disadvantages realized from the focus groups and interviews are consistent with and add to those identified from the literature review. While the words they used to describe their experiences with WET were quite different, the themes underlying the words were the same. The advantages of using WET, regardless of whether they pertained to work or nonwork, were all associated with the fact that WET gives employees greater control over when and where they worked and increases their ability to communicate with others. These two characteristics of WET can be seen as the reasons for the other advantages noted by focus group participants and interview respondents: increased flexibility, greater work-life balance, and a greater ability to work as a team.

Table 8. Perceived advantages and disadvantages of WET for the employer (Source: Interviews. Sample size = 30 low to moderate to users and 31 high users)

	Total Sample	High Dependence	Low Dependence
Advantages to Employer of WET			
More productivity - they get more work out of us	63%	71%	56%
More efficient - employees work smarter	39%	25%	46%
Employers are more able to stay in contact with employees	30%	29%	35%
Greater employee satisfaction	16%	17%	13%
Disadvantages to Employer of WET			
Cost of devices, updating, and training	39%	50%	29%
Loss of control - managers don't know where employees are, what they are doing	20%	17%	24%
Employees using WET for personal use during work hours	20%	13%	27%
Security problems	10%	8%	12%
Employee burnout, stress due to the high volume of e-mail	10%	22%	15%
Nothing, there are no disadvantages	10%	10%	10%

Table 9. Perceived advantages and disadvantages of WET: The employee (Source: Interviews. Sample size = 30 low to moderate to users and 31 high users)

	Total Sample	High Dependence	Low Dependence
Work-Related Benefits to Employee			
More efficient, more productive, better time management	34%	32%	36%
Flexibility, freedom, not tied to the desk	32%	28%	36%
Ability to work from home	25%	32%	21%
Being able to reach people; people can reach me when I am in transit/traveling/off-site	24%	20%	27%
Easier to get work done outside of normal working hrs	12%	12%	12%
Non-Work-Related Benefits to Employee			
Family can get in touch with me	26%	20%	32%
I can be at home (don't have to put in overtime in office)	22%	22%	19%
Family also can make personal use of WET	19%	25%	16%
I can work from home while meeting family commitments (sick kids)	19%	16%	20%
Less stressful - healthier - helps with work life balance	14%	26%	4%
Work-Related Drawbacks to Employee			
I am never off duty - I can be reached at any time, always expected to be accessible	40%	41%	40%
It increases work expectations re workload and turnaround time	20%	16%	23%
Lengthens the workday	19%	24%	12%
Limitations of device, malfunctions, lack of functionality	16%	20%	14%
No drawbacks	16%	21%	12%
Non-Work-Related Drawbacks for Employee or His or Her Family			
There are no drawbacks	38%	44%	34%
It takes away from my time with them and my personal time - time is given to work instead	38%	44%	33%
It is intrusive	19%	24%	16%

Unfortunately, the same features of WET that increase perceived control and facilitated communication also appear to be the source of many of the disadvantages. Respondents in both the high-dependence and low-dependence focus groups as well as those who participated in the interviews identified the following disadvantages of using WET: the feeling of being on call all the time (it is interesting to note that participants felt this type of pressure from work and family) and the lack of a clear boundary between work and home. Disadvantages unique to the work domain include increased workloads and increased work expectations (speed of response, how much work can be done in a given time period). The main disadvantage relating to the family was that less time was available to spend with the family, which may be a direct consequence of increased workloads and expectations as well as the increased ease with which one can work from home and elsewhere. Some respondents also felt that WET was intrusive. Interestingly, both focus groups expressed frustrations with the technology itself, with low users talking about frustrations associated with the limits of the technology and high users talking about frustrations associated with people not using the technology appropriately.

Finally, the focus group data suggest that dependence on WET and attitudes toward WET are highly interrelated. Employees in the heavy user group had a significantly more positive attitude toward the use of WET; users in the low dependence group were more ambivalent. The high-user focus group was unanimous that the advantages of using WET far outweighed any disadvantages, while the low user group could not make a decision with respect to this issue. It is difficult to determine from these data, however, if greater exposure to WET provides one with a more balanced view of technology or if cognitive dissonance means that people who make heavier use of such tools ignore the negative consequences of such actions.

Optimizing the Use of WET

Employees and employers who wish to realize greater benefits from the use of WET need to either reduce the perceived drawbacks associated with the use of this technology, enhance the perceived benefits of use, or both. Accordingly, to help us to formulate advice to organizations on how best to manage WET, we asked focus group and interview respondents to give us recommendations on the use of WET. Key suggestions offered by study participants are summarized in Table 10.

Employees recommended that their employers create appropriate policies to define what constitutes appropriate work use of the technology and how much work can be expected from employees in a given time period. They also wanted the organization to provide appropriate tools, tech support for their WET, and training on how to make good use of their technology. One in four of those with high dependence on WET felt that employers needed to deal with the culture within the organization. They noted that possession of the technology should not be viewed as an open invitation to be contacted at any time. Respondents also felt that work and nonwork boundaries must be clearly established and respected and that the development of an effective etiquette around proper use of WET should be considered a priority. Finally, adequate training on the use of WET and technical support for WET users were considered important steps to facilitate and encourage adoption.

Table 10. Enhancing benefits/reducing drawbacks of WET (Source: Interviews [Sample size = 30 low to moderate to users and 31 high users] and Focus Groups [eight low to moderate users and nine high users])

	Total Sample	High Dependence	Low Dependence
How employers could reduce drawbacks of WET			
Create policies for appropriate work use (i.e., discuss appropriate response time, how much work can be expected in certain time period)	26%	33%	21%
Make sure devices work or are appropriate (i.e., provide tech support, spam control)	12%	12%	12%
Give faster Internet connections for home computer use/pay for home Internet connection	12%	22%	3%
Provide training on the use of the devices	12%	0%	21%
Change culture - encourage work/life balance	12%	23%	0%
How individuals could reduce drawbacks of WET			
Discipline myself - set personal guidelines on use and then stick to them	47%	55%	38%
Take the initiative and get training on how to use the WET	10%	4%	20%
How employers could enhance benefits of WET			
Make more devices available to more colleagues, particularly for those who can use them	35%	45%	22%
Provide high-speed Internet connection for home computer use	22%	12%	30%
Update and maintain WET technology regularly	18%	16%	14%
Allow/encourage teleworking	15%	20%	10%
How individuals could enhance benefits of WET			
Limit my use, manage expectations	31%	37%	22%
Get training	16%	17%	15%
Maximize existing benefits	13%	4%	21%
Buy additional devices/lobby for more	11%	13%	25%

Solutions and Recommendations

This study indicates that although WET has the potential to benefit employees and employers, these benefits will not materialize if some basic constraints and rules on WET use are not developed and implemented. A number of authors (Dinnocenzo & Swegan, 2001; Gordon, 2001; Weil & Rosen, 1997) have written extensively on how employees can cope with the "anytime-anywhere work world" and "reclaim the boundaries between work life and personal life" (Gordon, 2001, p. 30). A review of some of the prescriptive literature in this area as well as the interview findings resulted in the following recommendations to enhance the usefulness of the technology and to minimize negative aspects. Employers should do the following:

- Set limits on the amount of usage of work-extending technology. For instance, if a person is issued a cell phone, the employer should develop effective policies such as rotating cell phones among on-call personnel so that the interruptions are more predictable and balanced.
- Be realistic about the amount of work one can do outside traditional work hours.
- Train employees on how to use existing applications properly. For instance, teach employees how to use the rules available in common applications such as Microsoft Outlook to manage their e-mail more effectively.

Employees also need to set limits to availability and to the use of WET. Examples of strategies that can be used by employees include the following:

- Do not give cell phone numbers out indiscriminately.
- Check e-mail and voicemail only at designated times.
- Set response limits (e.g., respond to a voicemail within 60 minutes during the week but the next day if picked up after hours).
- Create zones of separation (Gordon, 2001). Some periods are only for work, some are only for personal time, and some are when one allows work to intrude, if necessary.
- Set clear rules with co-workers and supervisors about when one is available and under what circumstances one can be contacted.
- Do as much work as possible within the office. Avoid procrastination and excessive socializing.

Future Trends

Our review of the literature indicates that the topic of work extension has received very little attention from academics and managerial writers. While there is a great deal of literature available on teleworking, the significant differences between teleworking and work extension imply that the applicability of the telework literature to the field of work extension is questionable. The existing literature also lacks any discussion of the impact of work extension on productivity, a key concern for many employers. We do not know, for example, if work extension is the result of overwork or if the technology itself increases work by increasing expectations and response time. The effectiveness, productivity, and health of the employee who is putting in longer hours using WET are also unexplored topics. Our aim in this section is to provide a general view of the areas that need to be explored if the subject of work extension is to be understood fully.

First, there is a lack of basic data about work extension. While this study is a start, it needs to be replicated. Data also need to be collected to allow researchers to identify key moderators of the relationship between dependence on WET and key outcomes. Variables that need to be examined include gender, age, education, organization type, industry, nature of the work, and place of abode (i.e., rural vs. urban).

Second, we are not presently in a position to talk with absolute certainty about why people make heavy or light use of WET. The direction of causality for increased dependence on WET is difficult to ascertain. Does having technology available outside the office encourage one to work long hours, or are employees with higher levels of stress, heavier workloads, and greater expectations with respect to availability and response time more likely to acquire WET so that they more easily can extend their workday to better manage their work? While this study unequivocally links higher workloads with increased dependence on WET, it is not possible to conclusively determine the direction of causality. Future studies with a longitudinal design should be undertaken to address this issue.

An associated issue is the effectiveness of work extension. Future research on work extension activities should look at this work arrangement through the productivity paradox lens. Several findings are behind this recommendation. First, we need to look further at the link between perceived productivity and the use of WET. While the fact that virtually no one in our sample perceived that the use of WET had lowered their productivity, what is troubling is that almost half of the respondents are not experiencing expected productivity gains from using WET. The results with respect to work-life balance also were inconclusive. Relatively few people reported that their use of WET had reduced their ability to balance work and life. This suggests that the potential for these devices to help control work-life balance is not being realized. As noted in this study, work extension is associated positively with longer hours, increased workloads, and higher levels of stress. Further research is needed to quantify the extent to which the efficiency and perceived productivity gains offered by WET are offset by declines in the physical and mental health of the WET user.

Third, an important topic for research is to identify benefits and problems for the family that are caused by work extension. Future research could look at how spouses and children feel when a work extender is checking e-mail during the family holiday or during the weekend and explore how family attitudes to work extension affects dependence on WET and attitudes toward WET.

Fourth, the advantages of using WET, regardless of whether they pertain to work or nonwork, all were all associated with the fact that WET gives employees greater control over when and where they worked and increases their ability to communicate with others. Unfortunately, the same features of WET that increase perceived control and facilitate communication also appear to be the source of many of the disadvantages such as the feeling of being on call all the time and the lack of a clear boundary between work and home. There is a need for research that determines empirically how organizations and employees can best deal with these competing issues.

Fifth, the data suggest that dependence on WET and attitudes toward WET are highly interrelated. Employees in the high dependence group had significantly more positive attitudes toward the use of WET; users in the low-dependence group were more ambivalent. The high dependence focus group was unanimous that the advantages of using WET far outweighed any disadvantages, while the low-dependence group could not make a decision with respect to this issue. It is difficult to determine from these data, however, if greater exposure to WET provides one with a more balanced view of technology or if cognitive dissonance means that people who make heavier use of such tools ignore the negative consequences. Again, longitudinal studies should be undertaken to address this issue.

Conclusion

Based on an extensive review of the existing literature, a survey of 845 Canadian knowledge workers, a national survey of 31,571 Canadians, two focus groups, and 61 in-depth interviews, this chapter provides new information on how work extension technologies are being used and links the use of WET to key organizational and individual outcomes. This chapter also has outlined ways in which employers and employees can use WET more effectively. The conclusions obtained from the four different data sources were very consistent and increased our confidence in the validity and generalizability of our findings. The following key conclusions can be drawn from this research:

- The use of WET to support work outside the office is widespread in Canada.
- Knowledge workers spend a high proportion of their workday engaged in technology-supported work outside the office.

- Employees have varying levels of dependence on WET. Dependence on WET is associated strongly with the type of tool used. Employees with low dependence on WET make relatively high use of a home PC with e-mail and a PDA, while those with moderate dependence are more likely to use a laptop with e-mail and a cell phone. Those with high dependence, on the other hand, rely more on cell phones and BlackBerry devices.

- Employees with different work requirements need different types of WET. Cell phones are essential work tools for employees who need to stay in touch with clients and colleagues when away from their desks. Laptops facilitate mobile work and are essential work tools for employees who want to work seamlessly away from the office. BlackBerry devices are used as e-mail management systems and are essential tools for employees who receive a large number of e-mails that require immediate attention. PDAs facilitate time management and are important tools for employees who have a lot of different demands on their time and need to stay organized. The home PC allows work from home and is an essential tool for an employee who wants to work at home on evenings and weekends.

- Using WET to provide flexibility with respect to when and where work is done is considered an appropriate use of the technology. Personal use and overuse (i.e., in meetings, being always available) is considered inappropriate. By their own admission, one in three in the sample makes inappropriate use of WET. Inappropriate use of WET is attributed to workload pressures, the need to be available, and a work culture that rewards accessibility and working long hours through career advancement.

- Study participants identified a number of work and nonwork advantages that they had experienced from their use of WET, including better teamwork, a decrease in the time spent commuting, increased flexibility with respect to when and where work is done, an increased ability to communicate with people, an increased ability to deal with family and personal emergencies, and a better balance between work and family needs.

- Respondents were unanimous that WET users experience the following problems: increased expectations with respect to availability for work and family, increased workloads, less time for family and personal activities, inappropriate use of the technology by others, and privacy issues regarding personal use of the technology.

- The impact of WET on key work and employee outcomes depends on job type; for some, WET is a stressor, for others it is a way to cope with multiple demands, and for others it serves as both a stressor and a coping mechanism. This suggests that it is how the technology is being used that makes a difference, not the technology itself.

- WET is not the panacea for many of the challenges experienced by today's workforce; the potential for WET to alleviate work-life conflict or increase productivity is not being realized. The findings from this study are consistent with the productivity paradox and suggest that the potential productivity gains offered by WET technology are offset by the human costs such as stress, burnout, and work-life conflict associated with the inappropriate or overuse of WET.

In summary, our research found that for most employees, the benefits of using WET far outweigh the costs. The use of WET tools has become institutionalised, and most people could not imagine going back to a world without access to these tools. For better or for worse, WET is here to stay. The challenge now is to use these tools effectively.

References

American Management Association. (2002). *2002 summer vacation plans survey*. New York: AMA Research.

Bassett, C. (2000). In the company of strangers: User perceptions of the mobile phone. In L. Haddon (Ed.), *Communications on the move: The experience of mobile telephony in the 1990s* (COST [European Organisation for Co-operation in the field of Scientific and Technical Research] Report 248, pp. 116-135). Tells, Norway: Farsta.

Bluedorn, A. C. (2002). *The human organization of time*. Stanford, CA: Stanford Business Books.

Bolan, S. (2001, November 16). Stress-free IT professionals. *Computing Canada*, pp. 8-9.

Bolino, M. C. (1999). Citizenship and impression management. *Academy of Management Review, 24*(1), 82-99.

Brown, B. (2002). Studying the use of mobile technology. In B. Brown, N. Green, & R. Harper (Eds.), *Wireless world: Social and interactional aspects of the mobile age*. London: Springer.

Canalsys. (2005). *Worldwide smartphone market*. Retrieved February 21, 2006, from http://www.canalys.com/pr/2005/r2005102.htm

Cohen, S., Kamarck, T., & Mermelstein, R. (1983). A global measure of perceived stress. *Journal of Health and Social Behaviour, 24*, 385-396.

Corbett, N. (1994). *The use of mobile offices: An explanatory analysis* (unpublished MMS thesis). Ottawa: Carleton University.

Diener, E., Emmons, R. A., Larsen, R. J., & Griffin, S. (1985). The satisfaction with life scale. *Journal of Personality Assessment, 49,* 71-75.

Dinnocenzo, D, & Swegan, R. (2001). *Dot calm.* San Francisco, CA: Berrett-Koehler.

Drucker, P. (1999). Knowledge worker productivity: The biggest challenge. *California Management Review, 41*(2), 79-94.

Duxbury, L. E., & Higgins, C. A. (2002). The 2001 national work-life conflict study: Report one. *Health Canada.* Retrieved February 21, 2006, from http://www.phac-aspc.gc.ca/publicat/work-travail/report1/index.html

Eriksen, T. H. (2000). *Tyranny of the moment.* London: Pluto Press.

Fairweather, N. B. (1999). Surveillance in employment: The case of teleworking. *Journal of Business Ethics, 22*(1), 39-49.

Frase-Blunt, M. (2001). Busman's holiday. *HR Magazine, 46*(6), 76-80.

Gant, D., & Kiesler, S. (2002). Blurring the boundaries: Cell phones, mobility, and the line between work and personal life. In B. Brown, N. Green, & R. Harper (Eds.), *Wireless world: Social and interactional aspects of the mobile age* (pp. 121-131). London: Springer.

Gartner Group Research. (2001). *Benefits and TCO of notebook computing.* Santa Clara, CA: Computers Inc.

Gartner Group Research. (2006). *Dataquest alert.* Santa Clara, CA: Computers Inc.

Gleick, J. (1999). *Faster.* New York: Pantheon Books.

Gordon, G. (2001). *Turn it off.* New York: Three Rivers Press.

Green, N. (2001). On the move: Technology, mobility and the mediation of social time and space. *The Information Society, 18,* 281-292.

IDC Research. (2001). *IDC mobile consumer survey.* Framingham, MA: IDC.

IDC Research. (2002). *Web interfaces deliver increased productivity to your virtual workforce.* Framingham, MA: IDC.

Internet World Stats (IWS). (2006). Internet World Stats Surfing and Site Guide. *Internet World Stats.* Retrieved February 21, 2006, from http://www.internetworldstats.com/surfing.htm

Ipsos Reid Research. (2001). *Analyzing the return on investment of a BlackBerry deployment* (Research study prepared for research in motion). Waterloo, ON: RIM.

Kennedy, S. (2002, June 10). Workers of the world unplug. *The Globe and Mail,* p. C1.

King, W. R. (2002). IT capabilities, business processes, and impact on the bottom line. *Information Systems Management. 19*(2), 85-87.

Maslach, C., & Jackson, S. (1986). *The Maslach burnout inventory*. Palo Alto, CA: Consulting Psychologists Press.

McGinn, D. (2002, April 29). I'll help myself. *Newsweek—Next Frontiers, 139*(17), 52-56.

Quinn, R., & Staines, G. (1979). *The 1977 quality of employment survey*. Ann Arbour, MI: University of Michigan.

Rizzo, J, House, R., & Lirtzman, S. (1970). Role conflict and ambiguity in complex organizations. *Administrative Sciences Quarterly, 15,* 150-163.

Salazar, C. (2001). Building boundaries and negotiating work at home. *Proceedings of the 2001 International ACM SIGGROUP Conference on Supporting Group Work,* Boulder, Colorado (pp. 162-170).

Schlosser, F. K. (2002). So, how do people really use their handheld devices? An interactive study of wireless technology use. *Journal of Organizational Behavior, 23*(4), 410-423.

Silver, C., & Crompton, S. (2002). No time to relax? How full-time workers spend the weekend. *Canadian Social Trends, 62,* 7-11.

Strom, G. (2002). Mobile devices as props in daily role playing. *Personal and Ubiquitous Computing, 6*(4), 307-310.

Weil, M. M, & Rosen, L. D. (1997). *Technostress*. New York: John Wiley & Sons.

Chapter XV

The Role of User Characteristics in the Development and Evaluation of E-Learning Systems

Dianna L. Newman
University at Albany / SUNY, USA

Aikaterini Passa
University at Albany / SUNY, USA

Abstract

This chapter presents a multi-phase cyclical model of designing, developing, and evaluating instructional technology (IT) learning systems based on inclusion of users' characteristics (experience with technology, familiarity with content, adaptability, learning style, gender, professional level). The model was developed and piloted over the course of seven years in more than 50 learning communities and has resulted in documentation of stages in which user variables interact with the process. Key elements of the model are presented in detail and supported by samples of development and related evaluation. The authors hope that the chapter contains excellent recommendations for the practice of designing and evaluating IT learning systems that meet the varied individual, cultural, and contextual needs of users.

The role of technology as a support to instruction and curriculum is now a major component of global educational systems; an increasing number of organizations are using technology, both in and out of traditional learning settings, as a means of transferring knowledge, skills, and abilities to students, employees, and consumers around the globe. As IT systems that support this transfer are being integrated and implemented, managers, designers, and evaluators are being asked to provide evidence of successful outcomes (Gallupe & Tan, 1999). Only limited efforts have been made to establish overarching evidence of valid ways to improve learning as it occurs or to document the impact of learners' cultures on the process (Nemeth, 2004). As managers and decision makers allocate an increasing amount of resources to IT development, implementation, and sustainability, there is a need to document successful efforts, especially in areas related to training and professional development (Gonyea, 2005).

Overview

The theory of information processing, or how "we convert information from stimuli into interpretations of what we are perceiving and what it means" may serve as an underlying model for investigating the interrelationship of users' learning schemata and technology use (Hill, 1997, p. 113). According to Mayer (1996), the best learning occurs when it is active; that is, when learners can select and attend to features of the environment, transform and rehearse information according to their own backgrounds, relate new information to previously acquired contextually based knowledge, and organize that knowledge in a way that makes it meaningful to their own context. According to Myers and Tan (2002), this requires looking at the process as one that is dynamic, interactive, and culturally based, bounded by temporal as well as emerging user needs. In this process, the learner can relate his or her existing knowledge and experience to the new knowledge, and as a result, retention will be enhanced.

This view of learning far exceeds the original concept of IT systems based on behavioral models in which learning occurred in sequential stages between receipt of an external stimulus and the production of a response (Skinner, 1954); instead, information processing and cognitive psychology emphasize the interaction of internal mental processes that result in learning. This approach recognizes and incorporates variations in acquiring, processing, storing, and retrieving information and how they are influenced by internal as well as external cultural stimuli (Greeno, Collins, & Resnick, 1996; Schunk, 2000). Instructional scaffolding or controlling the presentation of task elements so that learners can focus on and master those that are appropriate to their current level or strategy of learning is an important part of this process (Bruning, Schraw, & Ronning, 1995). Scaffolding is derived from

Vygotsky's (1978) theory that learning is embedded within cultural and contextual environments and that these must be taken into account as part of the controlled presentation if learning is to occur. Scaffolding assists in learning by identifying the stage at which a learner is functioning, recognizing concomitant user and environmental variables, and infusing this information into the design and development of the learning tasks. Preece, Rogers, and Sharp (2002) noted that scaffolding is a key part of designing and evaluating new technologies that will cross cultural, geographical, and user boundaries.

User Characteristics that Impact Learning

Learners' cultural, contextual, and personal variables all play a major role in how they interact with information; as noted above, a learner's orientation goes beyond externally controlled processes to encompass internal factors such as expectations, feelings, attitudes, values, interests, intentions about what to learn, perceptions about how new learning is acquired, and perceptions of its worth. Key user characteristics that, in turn, impact these factors include learning style, acceptance of the mode of presentation, identification with the content, and prior experiences with or knowledge of all three. As examples of their importance, Riding and Rayner (1998) noted that learners' perceptions of and acceptance of the structure of material (format and conceptual) impacted holistic integration of information, acceptance of the order of presentation, and flexibility in self-structuring. Additionally, Keller (1987) noted that learning was facilitated when instruction was appealing to students and, as a result, when students developed positive attitudes toward it. This occurred when instruction was identified positively with prior experiences, when it encouraged their active involvement, and when it facilitated persistence on task.

In a review of the research on the management, design, and evaluation of IT systems, Stone, Jarrett, Woodroffe, and Minocha (2005) noted that the role of instruction is to activate internal learning and that technology should be used to support this activation. They found that user characteristics such as age, gender, ethnicity, prior content and technology experience, motivation, and attitude had an impact on users' interactions with IT systems. Consequently, they noted that IT systems should make learners aware of how much they do know and help them in managing and monitoring their knowledge acquisition. In IT settings, learning should be managed both by the learner and by the learning agent with learners becoming increasingly responsible for their own processes in new settings (Butler & Winne, 1995; Roberts & Erdos, 1993). As a direct consequence, learners' cognitive abilities; their prior knowledge, skills and experiences; and their interest in a particular topic should become central to the design of the instructional process.

Cognitive psychologists and advocates of constructivist learning also have had a major impact on expectations of technology-supported learning (Lajoie, 2002); most

Table 1. User characteristics related to IT Design

Experience with Technology
Availability of Technology
Familiarity with Content
Professional Level (e.g., novice, expert, etc.)
Learning Style
Adaptability (e.g., user's physical and cognitive needs)
Student Role (e.g., student to student, student to teacher, group/individual)
Gender
Language
Geographical/National Background

IT managers, developers, and educators now expect technology to serve as a facilitator for active learning, not just as a resource agent (Perkins, 1991). It is expected that technology will facilitate learning by providing appropriate contexts in order for learners to explore and come to understand complex phenomena in a variety of subject areas for multiple types of users (Jonassen, 1991; Rieber, 1996). This is especially true when technology is used to support problem solving by engaging learners in real-world contexts (Duffy & Cunningham, 1996; Honebein, 1996). Accordingly, successful technology-based instruction now must be tied to integrated, culturally based, learner developed processes. IT systems must provide a means by which individuals can manipulate resources and make their thinking visible; it must be engaging and provide cognitive support for learners from different backgrounds, with different abilities, and with different ways of learning (Hanaffin, 1992; Lepper & Gurtner, 1989; Salomon, Perkins, & Globerson, 1991). In addition, the system must allow for various representations of knowledge, facilitate communication within and among different groups of learners, and enable learners to experiment with different ideas. A summary of key user variables that should be considered when planning IT systems is presented in Table 1.

A User-Centered Model for Development and Evaluation and Examples of Use

The design of information systems must be directed toward the end consumer, and nowhere is this more important than in the development of technology systems that support learning (Rowland, 2004). If learner characteristics are important factors in the development of technology-supported instruction, it is necessary that these characteristics be reflected in the design and evaluation of modules that support learning. The challenge for IT managers is to lead development teams in designing technology-supported systems of learning that will facilitate the acquisition of factual and structural knowledge and its meaningful integration into the cognitive frameworks for learners from different backgrounds and with different degrees of prior knowledge (Nemeth, 2004). As a result, the optimal goal is to structure the external conditions in a manner that will facilitate internal learning and information processing within each learner's zone of proximal development or scaffolding level. To meet this need, Newman (1998) developed a multi-stage process for team leaders to use in designing, validating, and documenting IT learning systems. Each stage of the model involves acknowledgment of user characteristics; their prior levels of knowledge, skills, and experiences; and the external, contextual variables that influence the formation of new learning systems. A summary of the user-centered model is presented in Figure 1.

Figure 1. Summary of the steps involved in the user-centered model

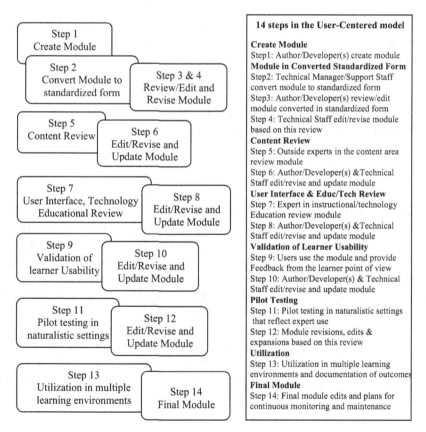

Over the course of the past decade, the user-centered model has been used to develop and validate more than 100 computer/Web-based modules of instruction in more than 50 settings. Multiple demonstrations of culture representing a nongeogrphic approach (Myers & Tan, 2002) were included within this process. Discipline cultures include mathematics, engineering, psychology, addictions, and leadership. Stages of learning cultures include elementary, secondary, post-secondary, and graduate education as well as workforce and professional development training. Users included persons of both genders with varied ethnic and ability backgrounds and differing levels of experiences and contexts. Following are brief summaries of some of the applications of the model.

Integrated Mathematics Curriculum at the Post Secondary Level

The major goal of this project[1] was to create a library of interactive multimedia hypertext modules that would link abstract mathematical constructs with practical real-life applications in engineering and the sciences. Cultures included educational settings as well as varied workforce settings, both genders and multiple ethnic groups, and variations in primary spoken and written language. The specific method adopted by project staff was the development of computerized Web-based modules that could be used as adjuncts to classroom instruction, facilitating the review of old material and the instruction and learning of new principles or as stand-alone materials for professional development purposes. To facilitate this process, teams of faculty representing the disciplines of mathematics, engineering, science, and technology and instructional design cooperatively determined content of the modules and jointly developed and reviewed the modules.

Engineering Curriculum

A second subset of projects[2] involved design, development, and evaluation of student-initiated, student-centered learning tools utilizing innovative Web-based curricula in the pre-engineering and engineering domain. The goal of the curriculum units was to assist in presenting knowledge in a manner that would allow for immediate transfer to practice in applied settings while motivating and rewarding learners as they progressed through the modules. Subcultures included educational level, workforce areas, geographic sites, ethnic groups, and variations in prior knowledge and experience. To facilitate the development of the modules, teams consisting of discipline-based faculty; professionals from the specific content field; and specialists in learning, instructional design and evaluation, and technical developers were created. A formative, cyclical process of development, assessment, and evaluation was established that allowed the team to share information on an ongoing basis, to use discussion and feedback to make decisions and refinements in the process and the products, and to share findings of intermediate outcomes.

Workforce Development in Developmental Disabilities

The purpose of this project[3] was to develop an online course that could be used for workforce development in a human service setting. The online course used a technology-supported online distance education program that subsequently was compared to a traditional cohort clustered classroom approach. The participants were human service workers with varying degrees of expertise, educational backgrounds, geographical settings, and primary languages, who served persons with disabilities. The content of the course prepared service coordinators to obtain personal service programs for individuals with developmental disabilities. The iterative process described in Figure 1 was used to provide formative feedback via technology reviews and usability procedures as well to pilot the online version of the course.

Elementary and Secondary Education

Several projects[4] supported the development of technology-supported curriculum for K-12 educational environments. These efforts ranged from professional development offered by and about technology, for pre-service and in-service teachers, and to modules and applets that supported direct and remediation instruction in the classroom or in tutorial settings. Learners varied by age, content and technology expertise, English language ability, and work setting. Teachers were involved directly in the development of curriculum units, and the focus was on supporting standards-based instruction and inquiry-based learning. Both external and internal reviews of curriculum were conducted, as were technology reviews and observations of real-time use. Content reviews in this setting focused on support for national and state standards as well as appropriate use of instructional techniques that would meet the needs of students with varying ability levels and districts with varying levels of technology.

Workforce Development in Addictions Education

This project developed two primary methods of instruction[5] to meet different user needs. The first was a traditional face-to-face, instructor-centered, on-site approach (the traditional classroom/workshop classroom environment). The second was a model of distance education based on technology-supported instruction that supported local, regional, national, and international users. Subcultures included variations in

English as the primary language, academic discipline and/or area of practice, education level, familiarity with technology and content, and geographical setting.

The Evaluation Model in Action

The model was used to develop and validate approximately 100 modules in multiple settings across multiple content domains for multiple types of users. Extensive notes on the process were developed for each stage and module. A meta-analysis of the process across these uses resulted in the identification of four phases involving assessment and inclusion of user characteristics. A summary of the reduced model as it pertains to user variables is presented in Figure 2.

The first phase identified by the meta-analysis was related to content review. Designing modules for learning systems typically requires use of experts in the field who are extremely knowledgeable about the content being created. Thus, the first phase recognizes cultural and contextual variables consistent with the biases of designer and developer; it consists of an external validation of the module content. In the user-centered model, the design team utilizes experts from two cultures representing theory and practice. Content area faculty and field-based professionals review the instructional content and the proposed use of the content in various applied settings. It is during this stage of review that cultural variations in definitions, formulae, and acceptable usage are noted. Specific domains covered by reviews should include the following:

- Importance and relevance of the module's instructional objectives to diverse users

Figure 2. The validation cycle and user-centered characteristics

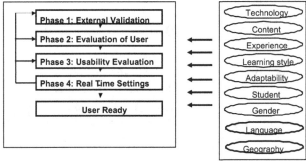

- The accuracy of the content (both applied and theoretical)
- The use of relevant real-life applications
- The level of difficulty and the appropriateness of material for learners with different goals
- The integration of concepts with practice and the relevance of the module in facilitating inquiry-based learning
- Appropriate presentation of facts and aids in a manner that facilitates user involvement in learning.

Results of these reviews should be provided to the module developers, project staff, and the technical manager on a periodic basis through written and in-person team reviews.

The second reoccurring phase found in the meta-analysis consisted of the need for an instructional technology review by experts in information systems design. The critical factor in these reviews is whether or not the technical design of the learning system acknowledges learner variables. Initial reviewers who are part of the design team should work with project staff on a regular basis, providing formative feedback on user-centered needs. In addition, external reviewers who are specialists in instructional design and technology should review and rate each system and/or module, assessing the relevance and flexibility of the presentation mode for different types of learners and the flow of instruction and learning. Specific questions that can be asked include the following:

- Does the IT module provide stimuli for obtaining, sustaining, and directing learners' attention to the important information?
- Can the IT module provide a concrete referent to explain an abstract concept, and does it guide users' learning to relate the elements and the relationships between concepts?
- Are images relevant to a variety of learner cultures such that users can extract the key concepts from verbal and pictorial content so that information will be rehearsed and remembered easily?
- Is the information presented in a way that acknowledges and is meaningful with consideration of learners' familiarity in specific areas?
- Does it allow for different variations in different levels of entrance and exit based on prior experience and level of organized culture?

Additional user-indicators checked by the reviewers should include presence or absence of culture and gender bias, usability by those with varied learning styles, and appropriateness of supporting materials. For example:

- Does the IT module stimulate emotional arousal for all learners?
- Does it stimulate all learners' interests so they gain enjoyment and generate positive attitudes toward the learning materials?
- For varying levels of ability, what types of information need to be conveyed, and how much information is to be given? Is it provided in a way that addresses different learners' needs?
- How much time is provided for conveying the information? Does the system allow for variation?

In general, each module should receive at least two sets of external reviews. Earlier reviews should focus on the process of using technology to deliver instruction, and later reviews should emphasize the instructional relevance of material to different types of users.

The final two stages identified by the meta-analysis address the adaptability of the learning system to diverse audiences in diverse settings. Usability evaluation (phase three) specifically assesses the IT modules for relevance across different learning styles, prior experience with content, and previous experience with the specific mode of technology-supported system that is being developed. At least four types of learners should be involved in this process: those who are novice or experts in the content area and those who are novice or expert in the technology mode. In the previous examples, individual learners utilized the module in lablike settings in which specialists in evaluation and learning theory directly observed the use. As learners interacted with the modules, changes in their knowledge structure were noted, and the modules were adjusted accordingly. In addition to being observed in the use of the module, participants also were encouraged to think aloud as they went through the module; that is, to verbalize freely what they were thinking throughout their experience. The goal was to get inside the learner's head and elicit what they were doing and why; this progression through the module along with the think-aloud comments were videotaped and used to determine sequences and patterns of module use by different types of learners and the ability of the module to be effective for varied types of users. Where possible, variations in learning style also should be recorded. In our research, at the initiation of the module use, participants were given a learning style inventory that allowed for assessment of global and sequential patterns of thinking; results indicated that this characteristic impacted the ability to use the system. Learning styles to consider when developing a usability design are summarized in Figure 3.

The fourth phase that consistently yielded relevant information about users' characteristics involved direct observation of module use in real-time settings. This phase provided project staff with formative feedback on the utility of the modules in varied learning settings and work contexts. Procedures included observation of the use of the module, surveys of users, interviews with randomly selected learners,

Figure 3. Setting up a usability design

| | | Experience with the Specific Mode of Technology ||
		Expert	Novice
Experience with the Content	Expert	*Experts in the content area and in the use of the technology that is being developed*	*Experts in the content area and novices in the use of the technology that is being developed*
	Novice	*Novice in the content area and experts in the use of the technology that is being developed*	*Novices in the content area and in the use of the technology that is being developed*

and follow-up interviews with instructional staff and developers. Variables assessed during this phase should include divergent users' perceptions of the IT module content, the format and setting of its use, the perceived relevance of the information, and the effects of module integration on learning. Divergent learners also should be asked their perceptions of the impact of the process on their overall learning and areas in which they perceive that module refinements should be considered. Observation and interview variables should be used to confirm these points as well as to document perceptions of the ability of the modules to meet the needs of both traditional and nontraditional instructional settings and students. This phase should include nondeveloper settings involving as many different user characteristics and environments as possible.

Recommendations for Developers

The projects already summarized resulted in a demonstrated effective means of designing, developing, and assessing technology-based learning systems that take into account variations in users' characteristics. A need to optimize the capabilities of both emerging technologies and learners is apparent. The use of multi-disciplinary-based teams coordinated by project staff and assisted by technical assistants proved to be an effective method of ensuring learning systems that could and would be used by faculty and students in discipline-based and cross-discipline-based learning settings. As this model was developed and refined, it was noted that greater emphasis was put on the co-creation of learner-centered real-life knowledge generation. Additionally, all team members reported a greater understanding of not only the importance of correct content and the application of the content but also on methods that would facilitate inquiry-based, collaborative learning reflective of higher-level career patterns that would cross discipline and regional and national boundaries. Within this

Table 2. Key components when developing user-centered modules

Need for content and instructional design/technology reviews intra- and interdisciplinary based.
Need for evaluation of 'real' time use with cross-sectional and longitudinal observations in natural settings.
Need to address user and instructional variables in evaluation with usability studies, controlled observations, videotaping, and think aloud techniques.
Need for continuous cyclical formative evaluations on all parts of module design and development.

multi-stage cyclical model for developing information systems, several components were noted as key to the development of sound instructional modules. A summary of these key components is found in Table 2 and discussed next.

- **Need for content and instructional design/technology reviews:** The use of formal content and instructional design/technology reviews was seen as fundamental to the process of developing multi-setting, user-centered curriculum. The information provided allowed developers and project staff to design material that would interface with multiple classrooms, multiple modes of instruction, and multiple types of users with confidence in the validity of content and its applications. These reviews also provided evidence that the material was not individual-instructor-based but was, in fact, intra- and interdisciplinary-based as well as relevant at multiple cultural layers within organizations and across geographic boundaries. In addition, the standardized form of these reviews allowed for comparison and design improvement across multiple modules in different settings at the same time.

- **Need for evaluation of real-time use:** Through the consistent use of cross-sectional and longitudinal observations in real settings, the evaluators and project staff were able to document technology, design, and instructional variations that impacted learning. For instance, it was through the observation of use in real classes and the subsequent interviews with learners and instructors that findings pertaining to the positive use of practice assignments that blended and reinforced content and applied problem solving were noted. As a result of these findings during the initial use of modules, subsequent designs automatically included this technique. Additionally, it was through observations and interviews with learners who used the modules in real-time settings that disparity between users on module work, graded outcomes, and applied usage were noted.

- **Need to address user and instructional variables in evaluation:** Usability studies, controlled observations, videotaping, and think-aloud techniques allowed for study and adaptation of the modules in ways that would support users with varied backgrounds in terms of learning styles, expertise with content, expertise with technology, and expertise with language. The direct observation of individual learners in controlled settings allowed the evaluators to document the need for designs that support both global and sequential learners. Additional findings indicated that female users perceived the motivation of the modules and their subsequent impact differently than did male users and that English language learners had difficulties with simultaneous video and audio presentations.

- **Need for continuous cyclical formative evaluation:** Early formative evaluation not only allowed for correction of specific modules but also altered the general design of the module and subsequent development for all modules (Newman, 1998). As this process continued, later stages of curriculum development were able to focus more on adaptation of the product to specific users and less on the process of adaptation. It also was noted that a teaming process was supported by the cyclical, formative nature with project staff that were technically oriented becoming more adept at content and instructional practices, while team members whose backgrounds were content in nature became more adept at the development and use of varied technology modes that supported different learners.

Discussion

Overall, the process of integrating technology-supported material proved to be successful when it drew learners' attentions, motivated interaction, and helped users to accomplish learning goals without confusion and fatigue. Learners and instructors reported more motivation for learning, more desire for real-life problem solving, and subsequently more ability at real-life decision making in broader contextual settings. Instructors noted that learners had a broader view of the need for cross-disciplinary transfer of knowledge and were more adept at the process. When the material was applied directly to application in different settings via virtual reality, learners acquired a greater relevance of the content and its use in practical settings, resulting in longer retention. User-centered development also allowed instructors to assist learners in transfer of knowledge in a way that met individual variations of students and instructional settings. Specific findings pertaining to the use of technology-supported learning include the following:

- **Impact of learning style:** Evidence indicated that the design of technology might either positively assist or negatively impact an individual's method of cognitive construction (Richardson & Newman, 2000). Users who are global or constructivist in nature are frustrated easily by linear technology-based instructional design and frequently will cease attempting to develop the cognitive structures necessary for task completion. In these projects, it was shown that IT systems could be designed to meet the needs of both global and sequential learners.

- **Impact of level of experience with technology and content:** Users' backgrounds, including familiarity with content and with the use of the specific mode of technology, interacted to impact perceptions of the relevance and usability of the IT system. Users who were novices in content and technology tended to prefer a linear method of learning and navigating through the system. Users who were more experienced in both the content and use of technology-supported learning had greater need for searching and seeking. In cases in which the module did not allow for individual pattern development, user frustration was high and frequently led to early termination without successful completion. This need for pattern searching indicates that either the type of learner must be more clearly identified by the developer or alternate methods of pattern searching must be included in the design.

- **Importance of user gender when studying cognitive growth:** These studies reinforced the need to continue the inclusion of gender differences when studying the use of IT systems (Richardson, Newman, Fama, & Bliss, 1999). Females initially reported a lower level of confidence in their abilities to learn material than did male users, but after completion, females reported higher levels of ability in problem solving, relevance of material, and motivation for learning than did male users. Female users also responded more positively to collaborative/cooperative IT learning methods than did male users and reported interaction with peers and/or team members to be more facilitative toward project completion. These findings were consistent across ethnic backgrounds.

- **Adaptability to varied   :** Findings also provided implications of the use of technology-supported curriculum in varied instructional settings. Although the initial assumption for development of many of these IT systems was that the instructor would be present to assist in learning, it was noted that the systems were adapted to other instructional environments. This included self-sustained review sessions, stand-alone homework assignments, instructor-guided lab settings, and professional learning environments. What is not documented, however, is the impact of the use of technology-supported modules on long-term problem solving.

- **Comparison to classroom-only instruction:** These studies add to the growing body of knowledge surrounding the utility and perceived quality of Web-based distance education programs by examining group differences among online and didactic participants' cognitive outcomes and appropriate covariates. Previous studies examining the utility of distance education programs suggested that participants in online and traditional didactic instruction formats do not differ significantly in their perceptions of course quality or in learning outcomes. In our research, however, cultural and contextual variables such as learning style and preknowledge were found to be potential covariates of growth in knowledge when IT systems were used to support learning.

- **Familiarity with technology:** The findings with regard to familiarity with technology are consistent with studies showing that online participants' self-confidence is directly related to outcomes. This supports the hypothesis that online learners rapidly are forming their own cultures, one that crosses geographic, ethnic, and organizational boundaries. Perceptions of support and self-confidence appear to be related to the learning process and perceived and actual outcomes.

This model and its supporting studies have important practical implications for those developing and managing global IT systems. IT managers, developers, and designers need to keep in mind user variables related to participants' knowledge, skills, and perceived effectiveness as well as the demographic variables that influence learning. The interaction of these variables form cultures of learning that may circle the globe or may be unique to an individual. Experiences, materials, and interactions delivered via IT systems create dynamic instructional contexts that serve multiple users and settings. The use of content and instructional technology reviews results in development of credible systems. Usability and real-time testing result in the development of transferable systems. The use of this multi-step iterative process will result in a learning system that meets the needs of users around the globe.

References

Bruning, R. H., Schraw, G. J., & Ronning, R. R. (1995). *Cognitive psychology and instruction* (2nd ed.). Upper Saddle River, NJ: Merrill.

Butler, D. L., & Winne, P. H. (1995). Feedback and self-regulated learning: A theoretical synthesis. *Review of Educational Research, 65*, 245-281.

Duffy, T. M., & Cunningham, D. J. (1996). Constructivism: Implications for the design and delivery of instruction. In D. H. Jonassen (Ed.), *Handbook of research for educational communications and technology* (pp. 170-198). New York: Simon & Schuster.

Gallupe, R. B., & Tan, F.B. (1999). A research manifesto for global information management. *Journal of Global Information Management, 7*, 5-18.

Gonyea, N. E. (2005). *Primary learner characteristics of recipients of professional development: A study of participants in the developmental disabilities field.* Unpublished doctoral dissertation, University of Albany, State University of New York, New York.

Greeno, J. G., Collins, A. M., & Resnick, L. B. (1996). Cognition and learning. In D. C. Berliner & R. C. Calfee (Eds.), *Handbook of educational psychology* (pp. 15-46). New York: MacMillan.

Hannafin, M. J. (1992). Emerging technologies, ISD, and learning environments: Critical perspectives. *Educational Technology Research and Development, 40*(1), 49-63.

Hill, W. F. (1997). *Learning: A survey of psychological interpretations* (6th ed.). New York: Longman.

Honebein, P. C. (1996). Seven goals for the design of constructivist learning environments. In B. G. Wilson (Ed.), *Constructivist learning environments: Case studies in instructional design* (pp. 11-24). Englewood Cliffs, NJ: Educational Technology Publications.

Jonassen, D. H. (1991). Evaluating constructivist learning. *Educational Technology, 31*(9), 28-33.

Keller, J. M. (1987). Strategies for stimulating the motivation to learn. *Performance and Instruction, 26*(8), 1-7.

Lajoie, S. P. (2000). *Computers as cognitive tools: No more walls, Volume II.* Mahwah, NJ: Lawrence Erlbaum Associates.

Lepper, M. R., Gurtner, J. L. (1989). Children and computers: Approaching the twenty-first century. *American Psychologist, 44*(2), 170-178.

Mayer, R. E. (1996). Learners as information processors: Legacies and limitations of educational psychology's second metaphor. *Educational Psychologist, 31*, 151-161.

Myers, M. D., & Tan, F. B. (2002). Beyond models of national culture in information systems research. *Journal of Global Information Management, 10*, 24-32.

Nemeth, C. P. (2004). *Human factors methods for design: Making systems human-centered.* Boca Raton, FL: CRC Press.

Newman, D. L. (1998). *Rensselaer Polytechnic Institute Project Links: Mathematics and its applications across the curriculum*, University at Albany/SUNY, Evaluation Consortium.

Perkins, D. N. (1991). Technology meets constructivism: Do they make a marriage? *Educational Technology, 31*(5), 18-23.

Preece, J., Rogers, Y., & Sharp, H. (2002). *Interaction design: Beyond human-computer interaction.* New York: John Wiley & Sons.

Richardson, J. C., & Newman, D. L. (2000). *Documenting cognitive patterns in computer supported learning: The validation of patterns of usability.* Paper presented at the Annual Meeting of the American Educational Research Association, New Orleans, LA, USA.

Richardson, J. C., Newman, D. L., Fama, L. D., & Bliss, L. A. (1999). *Gender difference on the impact of computerized instructional modules in mathematics, science, and engineering curriculum: Student perceptions of benefits and outcomes.* Paper presented at the Eastern Educational Research Association Annual Conference, Hilton Head, SC.

Riding, R., & Rayner, S. (1998). *Cognitive styles and learning strategies: Understanding style differences in learning and behavior.* London: David Fulton Publishers.

Rieber, L. P. (1996). Seriously considering play: Designing interactive learning environments based on the blending of microworlds, simulations, and games. *Educational Technology Research & Development, 44*, 43-48.

Roberts, M. J., & Erdos, G. (1993). Strategy selection and metacognition. *Educational Psychology, 13*, 259-266.

Rowland, G. (2004). Shall we dance? Designing for organizational learning and performance. *Educational Technology Research & Development, 52*(1), 33-48.

Salomon, G., Perkins, D., & Globerson, T. (1991). Partners in cognition: Extending human intelligence with intelligent technologies. *Educational Researcher, 20*(4), 2-9.

Schunk, D. H. (2000). *Learning theories: An educational perspective* (3rd ed.). Upper Saddle River, NJ: Prentice-Hall.

Skinner, B. F. (1954). The science of learning and the art of teaching. *Harvard Educational Review, 24*, 86-97.

Stone, D., Jarrett, C., Woodroffe, M., & Minocha, S. (2005). *User interface design and evaluation.* San Francisco: Morgan Kaufmann.

Vygotsky, L. (1978). *Mind in society: The development of higher psychological processes.* Cambridge, MA: Harvard University Press.

Endnotes

[1] For more information on the project, see http://links.math.rpi.edu/

[2] For more information on this project, see http://www.academy.rpi.edu

3   For more information on this, contact NY State Developmental Disabilities Planning Council at http://www.ddpc.state.ny.us or New York State Office of Mental Retardation and Developmental Disabilities at http://www.omr.state.ny.us to request information on New York State DEAL

4   For more information, see http://projectview.org

5   For more information, see http://www.attc-ne.org

Chapter XVI
A Model for Selecting Techniques in Distributed Requirement Elicitation Processes

Gabriela N. Aranda, GIISCo Research Group, Universidad Nacional del Comahue, Argentina

Aurora Vizcaíno, ALARCOS Research Group, Universidad de Castilla-La Mancha, Spain

Alejandra Cechich, GIISCo Research Group, Universidad Nacional del Comahue, Argentina

Mario Piattini, ALARCOS Research Group, Universidad de Castilla-La Mancha, Spain

Abstract

This chapter introduces a model based on techniques from cognitive psychology as a means to improve the requirement elicitation in global software development projects. Since distance negatively affects communication and control, distributed development processes that are crucially based on communication, such as requirements elicitation, have to be specially rethought in order to minimize critical situations. This chapter proposes reducing problems in communication by selecting a suite of appropriate elicitation techniques and groupware tools according to stakeholders' cognitive styles. It also shows how information about stakeholders' personalities can be used to make them feel comfortable and to improve their performances when working in a group.

Introduction

The development of software in scenarios in which stakeholders are in many geographically distanced sites increases day by day. One of the main reasons for such a growth is the possibility of counting on human resources from all around the world while travel costs are reduced to a minimum or do not even exist (Lloyd, Rosson, & Arthur, 2002).

As a consequence of working in a geographically dispersed manner, stakeholders must communicate with each other by means of specially designed technology called groupware. In doing so, members of a distributed requirement elicitation process have to deal not only with the normal challenges of a requirement elicitation process (Davis, 1993; Loucopoulos & Karakostas, 1995) but also with those derived from the lack of face-to-face interaction, time difference between sites, and the cultural diversity of stakeholders, which are typical of distributed environments (Damian & Zowghi, 2002).

There are several research areas that have attempted to find solutions to communication problems in workgroups. One of them is computer-supported cooperative work (CSCW), which focuses on providing technologies to enable communication and also analyzes human behavior when working in a group. Another is cognitive informatics, an interdisciplinary area that applies concepts from psychology and other cognitive sciences to improve processes in engineering disciplines, such as informatics, computing, and software engineering (Chiew & Wang, 2003; Wang, 2002).

Since our main goal is to enhance interpersonal communication in geographically distributed teams, concepts from both areas come together. On the one hand, people who are working at various geographic sites communicate with each other using groupware, which is part of the studies in CSCW. Examples of groupware used during multi-site developments are e-mails, forums, shared whiteboards, chat, instant messaging, and videoconferencing, among others (Damian & Zowghi, 2002; Lloyd et al., 2002). On the other hand, communication among people involves aspects of human processing mechanisms that are analyzed by the cognitive sciences. In our proposal, we are particularly interested in some techniques from the field of psychology, called learning style models, which may be useful to select groupware tools and elicitation techniques according to the stakeholders' cognitive styles.

Most of the related works that use learning styles in computer science concern educational purposes, such us their influence on learning (Bostrom, Olfman, & Sein, 1988; Thomas, Ratcliffe, Woodbury, & Jarman, 2002; Wu, Dale, & Bethel, 1998) and how to define frameworks for designing multimedia courses (Blank, Roy, Sahasrabudhe, Pottenger, & Kessler, 2003; Moallem, 2002). On the contrary, few related works use psychological techniques to solve problems in Software Engineering. One work in this field is the use of cognitive styles as a mechanism for software inspection team construction (Miller & Yin, 2004), which describes an experiment to prove that heterogeneous software inspection teams give a better performance than homogeneous ones in which the heterogeneity concept is analyzed according to the participants' cognitive styles. Although the concept of cognitive styles to classify people was used, our approach is not the same. As we have previously explained, we do not try to say which people seem to be more suitable to work together. On the contrary, we aim to give the best requirement elicitation techniques and groupware tools for an already chosen group of people.

Bearing this in mind, in the following section, we present some basic concepts about cognitive informatics and learning style models, and we introduce a general model based on fuzzy logic to select groupware tools and requirement elicitation techniques. The last sections present some motivating examples and address conclusions.

Why Cognitive Informatics?

Cognitive informatics relates cognitive sciences and informatics in a bidirectional way (Wang, 2002):

1. By using computing techniques to investigate cognitive science problems such as memory, learning, and thinking.
2. By using cognitive theories to investigate informatics, computing, and software engineering problems.

Our work is related to the second point of view, using concepts from cognitive psychology, which concern the way in which people pay attention to and gain information and how these information processing mechanisms affect human behavior (Chiew & Wang, 2003) to improve the requirement elicitation process.

Part of cognitive psychology theories is cognitive styles, which are based on Jung's theory of psychological types, published in 1921. Jung's theory classifies people's preferences about perception, judgment, and processing of information (Miller & Yin, 2004), which is why it has been used to analyze and understand differences in human behavior. As an extension of this, different instruments have been designed to measure human characteristics and to explain their differences. For instance, the learning style models (LSMs) classify people according to a set of behavioral characteristics pertaining to the ways in which they receive and process information, and their goal is to improve the way people learn a given task.

Even when LSMs have been discussed in the context of analyzing relationships between instructors and students, we think that it is possible to take advantage of this kind of model and to adapt it to virtual teams that deal with distributed elicitation processes. Since requirement elicitation is about learning the needs of the users (Hickey & Davis, 2003), and since users and clients also learn from analysts and developers (e.g., they learn how to use a software prototype, a new vocabulary, etc.), we propose an analogy between stakeholders and roles in LSMs, as shown in Figure 1.

The learning style model we have chosen as a basis for our methodology is the one proposed by Felder-Silverman (F-S) (Felder & Silverman, 1988), which classifies people into four categories, each of them further decomposed into two subcategories as follows: Sensing/Intuitive; Visual/Verbal; Active/Reflective; Sequential/Global.

The characteristics of each subcategory follow:

- *Sensing people* prefer learning facts and solving problems by well-established methods, while *intuitive people* prefer discovering possibilities and relationships and dislike repetition.
- *Visual people* remember best what they see (such as pictures, diagrams, flow charts, timelines, films, and demonstrations). On the contrary, *verbal people* get more out of words and written and spoken explanations.
- *Active people* tend to retain and understand information by doing something active with it (discussing or applying it or explaining it to others). In contrast, *reflective people* prefer to first think about information quietly.
- *Sequential people* tend to gain understanding in linear steps with each step following logically from the previous one, whereas *global people* tend to work in large jumps, absorbing material almost randomly without seeing connections and then suddenly "getting it."

Figure 1. Analogy between stakeholders and roles in learning models

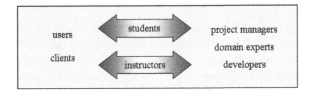

People classification is done by a multiple-choice test that returns a rank for each subcategory. Depending on the circumstances, people may fit into one category or another, being, for instance, sometimes active and sometimes reflective; so preference for each category is measured as strong, moderate, or mild. Only when there is a strong preference can a person be classified as a member of a certain group.

A Model to Support Personal Preferences in Distributed Requirement Elicitation

In order to support personal preferences toward groupware tools, in Aranda, Cechich, Vizcaino, and Castro-Schez (2004), a model based on fuzzy logic and fuzzy sets to obtain rules from a set of representative examples in the manner of patterns of behavior has been proposed. Those patterns tell us about the preferences of stakeholders in their daily use of groupware tools according to their classification in the F-S model. In a similar way, we propose finding a suitable set of requirement elicitation techniques according to the preferences for each category of the F-S model.

The input variables for both models are the four categories that correspond to the F-S model:

I = { Active-Reflective, Sensing-Intuitive, Visual-Verbal, Sequential-Global}

We have defined a domain (DDV) for each input variable by using the adverbs (and their corresponding abbreviations): Very (V), Moderately (M), and Slightly (S); these correspond respectively to strong, moderate, and mild in the F-S model, but we have changed their names to avoid confusion with respect to the use of the first letter.

We also have expressed the results of the F-S test with a negative sign for the categories that appear first on the presentation of the characteristics (sensing, visual, active, sequential) and with a positive sign for the latter ones (intuitive, verbal, reflective, global). By doing so, the domain for each category has been defined as shown in Figure 2(a), while the definition domain function for the Active-Reflective category (which has been defined similarly for the other three categories) as shown in Figure 2(b).

With respect to the output variable of our fuzzy model, in the first model, it represents the groupware tool a person chooses as his or her favorite:

O_1 = {Groupware Tool}.

Figure 2. Definition domain for input variables of our fuzzy model

	Very	Moderately	Slightly	Slightly	Moderately	Very	
Active	VAc	MAc	SAc	SRe	MRe	VRe	Reflective
Sensitive	VSe	MSe	SSe	SIn	MIn	VIn	Intuitive
Visual	VVi	MVi	SVi	SVe	MVe	VVe	Verbal
Sequential	VSq	MSq	SSq	SGl	MGl	VGl	Global

(a) Definition domain for F-S categories

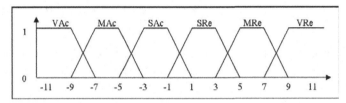

(b) Domain function for active-reflective category

and its domain would be all those groupware tools that can be used during a requirement elicitation process; for instance:

DDV Groupware Tool = {email, chat, videoconference, …}

On the contrary, in the second model, the output variable is the requirement elicitation technique a person would prefer:

O_2 = {Elicitation Technique}

and the definition domain would be a set of elicitation techniques that can be applied during a distributed requirement elicitation process:

DDV Elicitation_Technique = {interview, prototype, brainstorming, …}

Once both models are defined, we need to obtain examples to which the machine-learning algorithm can be applied. To do so, our first step is to ask people to fill in the test that will provide us with their learning styles (Phase 1 in Figure 3).

Later, we need to plan strategies to discover stakeholders' preferences when using groupware tools and requirement elicitation techniques. In the first case, we can ask people in a direct way which groupware tool they feel more comfortable using. This is simple, because people use e-mail, instant messaging, and chat quite normally in their lives, and even if they have never used videoconferencing or shared whiteboards, they easily can imagine how they would feel using them. For that, a simple question directed toward a group of stakeholders is enough for us to ascertain the output variable of our first model. On the contrary, knowing preferences about requirement elicitation techniques is not so easy. Analysts usually know a couple of techniques, and users and clients usually do not know any. So, in order to get a ranking of preferences, we need to develop experiments in which stakeholders receive some training in a set of requirement elicitation techniques, and then we can ask stakeholders about their experiences with each one.

Once the mechanisms for obtaining examples are defined, we can obtain a set of examples $\theta = \{e_1, e_2, ..., e_m\}$ in which each example would have the form $e_i = \{(x_{i1}, x_{i2}, x_{i3}, x_{i4}), y_i\}$. For instance, {(MIn, MVi, VAc, SSq), videoconference} would be a possible instantiation for the first model and {(SIn, VVi, VAc, VSq), prototype} for the second one. The process of obtaining examples for both sets is represented in Phases 2 and 3 of Figure 3.

Once we have an appropriate number of instances for both sets of examples, we can apply a machine-learning algorithm in order to generalize common features among examples. For instance, we have chosen the algorithm proposed in Castro, Castro-Schez, and Zurita (1999) that finds a finite set of fuzzy rules able to reproduce the input-output system's behavior as follows:

1. Convert each example in one rule.
2. Remove those rules that are the same from the initial set.
3. Analyze every initial rule to (where possible) extend it and generate a definitive rule.

Using this machine-learning algorithm over a set of examples that represent the preferences of many stakeholders, we expect to obtain rules such as *Ro: if X1 is VVi then Y1 is Instant Messaging* for the first model, or *Ro: if X1 is VAc then Y2 is Prototype* for the second model, which are interpreted respectively as, "If a user has a strong preference for the Visual subcategory, the groupware tool that this person would prefer is Instant Messaging," or "If a user has a strong preference for the Active subcategory, the elicitation technique that this person would prefer is Prototype." This is represented in Phase 4 of Figure 3.

Once we have obtained both sets of rules and we know the personal preferences of each person who works in a virtual team (Phase 5 in Figure 3), it is possible to choose the best elicitation techniques and groupware tools for that group of people

Figure 3. Phases in defining and analyzing personal preferences to choose appropriate technology in virtual teams

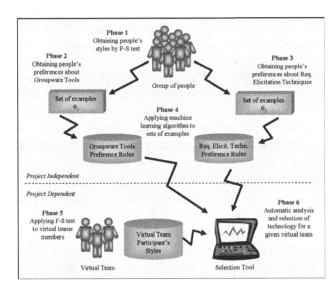

by analyzing the results with an appropriate tool that automates the process and by choosing those that have more adherents (Phase 6 in Figure 3).

By analyzing our model according to the framework of organizational learning proposed in Crossan, Lane, and White (1999), we have considered only two of the three levels of the framework—individual and group—while the organization level is not taken into account. This is because our model affects processes related to the learning of individuals that work in a group. As a consequence, we consider three processes: intuiting, interpreting, and integrating. We do not consider institutionalizing, which is typical of the organization level. However, even when institutionalization (the establishment of formal rules and procedures) is out of the scope of our model, its implementation depends on knowledge management activities that can work effectively when combined with our approach.

Similarly, our cognitive-based approach for technology selection could be used in many elicitation environments, including extreme programming (Beck, 2000), so that its combination may optimize communicational as well as management aspects.

Motivating Example

Once the preference rules are known, it is necessary to analyze different strategies for the last selection process. That means combining the information we know about the stakeholders to find the best suite of groupware tools and elicitation techniques for them. In Aranda, Vizcaino, Cechich, and Piattini (2005a) and Aranda, Vizcaino, Cechich, and Piattini (2005b), two different approaches are proposed: The first one is based on picking the technique that has more adherents in the group, and the second one takes into account those participants whose preferences are stronger.

To illustrate our approaches, let us consider the results of the test applied to three stakeholders:

S1 = (MAc, SSe, MVe, SGl); S2 = (SRe, VSe, VVi, VSq); S3 = (SRe, SSe, SVe, SGl)

Let us suppose that by the application of the rules that we have previously determined according to personal preferences we obtain a set of appropriate technologies for each stakeholder: E1, E2, and E3.

Assuming that S1 is the analyst, S2 and S3 are users, and {t} is the set of requirement elicitation techniques appropriate for the project and the stage in the life cycle, three scenarios are possible. These are shown in Figures 4 (a) and (b).

Case 1. Taking the Personal Preferences of all the Stakeholders into Account and Choosing the One that is Most Popular

In this case, techniques would be chosen from those that are repeated in most of the sets. Since S_1 and S_3 have some moderated and slight preferences for the verbal category, this possibly would cause them to prefer using a technique based mainly on words that would make stakeholder S_2 (who has a strong preference for the visual category) feel uncomfortable and, therefore, not committed enough to the collaborative task.

Case 2. Taking the Personal Preferences of all the Stakeholders into Account and Choosing One According to the Stakeholder Who Has Stronger Preferences

Since stakeholder S_2 has the strongest preferences, techniques would be chosen from the set E_2. Considering that the preferences of stakeholders S_1 and S_3 are moderated or slight, the selection of techniques would not negatively affect their performance but could significantly improve S_2's.

Figure 4. Strategies for technology selection based on stakeholders' learning styles

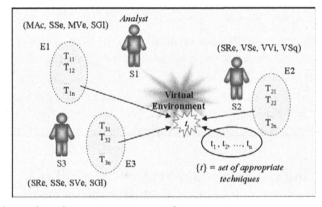

Figure 4(a). Selection driven by occurrences in personal sets

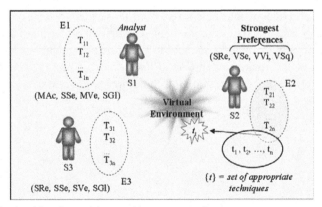

Figure 4(b). Selection driven by participants with strongest preferences

Conclusion

Organizations that develop software currently have the possibility of saving costs by adopting a distributed structure in which team members communicate through groupware tools. The selection of appropriate technology and requirement elicitation techniques in such environments is a subject of research, because when stakeholders feel comfortable with the technology and methodologies they use, information gathered during elicitation is expected to be more accurate.

Taking this into account, we have developed a methodology to select requirement elicitation techniques and groupware tools according to the learning styles of the members of a given virtual team. By developing our approach, we realized that although psychological techniques often are not applied to the improvement of software engineering processes, they could give people in charge of software development projects the chance to count on a widely proved theoretical basis, which also has the advantage of being understood easily by different types of stakeholders, whether or not they are professionals in informatics.

Acknowledgments

This work is partially supported by the MÁS project (TIC2003-02737-C02-02), Ministerio de Ciencia y Tecnología; the ENIGMAS project (PBI-05-058), Junta de Comunidades de Castilla-La Mancha, Consejería de Educación y Ciencia, both from Spain; the CompetiSoft project (CyTED 3789), and also by the 04/E059 project, Universidad Nacional del Comahue, from Argentina.

References

Aranda, G., Cechich, A., Vizcaíno, A., & Castro-Schez, J. J. (2004). Using fuzzy sets to analyse personal preferences on groupware tools. *Proceedings of the X Congreso Argentino de Ciencias de la Computación, CACIC 2004*, San Justo, Argentina (pp. 549-560).

Aranda, G., Vizcaíno, A., Cechich, A., & Piattini, M. (2005a). A cognitive-based approach to improve distributed requirement elicitation processes. *Proceedings of the 4th IEEE International Conference on Cognitive Informatics (ICCI'05)*, Irvine, California (pp. 322-330).

Aranda, G., Vizcaíno, A., Cechich, A., & Piattini, M. (2005b). Towards a cognitive-based approach to distributed requirement elicitation processes. *Proceedings of the WER 2005, VIII Workshop on Requirements Engineering,* Porto, Portugal (pp. 75-86).

Beck, K. (2000). *Extreme programming explained: Embrace change.* Addison-Wesley.

Blank, G. D., Roy, S., Sahasrabudhe, S., Pottenger, W. M., & Kessler, G. D. (2003). Adapting multimedia for diverse student learning styles. *The Journal of Computing in Small Colleges, 18*(3), 45-58.

Bostrom, R., Olfman, L., & Sein, M. K. (1988). The importance of individual differences in end-user training: The case for learning style. *Proceedings of the 1988 ACM SIGCPR Conference,* Maryland (pp. 133-141).

Castro, J. L., Castro-Schez, J. J., & Zurita, J. M. (1999). Learning maximal structure rules in fuzzy logic for knowledge acquisition in expert systems. *Fuzzy Sets and Systems, 101*(3), 331-342.

Chiew, V., & Wang, Y. (2003). From cognitive psychology to cognitive informatics. *Proceedings of the Second IEEE International Conference on Cognitive Informatics, ICCI'03,* London (pp. 114-120).

Crossan, M., Lane, H., & White, R. (1999). An organizational learning framework: From intuition to institution. *Academy of Management Review, 24*(3), 522-537.

Damian, D., & Zowghi, D. (2002). The impact of stakeholders geographical distribution on managing requirements in a multi-site organization. *Proceedings of the IEEE Joint International Conference on Requirements Engineering, RE'02,* Essen, Germany (pp. 319-328).

Davis, A. (1993). *Software requirements: Objects, functions and states.* NJ: Prentice Hall.

Felder, R., & Silverman, L. (1988). Learning and teaching styles in engineering education. *Engineering Education, 78*(7), 674-681.

Hickey, A. M., & Davis, A. (2003). Elicitation technique selection: How do experts do it? *Proceedings of the International Joint Conference on Requirements Engineering (RE03),* Los Alamitos, California (pp. 169-178).

Lloyd, W., Rosson, M. B., & Arthur, J. (2002). Effectiveness of elicitation techniques in distributed requirements engineering. *Proceedings of the 10th Anniversary IEEE Joint International Conference on Requirements Engineering, RE'02,* Essen, Germany (pp. 311-318).

Loucopoulos, P., & Karakostas, V. (1995). *System requirements engineering.* New York.

Miller, J., & Yin, Z. (2004). A cognitive-based mechanism for constructing software inspection teams. *IEEE Transactions on Software Engineering, 30*(11), 811-825.

Moallem, M. (2002). The implications of research literature on learning styles for the design and development of a Web-based course. *Proceedings of the International Conference on Computers in Education, ICCE 2002,* Auckland, New Zealand (pp. 71-74).

Thomas, L., Ratcliffe, M., Woodbury, J., & Jarman, E. (2002). Learning styles and performance in the introductory programming sequence. *Proceedings of the 33rd SIGCSE Technical Symposium on Computer Science Education,* Cincinnati, Kentucky (pp. 33-37).

Wang, Y. (2002). On cognitive informatics. *Proceedings of the First IEEE International Conference on Cognitive Informatics, ICCI'02,* Calgary, Alberta (34-42).

Wu, C. C., Dale, N. B., & Bethel, L. J. (1998). Conceptual models and cognitive learning styles in teaching recursion. *Proceedings of the Twenty-Ninth SIGCSE Technical Symposium on Computer Science Education,* Atlanta, Georgia (pp. 292-296).

Chapter XVII

Organizational Time Culture and Electronic Media

Cheon-Pyo Lee
Carson-Newman College, USA

Abstract

This chapter explains the electronic media diffusion process within organizations and provides a guideline to implement electronic media within organizations. The concept of Hall's (1976) time dimension culture, monochronic and polychronic, and two dimensions of media speed, production and interaction speed, are used to explain the media diffusion process within organizations. It suggests that the diffusion process and expected benefit of electronic media are significantly different, depending on national culture, organizational culture, and the characteristics of that medium. Therefore, careful examination and understanding of organizational time culture and the characteristics of media should be ahead of making a decision on electronic media adoption and implementation.

Introduction

Today's individuals in organizations frequently have access to new electronic media, such as short messaging service (SMS), mobile e-mail, and instant messaging (IM), in addition to stationary e-mail and more traditional communication media, such as telephone, memos, and face-to-face meetings. One of the noticeable changes in today's organizational communication pattern is that electronic media have supplanted traditional forms of communication media and have become common tools in organizations with their new capabilities, such as synchronous communication and storage and retrieval of communication (Carlson & George, 2004; Carlson & Davis, 1998). With these new capabilities, electronic media not only have improved communication quality within organizations but also have allowed communication that would not be possible using traditional media (Markus, 1994).

With the rapid deployment of new electronic media in the workplace and a wide array of choices, today's organizations and individuals often face a series of decisions about which medium to adopt and use in communicating with internal and external communication partners. Such decisions have shown a profound impact on individual and organizational performance and often have changed the organizational processes and structures (Straub & Karahanna, 1998). Thus, implementing an appropriate communication tool is an important mission for many organizations and has encouraged these organizations to introduce emerging electronic media as expeditiously as possible. However, while organizations generally introduce emerging media to improve effectiveness or productivity, the results are not always straightforward or beneficial at the individual or organizational levels.

Like in many other information systems (IS) studies, national culture articulated by Hofstede (1980) has provided an important foundation in explaining different performance and diffusion levels of electronic media. For example, in his cross-cultural study between the U.S. and Japan, Straub (1994) concluded that due to the characteristics of the written Japanese language and other national culture factors such as uncertainty avoidance, Japanese workers prefer to use fax machines rather than e-mail, whereas this preference was not noted in the U.S. However, even though Japanese cultures and languages have not changed much in the last 10 years, the usage rate of stationary and mobile e-mail in Japan far exceeds that of the United States (Kearney, 2004).

This trend simply implies that national culture heavily based on text-oriented culture cannot explain different electronic media diffusion and performance in different organizations effectively and that some other important factors are missing. Thus, by using the concept of Hall's (1976) time dimension culture—monochronic and polychronic—and two dimensions of media speed—production and interaction speed—this chapter explains the electronic media diffusion process within organizations and provides a guideline to implement electronic media within organizations.

Background

In his cross-cultural study between the U.S. and Japan, Straub (1994) identifies key variables in the media diffusion process. According to his paper, the chain of diffusion process begins with evaluating the characteristics of the medium, both in an overall sense and on a task-by-task basis, which ultimately leads to selection and use of that medium. He suggests that two important media evaluation criteria, social presence (SP) and information richness (IR), were used among employees in both countries. These two important media evaluation criteria, SP and IR, are the products of two dominant theories of media choice, social presence theory (SPT) and media richness theory (MRT) (Daft, Lengel, & Trevino, 1987; Williams & Christie, 1976), and are highly influenced by national culture.

Social presence (SP) is the degree to which people establish warm and personal connections with each other in a communication setting (Short & Christie, 1976), and the theory predicts that individuals will choose a medium based on the degree to which social presence is necessary for the particular communication task (Carlson & Davis, 1998). Information richness (IR) is one of the most widely studied models of media choices in management communication (Dennis & Kinney, 1998; Kahai & Cooper, 2003). IR proposes that media differ in the ability to facilitate changes in understanding among communicators (Daft et al., 1987). Thus, it predicts that richer media are used in highly uncertain and equivocal situations, and leaner media are used in more certain and unequivocal situations. Even though these two theories start from different efforts, the results of the two theories reach the same conclusion, since much of information richness theory is built on the presumption that increased richness is linked to increased social presence (Carlson & Davis, 1998; Zmud, Lind, & Young, 1990). Straub (1994) concludes that Japanese knowledge workers perceive e-mail to be lower in medium SPIR than U.S. workers do.

Like many earlier information systems (IS) studies that focused on work-related IS, Straub (1994) selects productivity benefits as a main performance outcome of media use. Therefore, in his model, media use is expected to lead to productivity benefits or drawbacks. He concluded that Japanese knowledge workers rate the productivity benefits of e-mail lower than U.S. workers do, due to two aspects of Japanese culture (uncertainty avoidance and complex written symbols), which slowed the diffusion of e-mail in Japan.

However, even though Japanese culture and language have not changed much in the last 10 years, the usage rate of stationary and mobile e-mail in Japan and other Asian countries far exceeds those of the United States (Kearney, 2004). Kearny (2004) reports that more than 70% of Japanese have an Internet enabled phone (IEP), and 77% of them actually use mobile e-mail; whereas 25% of North Americans have an IEP, and 27% of them actually use mobile e-mail. This trend simply implies that national culture heavily based on uncertainty avoidance and written symbols cannot explain different electronic media diffusion and performance in different organizations and that some other important factors are missing.

Understanding Electronic Media Diffusion and Performance

Electronic media have two different types of media speed, and the attributes related with these speeds may explain electronic media diffusion and performance in different organizations and countries. The two different types of media speed are production and interaction speed, and they are highly influenced not only by national culture but also by organization and individual culture.

Media Speed

According to Kahai and Cooper (2003), in electronic media, (media) production speed refers to the speed of transferring messages, such as speed of typing, reading, speaking, and listening. Some languages are difficult to compose using a keyboard, so the production speed of electronic media has been considered slower than traditionally richer media such as verbal communication. However, as new technologies are developed, production speed, especially typing speed, becomes as fast as or often faster than verbal communication.

The other type of media speed is (media) interaction speed. Interaction speed refers to the amount of time delay between the time information is sent and the time it is received (Carlson & George, 2004). Like traditional media, such as written letters or memos, some electronic media such as e-mail have perceptible delays ranging from a few seconds to a few days. However, new electronic media, such as IM and SMS, have capabilities similar to synchronous media and remove the disadvantage of interaction speed.

The result of Carlson and George's study (2004), which asked for the subject's evaluation of the medium, confirms the importance of interaction speed and its significant role in the diffusion of electronic media within organizations. In their research, when the subjects are asked to evaluate the perception of information richness and interaction speed of various media using a seven-point Likert scale, the mean of e-mail on the evaluation of media is the lowest among the five different media. Therefore, according to the evaluation of media richness, it may be concluded that e-mail will not be used as an effective communication medium within organizations. However,

Figure 1. Electronic media speed

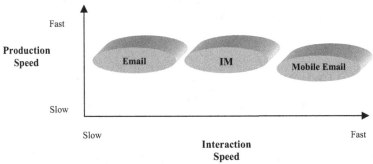

in terms of interaction speed, e-mail is ranked next to face-to-face communication, and it clearly explains the widespread use of e-mail in today's organizations.

Figure 1 shows the classification of recent electronic media by production and interaction speed. As noted, there are not many differences in production speed, since recently it takes almost the same amount of time for many people to compose a message either on stationary computers or mobile devices. However, the interaction speed of the three media is significantly different. Mobile e-mail and SMS allow individuals to send and review messages virtually anywhere and anytime, even while stuck in traffic or in a meeting. Therefore, the interaction speed of those media is as fast as verbal communication. However, even though instant messaging (IM) is a near-synchronous computer-based interactive communication media, its interaction speed is longer than mobile communication since it requires both parties to be in front of their computers. Finally, e-mail has delays ranging from a few seconds to a few days before the receiver checks the e-mail, so the interaction speed of e-mail is the slowest among the three media. This figure clearly shows that interaction speed is very important criteria that distinguish electronic media.

In sum, production and interaction speed of electronic media are very important determinants of electronic media selection. More importantly, even though production speed, which differs depending on each country's language and related culture, has provided important evaluation criteria of electronic media, the interaction speed can explain the different levels of electronic media performance more effectively.

Individual and Organizational Time Culture

The previous argument lead us to conclude that since the interaction speed of electronic media is related highly with the time dimension of culture, the understanding of culture related to the perception of time can be more suitable for explaining electronic media diffusion and performance. According to Hall (1976), individuals differ in the ways they approach time and in how they accomplish their goals. There are two different time dimension cultures: monochronic and polychronic. Monochronic time is characterized as linear, tangible, and divisible into blocks, consistent with the economic approach to time. Monochronic time use emphasizes planning and the establishment of schedules, with significant energy being put into the maintenance of established schedules. In contrast, polychronic time use occurs when two or more activities are carried out within the same clock block; switching among activities can be both desirable and productive.

Bluedorn, Kaufman, and Lane (1992) introduce a very good example of individuals in the two different types of time culture. "Two managers who are both planning to write a report in the morning. Both begin writing, and after thirty minutes, both managers receive a phone call. Manager *A* regards the phone call as an interruption and attempts to reschedule the call for time later in the day. Manager *B* answers the phone, has a complete conversation with the caller, and returns to work on the report after the call. Manager A is relatively monochronic because unplanned, unscheduled events are considered interruptions that should be minimized and not allowed to interfere with scheduled activities. Manager B is relatively polychronic because the unscheduled event was handled as a normal part of life, of equal or greater importance than planned activities" (p. 18).

There are degrees of polychronicity, ranging from people who tend to be very monochronic to those who are extremely polychronic, and an individual's polychronicity can used as a significant tool in understanding his or her personal approach to time management in the workplace (Kaufman-Scarborough & Lindquist, 1999). According to Conte, Rizzuto, and Steiner (1999), polychronicity is associated negatively with a preference for organization and related to organizational outcomes.

Like individuals, an organization also ranges from the one very monochronic to the other extremely polychronic. Bluedorn et al. (1992) provide a method for how an individual can measure his or her polychronic attitude. They also provide how an individual can measure his or her organization polychronic orientation scale. The result of the degree of polychronicity of individuals and organizations is a very important implication for organizations when implementing new electronic media. If the time culture of the organization or the individual within the organization doesn't match with the characteristics of new media, individuals may not produce the expected benefits from an implemented electronic media. More importantly, the implemented electronic media may produce significant negative impacts, such as information overload or digital depression (Farhoomnad & Drury, 2002).

In sum, as shown in Figure 2, electronic media diffusion and performance are influenced significantly by two dimensions of cultures: text and time. Especially, an

Figure 2. Electronic media diffusion and culture

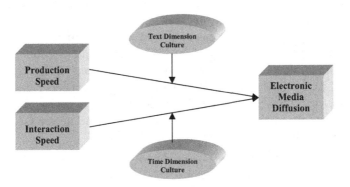

understanding of the monochronic/polychronic continuum can lead to better management of organizations and relationships by implementing electronic media.

Future Trends

In the future, it is expected that more electronic media will be implemented while existing electronic media continues to upgrade. As discussed earlier, the benefits of those media may not be delivered without a clear understating of organizational culture. Therefore, it is very important for organizations and managers to make their decision to implement after a careful examination of individual and organizational time culture in addition to national culture.

Also, it is very important for organizations to measure the outcome or impact of electronic media correctly, given a wide array of choices. It is often difficult to measure the outcome of IS based on a single construct, so various constructs, including productivity, decision quality, and efficiency, have been used widely as the major outcomes of IS (DeLone & McLean, 2003; Walther, 1995). In the case of electronic media use, it is often unclear how performance should be defined (Kahai & Cooper, 2003), but decision quality and decision efficiency may be used as major outcomes of electronic media use in the future (Carlson & George, 2004; Kahai & Cooper, 2003; Walther, 1995). In addition to those performance improvements, due to its somewhat hedonic nature, new electronic media often are selected and used based on the extent to which fun can be derived from using the media (van der Heijden, 2004). Thus, increasing user satisfaction driven by intrinsic motivation also can be a significant outcome of media use.

Conclusion

The diffusion process and expected benefit of electronic media are significantly different, depending on national culture, organizational culture, and the characteristics of that medium. Therefore, careful examination and understanding of organizational time culture and the characteristics of media should be ahead of making a decision on electronic media adoption and implementation. Also, when training or installing electronic communication media in an organizational context, the organization and managers should consider multiple factors related to electronic media, such as the speed of media and expected benefits.

References

Bluedorn, A. C., Kaufman, C. F., & Lane, P. M. (1992). How many things do you like to do at once? An introduction to monochronic and polychronic time. *Academy of Management Executive, 6*(4), 17-26.

Carlson, J. R., & George, J. F. (2004). Media appropriateness in the conduct and discovery of deceptive communication: The relative influence of richness and synchronicity. *Group Decision and Negotiation, 13*(2), 191-210.

Carlson, P. J., & Davis, G. B. (1998). An investigation of media selection among directors and managers: From "self" to "other" orientation. *MIS Quarterly, 22*(3), 335-362.

Conte, J.M., Rizzuto, T.E., & Steiner, D.D. (1999). A construct-oriented analysis of individual-level polychronicity. *Journal of Managerial Psychology, 14*(3/4), 269-287.

Daft, R. L., Lengel, R. H., & Trevino, L. K. (1987). Message equivocality, media selection, and manager performance: Implications for information systems. *MIS Quarterly, 11*(3), 355-366.

DeLone, W. H., & McLean, E. R. (2003). The Delone and Mclean model of information systems success: A ten-year update. *Journal of Management Information Systems, 19*(4), 9-30.

Dennis, A. R., & Kinney, S. T. (1998). Testing media richness theory in the new media: The effects of cues, feedback, and task equivocality. *Information Systems Research, 9*(3), 256-274.

Farhoomnad, A., & Drury, D. (2002). Managerial information overload. *Communications of the ACM, 45*(10), 127-132.

Hall, E. T. (1976). *Beyond culture*. New York: Doubleday.

Hofstede, G. (1980). *Culture's consequences: International differences in work-related values*. Newbury Park, CA: Sage.

Kahai, S. S., & Cooper, R. B. (2003). Exploring the core concepts of media richness theory: The impact of cue multiplicity and feedback immediacy on decision quality. *Journal of Management Information Systems, 20*(1), 263-300.

Kaufman-Scarborough, C., & Lindquist, J. D. (1999). Time management and polychronicity. *Journal of Managerial Psychology, 14*(3/4), 288-312.

Kearney, A. T. (2004). *Mobinet 5*. Retrieved February 18, 2004, from http://www.atkearney.com/main.taf?p=5,4,1,60

Markus, M. L. (1994). Electronic mail as the medium of managerial choice. *Organization Science, 5*(4), 502-527.

Short, W. E., & Christie, B. (1976). *The social psychology of telecommunications*. Chichester, UK: John Wiley & Sons.

Straub, D., & Karahanna, E. (1998). Knowledge worker communications and recipient availability: Toward a task closure explanation of media choice. *Organizations Science, 9*(2), 160-175.

Straub, D. W. (1994). The effect of culture on it diffusion: E-mail and fax in Japan and the U.S. *Information Systems Research, 5*(1), 23-47.

van der Heijden, H. (2004). User acceptance of hedonic information systems. *MIS Quarterly, 28*(4), 695-704.

Walther, J. B. (1995). Relational aspects of computer-mediated communication: Experimental observations over time. *Organization Science, 6*(2), 186-203.

Williams, E. S., & Christie, B. (1976). *The social psychology of telecommunications*. Chichester, UK: John Wiley & Sons.

Zmud, R. W., Lind, M. R., & Young, F. W. (1990). An attribute space for organizational communication channels. *Information Systems Research, 1*(4), 440-457.

Chapter XVIII

Global Organizational Fit Pyramid for Global IT Team Selection

Richard S. Colfax
University of Guam, Guam

Karri T. Perez
TeleGuam Holding, LLC, Guam

Abstract

Global organizations (GOs) rely on global IT teams to remain competitive and link global operations. The Global Organizational Fit Pyramid can facilitate decision making when selecting global IT team members and GO personnel for international team assignments. The Global Organizational Fit Pyramid provides a hierarchy of five decision levels to consider when putting together global IT teams. Level one: global status factors involves decisions regarding large-scale issues such as nationality, religion, and legal rules. Level two: social rank factors involves social background or history issues. Level three: experience factors involves experiences including work history. Level four: credential factors involves issues such as edu-

cational history and professional credentials. Level five: individual factors involves individual characteristics including physical, personality, and even emotional traits. The Global Organizational Fit Pyramid can be useful for selecting global IT team members. The Pyramid also may be applicable to other global team and HR-related decision-making issues.

Introduction

Businesses today are moving into the global arena and changing their operational structures to fit a new global business model. They are becoming global organizations (GOs) with the headquarters office in one location and many other operations and activities located in different nations around the globe.

GOs rely on global information technology (IT) teams in order to remain competitive. Therefore, the selection of global IT teams is important to the GO. Selecting global IT teams is a challenge because traditional team member selection practices may not meet the needs of GOs. Those who make decisions in global IT team selection need to use a new set of considerations so that selection procedures and practices are appropriate to the demands and conditions found in global team operations.

This chapter introduces the Global Organizational Fit Pyramid, which, when used with available organizational, HR, and IT tools, may assist and facilitate decision making by GO leaders and managers, especially human resource (HR) managers, when selecting global IT team members and GO personnel for international team assignments.

Background

Global organizations (GOs) require different configurations of personnel and human resource management practices as those GOs evolve. The global IT team is one aspect of the GO that is changing with the development of GO operations. However, as Ulrich (n.d.) noted, organizations continue to rely on existing frameworks and processes rather than revolutionizing the management structures and processes that have been applied for the last century.

Teams of IT professionals and experts are essential to the operation of GOs (Harris & Moran, 1991; Douglas, 1985) These teams offer expertise and enable organizations to become and remain competitive. Global IT teams usually are separated from the GO headquarters by vast distances and multiple time zones. Further, such global IT teams generally are composed of team members with different cultural, national, and educational backgrounds. These global IT teams may even work in conditions in which the global team members never meet face-to-face but rather conduct their portion of the business operations at a distance using the technology available (Gayle, 2005; SHRM Global Learning System, 2003; Odin, 1997).

Therefore, global IT team selection is an area that needs attention because traditional human resource (HR) or personnel selection practices may not meet the needs of GOs (Andors, 2005; Toffler, 1981). Those who take part in global IT team selection should use a new set of decision-making tools so that selection is more appropriate to the demands and conditions found in global team operations (Trompenaars & Hampden-Turner, 1998). Selection of global IT team members and leaders is still generally the responsibility of the human resource management (HRM) area of the organization, and HR operations usually are located at headquarters. It is equally important to understand that other headquarter executives and global IT team leaders also may be involved in global IT team selection.

Global Organizational Fit Pyramid for Global IT Team Selection

Efficient GO decision making includes global IT team composition or selection decisions. Therefore, a decision tree or pyramid of decision hierarchies will assist HR and other GO leaders to achieve the best global IT team selection processes in global operations. Such a decision hierarchy also can be applied to other aspects of global IT team management, including firing, discipline, evaluation, motivation, quality control, and job assignments. However, as noted earlier, the focus of this chapter is on decisions that impact global IT team selection.

A hierarchy of selection criteria and considerations should be utilized when considering potential global IT employees and team members. The Global Organizational Fit Pyramid (Figure 1) has been developed to facilitate decision making by managers, especially for global IT team member selection applications. Initially, this hierarchy was designed to assist global executives, HR professionals, and global team leaders with the selection process. Through practice and application, this Global Organizational Fit Pyramid should prove valuable in various other global decision-making activities as well.

The Global Organizational Fit Pyramid offers a progressive set of criteria and factors that may assist executives and managers who are involved in creating global IT teams that will achieve GO goals. Organizational leaders and IT team selection decision makers should consider each potential global IT team member and possibly provide appropriate training based on this hierarchy (Kleymeyer, 2005).

Decisions made by GO managers today rely almost totally on computer systems, databases, and information accessible throughout the GO rather than on the conditions that exist in the global workplace. So, global managers and global IT team leaders as well as individual team members need to have computer and technological competencies. Applications include all operational areas in which managers need to make decisions, not just team selection decisions. However, for this chapter, we limit our discussion to global IT team member selection applications.

Decision-Making Levels

There are five levels of decisions in the Global Organizational Fit Pyramid that are recommended be considered by global executives and leaders who are putting together global teams. While many organizational leaders make team creation decisions without consciously following the hierarchy of programmed decisions, it is advisable to use a programmed decision process when possible. Not doing so can lead to some extra issues and problems that can hamstring the whole team, sabotage the team's efficiency, and make project success a near impossibility.

Figure 1. Global Organizational Fit Pyramid: A decision-making hierarchy for selecting global IT team members and other global organization employees

Copyright © 2007, Idea Group Inc. Copying or distributing in print or electronic forms without written permission of Idea Group Inc. is prohibited.

Managers should consider possible global team members starting at the bottom, or level 1, of the hierarchy. If the issues and factors at that level show that a team member is acceptable for the team, then the manager moves to consider the next level factors. This process is continued until either the team member satisfies all factors and is added to the global team or is dropped from consideration when a conflict is identified at one of the five levels. While team members may be included despite issues at a given level, the global team manager and selection manager should be aware of the potential problems that may arise as a result. The potential problems probably will be rooted in the factors related to the level at which the global member-team fit failed.

Level 1: Global Status Factors

The first decision level, global status factors of the Global Organizational Fit Pyramid, includes decisions that must be made regarding issues related to the world as a whole. At this level, consideration is given to large-scale issues that relate to the global team and to individual team members.

The nationality, religion, and even color or race of potential team members can be an issue. Legal issues as well as stereotyped biases and prejudices should be considered in making team assignments. It is important to remember that a country's laws do not necessarily apply, even to its own citizens, when these citizens work abroad. So, team selection procedures that are illegal in one country may be legal in another, and vice versa. Of course, appropriate legal consultation is necessary and recommended to ensure that proper legal procedures are followed for both the headquarters' home country and the global operation site host country's laws.

Selection decisions related to creating IT teams also will be bound by national policies to protect local business operations, industries, and developments. This protectionism may be strong or weak, depending on the industry and government policies to build stronger local businesses either through efforts to limit foreign investors and products or by subsidizing local operations. Baker (2005) suggests that net exporters such as the U.S. can benefit by investing in local cultures, while net importers such as Brazil or India may find it beneficial to enforce weaker protectionism barriers.

All of this requires understanding local laws and how trade, investment, and other business activities are regulated by different governments (Xinhua News Agency, 2002) It is important to remember that legal restrictions and restraints are not limited only to business activities. In many countries of the world, travel to or from as well as extensive residency in a country may be restricted, limited, or even forbidden by the local government.

Another area to be considered involves the applicable laws of each nation. With the recent issues of business ethics applying to U.S. organizations, the Sarbanes-Oxley Act (SOX) has changed the way U.S. businesses are supposed to conduct and report on operations, all of which applies to HR roles and responsibilities (Grossman, 2005a, 2005b). It is now mandatory that firms "have an ethics code designed to deter wrongdoing, including a statement promoting financial integrity that clearly applies to senior financial officers" (Grossman, 2005a, p. 57). However, the ethical codes that are required of U.S. business leaders are not necessarily those of local governments or nationals. Global HR managers need to ensure that all senior officers, regardless of their nation of origin, understand and comply with these regulations when the organization operates in the U.S.

One other item to consider is the perception of social ranking of nations or cultures of team member origin. Different nationalities and cultures often are viewed as having either higher or lower social rank among the nations of the world. Thus, the nationality and cultural origin of a potential member could impact the ranking or desirability of certain team members, regardless of their actual qualifications or potential contributions to the global team. As a result, individuals' particular countries may be seen as having a higher or lower status in the minds of other team members. For example, team members from countries such as Japan, Germany, and the United States may be considered to have more knowledge and value than team members from less economically developed countries.

Another particularly important concern is the legal systems of the countries in which a GO operates. Conflicting legal systems pose dilemmas for the GO and global teams. One example is the U.S. restriction on giving or accepting bribes or gifts. U.S. nationals are not permitted to give or receive monetary gifts for favorable consideration in business. However, in some countries, the practices of providing or accepting gratuities to facilitate processing or handling in bureaucracies are the norm. In fact, failure to provide these gratuities may ensure that the paperwork and processing does not proceed. One reason for these fees may be more than the custom or normal practice. It very well may be that the officials we are dealing with are expected by local custom to obtain their normal income from these facilitation fees rather than a set salary. Once again, cultural, national and legal expectations and restrictions are factors that global IT team selection managers need to understand and be able to work with (SHRM Global Learning System, 2003).

Decision makers must remember and consider the fact that global IT team members will not only work together but also will communicate and depend on each other to complete the organization's tasks. They also may be forced to reside in global locations that may put them and their families at risk. So, although stereotypes may be overcome, the process of making international (global) assignments should include evaluating global factors as a team located at a site where global IT team members would be politically, religiously, racially, and even gender unwelcome.

Level 2: Social Rank Factors

The second decision level, social rank factors of the Global Organizational Fit Pyramid, includes decisions that must be made regarding issues related to the social background or history of each IT team member. These social rank factors involve the history and social roles that individual team members bring with them to the team.

Again, GO managers and leaders must have immediate access to information about local cultures and practices. At this point in time, information about cultural issues of social rank factors is limited because this is an unpopular issue in the global picture. No culture or nation wants to admit that different groups of citizens are treated unequally or discriminated against. The international reaction to such issues has been very apparent in cases of child labor or below-living-wage issues in offshore and outsourced investments and activities.

Class and caste can be issues during global IT team composition and IT team member selection. An example is a male IT manager with a college education in business and years of experience who is unable to get a female secretary to complete necessary office work. Further investigation reveals that both are from the Micronesian Region and even from the same island nation. (Micronesia is a region in the Pacific Ocean that includes more than a thousand islands and more than 10 independent nations.) On this surface, this sounds like simple insubordination or refusal to work. However, when looked at through a sensitive global perspective, it is found that the male IT manager comes from a lower clan on the island, while the female secretary is the daughter of the island chief. In this clan system, lower-level clan members cannot even talk to members of the chief's family (Perez, 2003).

However, Western business also places value on and is influenced by name and status. For example, European executives historically have been selected from an elite group or caste. As such, these executives will expect to interact with personnel at the same or a similar level. Furthermore, these executives generally are related in some way through family, school, club, or discipline. However, while the egalitarian U.S. system may have favored such a system with the old Ivy League system of the past, it does not necessarily favor such a system today.

Of course, there is the universal practice of nepotism found in all nations and cultures. Employees are treated deferentially if they are related to the boss, if their families are connected politically or financially, if their families are blue stock, or if they hold important community or social roles. It is just a different set of factors and values that is being applied to global IT team selection and needs to be understood in the global IT team selection process.

Global IT teams may include members selected from a global pool of qualified potential members. Selection managers in these situations should look beyond the résumé of the applicant when effective team interaction and performance is expected.

Level 3: Experience Factors

At the third decision level, experience factors of the Global Organizational Fit Pyramid, global managers consider issues related to the experiences of the potential global IT team members. These experience factors include evaluation of the whole work and related history of potential IT team members. This is the set of factors with which most global team selection managers traditionally have started when considering potential members for global IT or other team assignments. This is because it is often the easiest set of factors about which to gather information. The information tends to be factual and supportable with evaluation reports and hard data.

Information about potential global IT team members is generally factual. As a result, many HR departments are making use of available human resource information systems (HRISs) in order to access global IT team member applicants. This information generally includes work history items such as experience in the industry and company. The traditional work résumé or application attempts to summarize information related to these factors. However, the utilization of this information is now called into question when selecting global IT team members and making global assignments.

An example of poor selection for a GO management assignment is the American technical expert in computer systems design who is sent on an international assignment to set up an IT operation. The operation is in a country in which technology is fairly well developed. However, the languages are different. Further, the local culture can be characterized using Hofstede's categories as having a preference for high collectivism, low uncertainty avoidance, low assertiveness, and medium to high long-term orientation (Ting-Toomey, 1999). This is nearly the opposite of the typical U.S. orientation, which this manager represents and is used to employing (Moorhead & Griffin, 1998). As a result, local employees might feel insulted by the U.S. technical expert's insensitivity to the need for collaborative team decision making and interdependence. Communication among the IT team members could become cliquish with the classic "in" group development that fractures the cohesiveness that is needed to complete the project. Further, all the local IT experts who were part of the team at startup may leave for other opportunities both locally and elsewhere. As a further result, the overall project could stall and then fail to meet the GOs' expectations. The U.S. technical expert is evaluated by the headquarters' evaluation team as ineffective and a poor global IT manager. However, this manager was just not suited to the global environment and social needs of the assignment.

Global IT team selection managers may need to consider the language competencies, including language levels and structure of potential members from a given national or cultural background. Here, the language capabilities also may be appropriate selection criteria. A good example is the language capabilities of the Japanese applicant for an IT team position. These have not necessarily been mastered by all potential

Japanese IT team members. However, the uninformed global IT team selection manager might assume that a potential IT team member from Japan would be fluent in all forms of Japanese. This is not necessarily true, as the Japanese language has a number of structured levels that need to be used when conducting business in that language.

Another example is the different types of English that are used throughout the world. There are British, American, Australian, and many other localized forms of English. Each of these has different terminologies and even grammatical structures, not to mention pronunciation standards.

Clearly, these are areas in which global IT team selection practices and criteria regarding language experience and expertise need to be carefully established and applied by knowledgable selection managers.

Level 4: Credential Factors

The next decision level of the Global Organizational Fit Pyramid involves credential factors. Here again, the issues and factors considered are among those most usually looked at when selecting global IT team members. These include the educational history and professional credentials of each IT team member. Global IT team members often are considered and selected based on the schools (name and location) attended, subjects or majors studied, and degrees conferred. Professional credentials, of course, are important in professional fields.

Once again, the HR departments in GOs use existing HRISs to gather personal data and history information. As with Level 3, experience factors, the information needed at Level 4 by HR departments and global managers in the IT team selection decisions is factual. This can make it easier for global IT team selection leaders to sort, sift, and rank potential global IT team members when qualifications and histories are considered against the factors at this level.

However, there are other credential issues that global IT team selectors also should consider when evaluating these factors. A few to consider are listed as follows:

- Is the (bachelor's) degree from College A equivalent to that of College B?
- Is College A accredited in its home country? In other locations?
- Is the program of study equivalent to that offered at other schools?
- Does the rigor of the course of study meet company needs?
- Do graduates have the requisite preparation for company needs?
- Are professional credentials suitably stringent?

Some potential global IT team members may have paper qualifications that will not withstand serious comparative scrutiny.

Unfortunately, with the proliferation of online study programs and degrees, there also has been erosion in the dependability of a formal degree. While most programs, schools, and certifying agencies are legitimate, there are enough fly-by-night and buy-a-degree offerings available to cast doubt on the equivalence or acceptability of a potential team member's credentials. This does not begin to consider the corporate university or in-house training programs that now span the globe (Tucker, 2005).

Confirming credentials can be a challenge when considering potential global IT assignees and employees who come from countries other than the headquarters country. The legal issues within the jurisdiction of each nation in which a GO operates need to be understood and complied with by all members of management. However, this can be quite a challenge when IT and HR managers operate across national boundaries, especially if mixed global teams are being utilized. Among the many issues that need to be understood when selecting potential global IT employees is the restrictions that must be met for selection of EU applicants and employees with regard to background checks of international applicants. As Babcock (2005) notes, "(t)he United Kingdom doesn't allow third parties such as background checking firms to have direct access to criminal records held by local police. ... (and) it can take up to 40 business days to get information back" (p. 92). Further, more than 25 EU member nations have restrictions on information that an organization can obtain about and from employees or applicants. This includes sending information to others as well as requesting it. HR managers and other global IT team selection officers based out of the U.S. need to be compliant with the Safe Harbor arrangement in order to remain legal (Babcock, 2005; SHRM Global Learning System, 2003).

There is also the issue of equivalence when it comes to different educational systems. It should apply to both academic and professional credentials when conducting background and qualification checks. Degree or professional titles are not necessarily universal.

Level 5: Individual Factors

The fifth and final decision level of the Global Organizational Fit Pyramid involves personal or individual factors. At this level, consideration is given to the individual characteristics of the specific potential global IT team member. Here, physical, personality, and even emotional traits can be considered.

Individual factors are not generally included in global IT team applications and considerations from either the HR or IT perspectives. This is often because it is not acceptable to ask for or gather personal information, especially with the HR practices and legal restrictions in the U.S. regarding the gathering of personal information during the selection or employment application process.

The creation of a global IT team generally involves team member interactions at various levels and intensities. Consequently, members need to have a degree of compatibility, if at all possible. As the emergence of global IT team cohesion will take time (Moorhead & Griffin, 1998), global IT teams need all the help they can get to coalesce more quickly. Rapid development of global IT team cohesion will suffer if individual team members bring personality issues and conflicting behavioral traits and are not comfortable with other members. Some people just do not get along, no matter how professional they are or what is needed for the project at hand. There are personality clashes and conflicts that supersede the requirements of the global team that can seriously damage and even cripple a global team. The global IT team can polarize around the two or more factions and focus on the personal conflicts rather than the work and the GO itself. This is not productive nor the goal of the global IT team itself.

While it may not be politically or legally correct in the U.S., physical attributes may play a role in selection to a global IT team. Thus, global IT team member selection criteria may include physical, background, and other attributes that will help contribute to an effective global IT team. The amount of information available may vary, depending on the applicant's citizenship or present place of employment.

Summary

The role of managers has evolved to the point where managers no longer spend most of their time gathering and assimilating information to prepare for making decisions regarding global IT team composition. Managers in GOs today must accurately define the global team's needs and then sift to the level needed to identify suitable global IT team members around the world. Selecting and managing global IT teams in the high-tech environment of today do not fit traditional patterns (Puffer, 1996).

Decisions related to global IT team selection are vital to the success of these teams. As noted, the composition of global IT teams will impact the success rate of project completion in GOs, especially when different global sites and situations are necessary for the completion of a project (Lloyd, 2005).

The decision-making process is shared between HR managers and global IT managers. To assist global managers in this process, the Global Organizational Fit Pyramid provides a useful hierarchy for making global IT team selection decisions in the ever-evolving environment of global business today. "The complexities of a global environment are everywhere, whether across the world or in the same office. The need to maximize the value of diverse human capital is a global not just a domestic issue" (SHRM Global Learning System, 2003, Book 1, p. 14).

Once the information is reviewed and assessed, the global IT team selection decision maker can then evaluate the team and its goals. Based on this evaluation, the decision maker can determine the best individual global IT team member fit.

As a result, global IT managers should not only manage the tasks of the team but also contribute effectively to the global IT team selection processes. The Global Organizational Fit Pyramid provides five levels or guides that global HR managers and IT team leaders as well as global executives can utilize during all global IT team selection and other GO decision-making processes. It provides a hierarchy of decisions that if made progressively and if all factors meet the criteria of all levels, then global IT team cohesion and success can be expected. The degree of success of an IT team in the GO may depend on the degree to which these criteria are met.

Finally, we wish to emphasize that global IT team selection is not the only area in which the Global Organizational Fit Pyramid may be useful. Global managers may find the Global Organizational Fit Pyramid factors useful in operational, hiring, performance evaluation, disciplinary, promotion, and other management decisions. The use of this Global Organizational Fit Pyramid as a hierarchy of decision criteria to screen and evaluate decisions, whether related to programs, personnel, property, or locations, should prove beneficial to decision makers throughout the global organization.

References

Andors, A. (2005). Tech smarter. *HR Magazine, 50*(10), 67-72.

Babcock, P. (2005). Foreign assignments. *HR Magazine, 50*(10), 91-98.

Baker, E. (Ed.). (2005). *Scholarship* [speech]. Philadelphia: University of Pennsylvania Law School.

Douglas, M. (1985). Introduction. In J. L. Gross, & S. Rayner (Eds.), *Measuring culture: A paradigm for the analysis of social organization*. New York: Columbia University Press.

Gayle, A. (2005, October 17). Personal Communication.

Grossman, R. J. (2005a). Are you clear? *HR Magazine, 50*(10). 55-58.

Grossman, R. J. (2005b). Demystifying section 404. *HR Magazine, 50*(10), 47-53.

Harris, P. O., & Moran, R. T. (1991). *Managing cultural differences* (3rd ed.). Houston, TX: Gulf Publishing Company.

Kleymeyer, C. (2005). U.S. Army struggles to grasp foreign cultures. *Culture and Development Cyber-Library*. Retrieved October 25, 2005, from http://topics.developmentgateway.org/culture/rc/ItemDetail.do~1042889

Lloyd, T. (2005, October 25). We've got your JOB: Off-shoring. *Daily Yomiuri*, p. 13

Moorhead, G., & Griffin, R. W. (1998). Organiztaional behavior: Managing people and organizations. New York: Houghton Mifflin Company.

Odin, J. K. (1997). *Computers and cultural transformation*. Retrieved October 26, 2005, from http://www.hawaii.edu/aln/cul.htm

Perez, K. T. (2003). *Pohnpeians at work: From monocultural competence to bicultural fluency* (1st ed.). Ann Arbor, MI: UMI.

Puffer, S. M. (1996). *Management across cultures: Insights from fiction and practice* (1st ed.). Malden, MA: Blackwell Publishers.

SHRM Global Learning System. (2003). *The SHRM global learning system*. Alexandria, VA: The Society for Human Resource Management.

Ting-Toomey, S. (1999). Communicating across cultures. New York: The Guilford Press.

Toffler, A. (1981). *The third wave*. London: Pan Books.

Trompenaars, F., & Hampden-Turner, C. (1998). Riding the waves of culture: Understanding cultural diversity in business. London: Nicholas Brealey Publishing.

Tucker, M. A. (2005). E-learning evolves. *HR Magazine, 50*(10), 75-78.

Ulrich, W. M. (n.d.). *Collaborative business & IT infrastructures: Creating a culture of reuse*. Retrieved October 20, 2005, from http://www.flashline.com/Content/Ulrich/reuse_culture.jsp?sid=1160979784974-1123631859-101

Xinhua News Agency. (2002). Top news: China's e-banks set for golden age: Survey. *China Through a Lens*. Retrieved October 22, 2005, from www.China.Org.CN

About the Authors

Wai K. Law is a professor of business strategy and information systems at the University of Guam, USA. Dr. Law received his management PhD and computer science master degrees from Michigan State University. He has broad experience in information systems planning, quantitative modelling, computer simulations and database construction. His research interests are in business strategy and policy, logistics, IS strategy, multi-cultural IRM, e-business and innovative IT education. He has published in numerous journals and refereed conference proceedings. He also authored several books in the areas of strategic simulation and problem-based instruction. He is member of Beta Gamma Sigma.

* * *

Gabriela Aranda is an assistant professor and a member of the Research Group on Software Engineering (GIISCO: http://giisco.uncoma.edu.ar) at the University of Comahue, Argentina. In addition, she is a member of the MAS (Agile Systems Maintenance) Research Project, at the University of Castilla-La Mancha, Spain. Her interests are centred on improving the requirements elicitation process in distributed environments, and developing strategies for selection of groupware tools and elicitation techniques according to the cognitive aspects of stakeholders. She is currently a PhD student at the University of Castilla-La Mancha, Spain.

Copyright © 2007, Idea Group Inc. Copying or distributing in print or electronic forms without written permission of Idea Group Inc. is prohibited.

About the Authors

Saulo F.A. Barretto is currently researcher at the Research Institute for Information Technology, Brazil. He received his BS in civil engineering from Universidade Federal de Sergipe in 1986 and his MS in finite element methods from Universidade de São Paulo in 1990. He received his doctoral degree in boundary element methods from Universidade de São Paulo in 1995. In 1999 he moved his research interests to work with Web learning environments and Digital Culture and since then he has been conducting research projects funded by Brazilian funding agencies (FAPESP and CNPq) and European Community.

Amel Ben Zakour has a master's of human resource management at ISG, Tunis, Tunisia. She was a Fulbright visiting scholar at the University of Georgia in 2003-2004. She is currently a teaching assistant of management at the FSJEGJ, Tunisia. She is also a researcher at E.T.H.I.C.S research unit at the ESSEC, Tunis. Her research interests include cross-cultural IT issues and cross-cultural aspects of human resource management and leadership. She has presented her work in conferences such as the SAIS conference and the MCIS in 2004. She is finishing her doctorate at the University of Toulouse and that of Tunis.

Ioannis Bougos is a lawyer and a member of the Athens Bar Association. He has gained a long experience, as a lawyer-specialist for telecommunications matters, with major emphasis given to European regulatory policy issues, for the development and the promotion of modern information society applications and related facilities, in cooperation with national and European Authorities. After an extended involvement in a great variety of legal, technical and business affairs in the scope of modern telecommunications activities, he currently works as the head of the Department for Regulatory Issues, in the General OTE's (Hellenic Telecoms S.A.) Directorate for Regulatory Affairs.

Alejandra Cechich is a European PhD in computer science from the Castilla-La Mancha University, and an MSc in computer science from the University of South, Argentina. She is an adjunct professor and head of the Research Group on Software Engineering (GIISCO: http://giisco.uncoma.edu.ar) at the University of Comahue, Argentina. Her interests are centered on object and component technology and their use in the systematic development of software systems; quality assurance and quality process improvement .

Tom S. Chan is an associate professor at the Information Department, Southern New Hampshire University at Manchester, New Hampshire, USA. He holds a EdD from Texas Tech University, and MSCS from the University of Southern California. Prior to SNHU, he was an assistant professor at Marist College, and as project manager and software designer specialized in data communication at Citibank. He has published works in the areas of instructional design, distance learning, technology adaptation, information security and Web design.

Ioannis P. Chochliouros is a telecommunications electrical engineer, graduated from the Polytechnic School of the Aristotle University of Thessaloniki, Greece, holding also a MSc and a PhD from the University Pierre et Marie Curie, Paris VI, France. He possesses an extreme research and practical experience in various matters for electronic communications. Dr. Chochliouros currently works as the Head of the Research Programs Section of the Hellenic Telecommunications Organization S.A. (OTE), where he has been involved in different national, European and international projects and activities. He has published numerous scientific and business papers and reports, especially for technical, business and regulatory options arising from innovative e-infrastructures and e-services. He also works as a lecturer, in the Department of Telecommunication Science and Technology of the University of Peloponnese, Greece.

Stergios P. Chochliouros is an independent consultant, specialist for environmental studies, and holding a PhD from the Deptartment of Biology of the University of Patras, Greece, and a university degree as an agriculturist. Dr. Chochliouros has gained enormous experiences as an academic researcher and has been involved in various issues, also including use of modern technologies. In particular, he has participated, as an expert, in many European projects, relevant to a variety of environmental studies. Moreover, he has gained significant experience both as educator and advisor, while he is author of several papers and reports.

Richard S. Colfax, PhD GPHR, has more than 30 years of international business experience in the Pacific-Asia Region, including Japan, Guam & Micronesia. A fluent Japanese speaker and President of Colfax, Inc., he provides international HR/business consulting, facilitation and mediation. With a PhD from the Fielding Graduate University, he is an associate professor of HR and heads the HR program at the University of Guam. Professional qualifications include: Guam's first Global Professional in Human Resources (GPHR) certified by SHRM & HRCI; and certified Myers-Briggs Temperament Indicator (MBTI) Typewatching Trainer and 1999 CASE Carnegie Professor of the Year.

Linda Duxbury is a professor at the Sprott School of Business, Carleton University, Ottawa, Canada. Dr. Duxbury has published extensively in both the academic and practitioner literatures in the areas of work-life balance and stress, managing the new workforce, supportive management, generational differences in work attitudes and values, and managing change. Her most recent books (with Christopher Higgins) include "Who is at Risk? Predictors of Work-life Conflict," "Voices of Canadians: A View from the Trenches," and "Work-Life Confict within the New Millennium: A Status Report." Within the School of Business, Dr. Duxbury teaches the MBA and PhD courses on managing change.

Laurel Evelyn Dyson, BSc (Hons), BA (Hons), PhD, CELTA, GradDipABE, GradDipInfTech (Distinction), CCNA, CCAI, MInfTech, is a lecturer in Information Technology at the University of Technology, Sydney, Australia, where she teaches computer ethics and information systems courses. Her many years of teaching experience have included computer education programs for Indigenous Australians, senior citizens, adult literacy students and prisoners Her main research interests are in the fields of Australian cultural studies, and the cultural, ethical and design issues which impact on Indigenous people's adoption and use of information technology.

Xiuzhen Feng is a professor in the Department of Information Systems at Beijing University of Technology (BJUT), China. Dr. Feng graduated with a master's degree in management science at Xi'an Jiaotong University, China, and received her PhD in IS from Eindhoven University of Technology, The Netherlands. She also worked as an assistant professor in the department of business information systems at the University of Twente, The Netherlands. She is now involved in both teaching and researching of information systems at BJUT. Her current research interests include global information systems, cross-cultural information management, information service, information portals, and enterprise information systems.

Michel Grundstein is a consulting engineer and associated researcher at LAMSADE (Laboratory focused on analyzing and modeling decision aid systems) Paris Dauphine University. Formerly, he was corporate advisor, responsible for innovative methods and applications in the field of information technology, within a French nuclear power plant company. Thus, he had to handle successively: the introduction of computer-aided design, the passage from mainframe computer to departmental computers and personal computers, and artificial intelligence and knowledge-based systems deployment. His main research topic is knowledge management. He has initiated the so-called GAMETH® Framework, in order to locate crucial knowledge for business and running processes. He is the founder, with Camille Rosenthal-Sabroux of the SIGECAD Group, which domain topics are information system, knowledge management and decision aid. He has contributed to numerous books as co-author and published several articles (see www.mgconseil.fr Web site).

Chris Higgins is a professor at the Ivey School of Business, The University of Western Ontario, London, Canada. Higgins' research focuses on the impact of technology on individuals, including such areas as computerized performance monitoring in the service sector; champions of technological innovation; alternative work arrangements; and, most recently, work and family issues and their impact on individuals and organizations. Higgins has published articles in several top journals including *The Journal of Applied Psychology*, *Communications of the ACM*, *Administrative Sciences Quarterly*, *Sloan Management Review*, *Information Systems Research*, and *Management Information Systems Quarterly*. He is a former associate editor for *Information Systems Research*. Three of Higgins' doctoral students (Rebecca Grant, Betty Vandenbosch, and Deb Compeau) have won major awards for their dissertation research.

Don E. Kash is the Hazel Professor of Public Policy at George Mason University. Previously, he served as George Lynn Cross Research Professor of Political Science at the University of Oklahoma. Dr. Kash's research has focused on the interaction of technology, public policy and society. Among his recent books are *The Complexity Challenge: Technological Innovation for the 21st Century* (with Robert W. Rycroft) and *Perpetual Innovation: The New World of Competition*. He has authored numerous articles published in journals ranging from *Science* to *Research Policy*.

Cheon-Pyo Lee is an assistant professor of computer information systems at Carson-Newman College, USA. Dr. Lee received his PhD from Mississippi State University and his MS/CIS degree from Georgia State University. His research interests include IT adoption, mobile commerce, and business value of IT. He has authored several articles, books, and chapters, including *Communications of the Association for Information Systems (CAIS)*, *Information Technology and People*, and *Journal of Internet Banking and Commerce (JIBC)*. He has also presented at conferences such as the Americas Conference on Information Systems (AMCIS), Decision Sciences Institute (DSI) Conference, and International Resource Management Association (IRMA) Conference.

David Lewis is a professor of operations and information systems at the University of Massachusetts Lowell. He specializes in quality control, operations management, and international management. Dr. Lewis has published over thirty-five refereed articles, with his most recent works appearing in *Production and Inventory Management*, *Operations Management Review*, and the *International Journal of Management*. Dr. Lewis is currently doing research in business education, total quality control, and distance learning. Dr. Lewis holds a PhD from the University of Massachusetts Amherst, and is certified in Production and Inventory Management (CPIM), Quality Engineer (CQE), Purchasing (CPM), and Integrated Resource Management (CIRM).

Copyright © 2007, Idea Group Inc. Copying or distributing in print or electronic forms without written permission of Idea Group Inc. is prohibited.

Huixian Li is currently an ERP implementation consultant in Accenture Consulting Company. She received a bachelor's in management information system from Tian Jin University, China in 1999 and a master's of computing from National University of Singapore in December of 2003, with primary research focus on critical success factors of IS implementation. Her paper was published in 9th European Conference on Information System in 2003. Ms. Li has also obtained extensive industry experiences in IT and resources industry. Since she joined Accenture in Jan, 2004, she has participated actively in BPR, ERP planning and implementation, and CRM planning.

Ning Li is an assistant professor of public administration at University of Guam's School of Business and Public Administration. Dr. Li's current research interests are in public administration, technology policy and information management. He has numerous publications in Chinese and has authored or co-authored book chapters and journal articles in English. His English publications appear in such journals as *Administrative Studies*, *Technological Forecasting and Social Change*, and *Geographical Analysis*.

Ying Liang is a senior lecturer in the School of Computing at the University of Paisley in the UK. She worked as a lecturer at Wuhan University in China before she moved to the UK. in 1989. Her research has been in software engineering, object-oriented modeling and design, object-oriented system development, information systems, Web-based system development, and e-business. She has authored many publications in these areas. She has been the program committee member and the reviewer for a few international conferences and publishers.

John Lim is an associate professor in the School of Computing at the National University of Singapore. Concurrently, he heads the Information Systems Research Lab. Dr. Lim graduated with First Class Honors in Electrical Engineering and a MSc in MIS from the National University of Singapore, and a PhD from the University of British Columbia. His current research interests include e-commerce, collaborative technology, negotiation support, IT and education, and IS implementation.

Dalton Martins is currently researcher at the Research Institute for Information Technology, Brazil. He received his BS in electrical engineering from Universidade Estadual de Campinas in 2002 and his MS in P2P networks from the same university in 2004. Since 2003 he is engaged in researching social technologies and MetaReciclagem project.

Copyright © 2007, Idea Group Inc. Copying or distributing in print or electronic forms without written permission of Idea Group Inc. is prohibited.

Dianna Newman is an associate professor at the University at Albany/SUNY and Director of the Evaluation Consortium at Albany. Dr. Newman has served as an evaluator for multiple federal and state funded technology-based curriculum integration grants and currently is developing an innovative model of evaluation that will document systems changes resulting from technology-based curriculum integration educational settings. Dr. Newman is widely published in the area of technology innovation and k-12 curriculum practices. Dr. Newman has served on the Board of Directors for the American Evaluation Association and assisted in writing the Guiding Principles for Evaluators, the professional guidelines for practice.

Indrawati Nataatmadja joined the Faculty of Information Technology at University of Technology, Sydney, Australia, after completing her MBA and many years experience in systems development and implementation. Her previous involvement in training and development in the area of international banking as well as her experiences working and studying in different countries led to her interest in cultural diversity issues in organizational settings. She has published in this area and has been awarded a grant to undertake a study of cultural factors affecting student class participation in tertiary education.

Aikaterini Passa is currently a joint degree graduate student in the Department of Educational and Counseling Psychology and the Department of Biometry and Statistics at the University at Albany/SUNY. She also has an MS in educational psychology and Methodology with a concentration in research design and statistical analysis. She is currently working as a research project assistant in the Evaluation Consortium at the University at Albany, conducting studies of federal and state funded grants. She also serves as a graduate school representative on the leadership board of the Academy for Character Education at Sage College's School of Education.

Karri T. Perez, PhD, SPHR, is VP for human resources and administration for GTA, a Guam telecommunications company. Previous positions include Vice president/HR manager for Bank of Guam, senior HR director for Starwood Resorts Worldwide, and HR management in retail, hotel and manufacturing. Perez has a PhD from Fielding Graduate Institute, and the senior professional (SPHR) and global professional (GPHR) in HR designations from the Society for Human Resource Management. Karri is an adjunct professor, a consultant and trainer, co-authored an article on human and IT systems integration, and writes for Directions Magazine. She has lived and worked in Guam, Virginia and Japan.

Mario Piattini has an MS and a PhD in computer science from the Politechnical University of Madrid, and he has an MS in psychology from the UNED. He became a certified information system auditor and certified information security manager by ISACA (Information System Audit and Control Association). He is a professor at the Department of Computer Science at the University of Castilla-La Mancha, in Ciudad Real, Spain, and the author of several books and papers on databases, software engineering and information systems. He leads the ALARCOS research group specializing in information system quality. His research interests are software quality, advanced database design, metrics, software maintenance, information system audit, and security.

Renata Piazzalunga is currently the president of the Research Institute for Information Technology, Brazil. She received her BS in architecture and urbanization from Universidade de São Paulo (USP) in 1991 and her MSc degree in urban design, also from USP in 1998. She received her doctoral degree in cyberspace architecture at USP. She is currently engaged in researching how the information society can influence the way of creating spaces and representations in architecture. She is also engaged in researching interactive computer-based learning environments involving cognition systems development for the Web.

Claudio Prado is currently the coordinator of digital policy of the Ministry of Culture of Brazil. He is a self taught anti-specialist in holistic (trans-disciplinary) matters who believes that the DNA of the human adventure of the 21st century cannot be understood from any disciplined angle.

K. S. Raman is an adjunct professor in the Department of Information Systems (IS) at the National University of Singapore. Prior to this, he has served as senior fellow and coordinator of the Information Program and adjunct associate professor with the department. His research interests cover cross-cultural issues in IS, group support systems, IT in small enterprises, and IT policies in Asia-Pacific countries, and IT in developing countries. He has published widely on these topics in international journals and conferences and served on the editorial boards of journals, including the MIS Quarterly. Raman has extensive industry experience in planning, design, implementation, and management of Information Systems, including ERP systems. He has led consulting and research projects for large organizations and government agencies in Asia-Pacific and the World Bank.

Fjodor Ruzic is doctor of information sciences at University of Zagreb where he is lecturer in new media, and interactive multimedia systems. His recent research activities are covering the integration of information content and integration impacts on development of information theory and practice. He is in information sciences from 1975, and he is member of many national and international bodies relating to telecommunications systems integrity, information resources management and multimedia system environment. He was working on both research and implementation sides of networked databases, educational material, and digital media. He published over 125 scientific and research papers in various international journals and he is author of several books dealing with graphical user interfaces, multimedia and Internet.

Timothy P. Shea is an associate professor of management information systems in the Charlton College of Business at the University of Massachusetts Dartmouth. He received his DBA in management information systems from Boston University, his MBA from Indiana University, and his BS in operations management and computer science from Boston College. Dr. Shea has had numerous publications in the research areas of e-commerce, Web-based learning, ERP readiness and assessment, and communities of practice, with his most recent works appearing in *The International Journal of Business Information Systems*, *The Review of Business Information Systems*, and *The International Journal of Enterprise Information Systems*. He has also written a number of software training manuals.

Anastasia S. Spiliopoulou is a lawyer, LL.M., and a member of the Athens Bar Association. During the latest years, she had a major participation in matters related to telecommunications and broadcasting policy, in Greece and abroad, within the framework of the Information Society. Mrs. Spiliopoulou has been involved in current legal, research and business activities, as a specialist for e-commerce and e-business, electronic signatures, e-contracts and e-procurement, e-security and other modern information society applications. She has published numerous scientific papers, with specific emphasis given on regulatory, business, commercial and social aspects. She currently works as an OTE's (Hellenic Telecoms S.A.) lawyer for the Department of Regulatory Issues, of the General OTE's Directorate for Regulatory Affairs.

Ramesh Subramanian is the Gabriel Ferrucci Professor of Computer Information Systems at the School of Business, Quinnipiac University, Hamden, Connecticut. Dr. Subramanian received his PhD in computer information systems (1992) and MBA from Rutgers University, NJ. He also holds a post graduate diploma (honors) in management from XLRI, Jamshedpur, India, and a Bachelor of Applied Sciences from Madras University, India. Dr. Subramanian's research interests include information systems strategy, cross-cultural issues, security, historical and philosophical underpinnings of IS/IT, digital asset management, e-commerce, peer-to-peer computing and IT education.

John Ajit Thomas is a PhD candidate in information systems at The Eric Sprott School of Business, Carleton University Ottawa. His research interests are in how companies can use dynamic capabilities to create sustainable and enduring sources of competitive advantage. He works as a technical architect and business intelligence consultant with Cognos Inc. He resides in Ottawa, Canada.

Ian Towers is a PhD candidate at the Sprott School of Business, Carleton University, Ottawa. His thesis is on organizational change at the Canadian subsidiary of a multinational pharmaceutical company. He is investigating the nature of change as a process, as part of a multilevel, contextual analysis. His research interests include middle management, identity and discourse.

Célio Turino is currently secretary of cultural programs and projects of the Ministry of Culture of Brazil. He received his BS in history and his MS in history from Universidade Estadual de Campinas, Brazil. He is the person responsible for the Livings Culture (Cultura viva) Program.

Aurora Vizcaíno is an associate professor in the Computer Science Department of the University of Castilla-La Mancha, Spain. She has a European PhD in computer science, a BSC in computer science from University of Castilla-La Mancha, 1995, and a graduate degree in computer science from the University of Granada, 1997. Her research interests include software engineering, agents, and knowledge management. She has numerous publications in important International Conferences and is part of the program committee of many of them.

Yin Ping Yang is currently a PhD candidate in the Department of Information Systems, School of Computing at the National University of Singapore. She has received Bachelor of Computing (honors) in June 2003. Her papers have appeared in IFIP TC8/WG8.3 International Conference on Decision Support Systems, Hawaii International Conference on System Sciences, Group Decision and Negotiation Conference, and *Journal of Global Information Management*. Her research interests include negotiation support systems, culture and negotiation, electronic commerce, and e-marketplace in China.

Index

A

active people 355
addictions education 339
ad hoc infrastructures for KM implementation 252
anachronism 64
anthropology 27
ASICs (see application-specific integrated circuits)

B

bazaar model 149
behavioral patterns 30
Bhabha, Homi 133
BIR (see business integration and restructuring)
BlackBerry devices 305
BPR (see business process re-engineering)
Brazilian government 147
business contexts 170
business integration and restructuring (BIR) 216
business intelligence 265
business process re-engineering (BPR) 77

C

captology 69
caste 379
cell phones 305
chat rooms 293
chief information officer (CIO) 84
CIO (see chief information officer)
class 379
classroom-only instruction 346
cognitive frameworks 337
cognitive growth 346
cognitive informatics 353
cognitive styles 353
collaborative tools 265
collecting information 8
collectivism (see also individualism) 5
color 197
communicational requirements 168, 169
communities of practice (CoPs) 265
community 265
competitive advantage 213
complex technologies 212
computer-based information technology 56
computer-based MIS 43
computer-integrated manufacturing 217

computer-supported cooperative work (CSCW) 352
computing 352
conceptual framework 27
conceptual systems 214
Confucian teachings 37
Confucian work dynamism 37
consensual relationships 31
contemporary democracy 115
content management 265
context-specific CSF 76
continuous cyclical formative evaluation 345
convergence 54
Conversê 156
CoPs (see communities of practice)
corporate governance 247
credential factors 381
critical success factors (CSF) 76, 77
cross-cultural differences 267, 288
cross-cultural environment 1
cross-cultural information systems 2
cross-cultural psychology 27
cross-cultural studies 33
CSCW (see computer-supported cooperative work)
CSF (see critical success factors)
cultural diversity 60, 287
cultural expression 64
cultural homogenization 60
cultural literacy 61
cultural shift 56
cultural spot 155
culture 218, 243
culture-centric Web site 192
culture theory 25, 30
currency 201
curriculum 334
customer service 129, 217
cybercafé 194
cyberspace 70

D

database management 95
data management system 11
data processing (DP) 216

data resources sharing 217
date format 201
decision-making levels 376
decision-making tools 375
Deming's cycle by quality management practitioners 251
dialogue act diagram 173
dialogue act model 169
dialogue act modeling 164
dialogue act modeling approach 166
dialogues 170
digital 55
digital content 116
digital culture 155
digital depression 369
digital e-culture 54
digital information 58
digital information system 256
digital libraries 117
digital literacy 1, 76, 113, 146, 162, 305, 386
digital media 58
digital rights management (DRM) 119
discussion forums 292
distance communication, 58
domain 265
DP (see data processing)
DRM (see digital rights management)
dynamic digital content 116

E

e-culture 55, 62
e-culture literacy 55
e-mail 292
e-technology 55
ease of use 66
EIS (see executive information systems)
electronic government (e-government) 113
emic perspective 27
employee effectiveness 288
Estúdio Livre 156
etic perspective 27
evolutionary economics 213
example-based systems 195
executive information systems (EIS) 45
experience of reading 386

F

face-to-face meetings 365
family conflict 311
family pattern 31
feedback 171
feminine cultures 45, 199
femininity (see also masculinity) 5
first-hand support 98
first-of-a-kind technology 215
flexibility 66, 290
formal information network 239

G

gender roles 5
global educational systems 334
global environment 1
global information society 108
globalization 26, 60, 129, 192
Global Organizational Fit Pyramid 373
global organizations (GOs) 373
global people 355
global space 238
global status factors 377
global users 165
GOs (see global organizations)
government policy 91
groupware 293
guanxi 90

H

hacker 153
hierarchical structure 31
hierarchy 34
hierarchy/egalitarianism 34
high-level social system 3
high/low context 25
high and low context 290
high context/low context of communication 37
HRISs (see human resource information systems)
HRM (see human resource management)
human resource information systems (HRISs) 380

human resource management (HRM) 375
human sexuality 199
hybridity 131
hyperspatial 64

I

I/C (see individualism/collectivism)
ICT (see information and communication technology)
IEP (see Internet enabled phone)
IIS (see information innovation and support)
illocutionary acts 172
IM (see instant messaging)
image 198
impression management 310
individualism (see also collectivism) 5
individualism/collectivism (I/C) 5, 25, 32, 34, 285, 290
informal information network 239
informatics 352
information-communications systems 54
information and communication technology (ICT) 109, 237, 284
information appliance 65
information behavior 69
information economy 63
information innovation and support (IIS) 216
information management 216
information overload 369
information presentation 1
information procurement 1
information resource management (IRM) 1, 263
information resources management 236
information richness (IR) 366
information richness theory 366
information society 55, 108
information systems (IS) 14, 26, 77, 237, 365
information systems management (ISM) 14
information technology (IT) 26, 57, 77, 374
information technology practitioners 306
innovation network 216

innovation patterns 212, 214
instant messaging (IM) 365
instructional design 344
instructional settings 346
instructional technology (IT) 333
integration 337
intellectual property rights (IPR) 119
intelligent networks 58
interactive communication 169
interactivity 290
internal communication 90
internationalization 297
Internet enabled phone (IEP) 366
interpersonal communication 352
intranet 293
intuitive people 354
invisible 65
invisible culture 62
invisible e-culture 66
IPR (see intellectual property rights)
IR (see information richness)
IRM (see information resource management)
IRM (see information resources management)
IS (see information system)
IS implementation 89
ISM (see information systems management)
IS success measures 80
IT (see information technology; instructional technology)
IT adoption 41
IT management 72

J

just-in-time inventory management 217

K

KA (see knowledge acquisition)
KBS (see knowledge-based systems)
KD (see knowledge documentation)
key user involvement 92
KM (see knowledge management)
KM (see knowledge mangement)
KMS (see knowledge management system)
knowledge-based systems (KBS) 253
knowledge-related activities 264

knowledge acquisition (KA) 264
knowledge documentation (KD) 264
knowledge management (KM) 216, 237, 262, 287, 288
knowledge management governance 247
knowledge management system (KMS) 238, 243
knowledge sharing (KS) 262, 264
knowledge worker 256
KS (see knowledge sharing)
KS (see knowlege sharing)

L

language accommodation 296
learning 337
learning style 345
learning style models (LSMs) 354
Living Culture (Cultura Viva) 148
localization 297
local space 238
long-term orientation 5, 37
long-term vs. short-term orientation 285
LSMs (see learning style models)

M

magnetic resonance imaging (MRI) 226
management information system (MIS) 2, 43, 216
managerial guiding principles for KM 251
manufacturing requirements planning (MRP) 77
MAS (see masculinity/femininity)
masculine societies 199
masculinity (see also femininity) 5
masculinity-femininity index 8
masculinity/femininity (MAS) 25, 32, 35, 285
mechatronics 216
media richness theory (MRT) 366
media speed 367
MGKME (see model for global knowledge management within the enterprise)
mimicry 131
Ministry of Culture 155
MIS (see management information system)

MIS (see management of information system)
mobile e-mail 365
mobile technologies 294
model for global knowledge management within the enterprise (MGKME) 236, 249
monochronic 369
movement 198
MRI (see magnetic resonance imaging)
MRP (see manufacturing requirements planning)
MRT (see media richness theory)
multidimensional 64
multilingual 202

N

Napster 152
national cultural differences 1
national culture 41
Nehru, Jawaharlal 135
neoclassical growth model 213
networking 64
New Education Policy 134
new electronic media 365
new growth theory 213
new market 114
new media 57
new services 121
nonfunctional requirements 169
normal innovation pattern 215
number format 201

O

object modeling technique 181
OECD (see Organization for Economic Co-operation and Development)
offshoring 129
online games 118
open source software 157
open sources software 71
organizational culture 270
organizational leaders 375
organizational learning 251, 287
organizational learning processes 255
organizational time culture 368

Organization for Economic Co-operation and Development (OECD) 247
organization polychronic orientation scale 369
organizing information 8

P

page design 200
paternalism 31
pattern matching 195
PBL (see problem-based learning)
PDA (see personal digital assistant)
PDI (see power distance) 34
peer-to-peer content 116
peer-to-peer file-sharing systems 148
peer-to-peer file transfer 118
PEOU (see perceived ease of use)
perceived ease of use (PEOU) 38
perceived usefulness (PU) 38
perlocutionary acts 172
person's name 201
personal digital assistant (PDA) 111, 294, 307
personal preferences 359
personal time and space 310
polychronism/monochronism 25
portable computers 305
post-colonial social theory framework 130
post-colonial theory 130
post-independent India 140
postal address 201
power distance (PDI) 25, 32, 34, 285
pplication-specific integrated circuits (ASICs) 226
practice 265
prepositional acts 172
prganizational learning 218
problem-based learning (PBL) 277
production speed 367
productivity 320
project management 95
PSI (see public sector information)
PU (see perceived usefulness)
public domain 149
public sector information (PSI) 114
pyramid pattern 31

Q

quality management system 251

R

R&D (see research and development)
readers 386
real-time use 344
reflective people 355
rehearsability 290
reprocessability 290
research and development (R&D) 113, 213
richness 290

S

salutation 201
scaffolding 334
scanners 1, 25, 54, 76, 107, 128, 146, 192, 211, 236, 262, 305, 333, 351, 364, 373, 386
self-determination 56, 66
self-enhancement/self-transcendence 34
self image management 310
semantic and syntactic approaches 195
sensing people 354
sequential people 355
short-term orientation 37
short messaging service (SMS) 365
SIN (see system integration and networking)
sly civility 131
SMS (see short messaging service)
social influence 46
social presence (SP) 290, 366
social rank factors 379
social responsibility 54
software 69
software development 185
software engineering 352
SP (see social presence)
space of influence 238
speckled computing 65
speech act theory 172
standardization 15
stereotype 131
storing information 8

stress 319, 320
stress levels 320
symbol 198
system integration and networking (SIN) 213

T

tacit knowledge 219
TAM (see technology acceptance model)
Tata, J. R. D. 138
technical infrastructure 165
technological advancements (see also technological innovation) 68
technological innovation (see also technological advancements) 150
technological progress 112
technology-supported instruction 339
technology-supported systems of learning 337
technology acceptance model (TAM) 25, 27, 38
technology development 60
text sorting sequence 201
theory of reasoned action (TRA) 39
time format 201
time orientation 36
tools-services-content triangle 54, 55
TRA (see theory of reasoned action)
transformational 214
transformational innovation 218
transformational pattern 224
transitional innovatio 215
triple convergence 59

U

ubiquitous e-technologies environment 66
UCS (see Universal Muliple-Octet Coded Character Set)
uncertainty avoidance 25, 32, 36, 285, 290
Unicode Consortium 203
Universal Multiple-Octet Coded Character Set (UCS) 204
usability evaluation 342
user-centered model 337
user interface design 193

utterance acts 172

V

verbal people 354
video-over IP 300
videoconferencing 294
village market pattern 31
virtual computer 64
virtual European library 117
visible Culture 61
visualization of interaction 184
visual people 354
Voice-over IP (VoIP) 300
VoIP (see Voice-over IP) 300
volatile development process 165

W

WBIS (see Web-based information systems)
Web-based information systems (WBIS) 162
Web developers 169
Web engineering 183
well-oiled machine pattern 31
Western scientific education 135
WET (see work extending technologies)
WET survey 312
word-for-word translations 195
work-extension technologies 305
work-life balance 319
work extending technology (WET) 306, 308
workforce development 339
workload 319

Single Journal Articles and Case Studies
Are Now Right at Your Fingertips!

Purchase any single journal article or teaching case for only $25.00!

Idea Group Publishing offers an extensive collection of research articles and teaching cases in both print and electronic formats. You will find over 1300 journal articles and more than 300 case studies on-line at **www.idea-group.com/articles**. Individual journal articles and cases are available for only $25 each. A new feature of our website now allows you to search journal articles and case studies by category. To take advantage of this new feature, simply use the above link to search within these available categories.

We have provided free access to the table of contents for each journal. Once you locate the specific article needed, you can purchase it through our easy and secure site.

For more information, contact cust@idea-group.com or 717-533-8845 ext.10

- Databases, Data Mining & Data Warehousing
- Distance Learning & Education
- E-Commerce and E-Government
- E-Government
- Healthcare Information Systems
- Human Side and Society Issues in IT
- Information Technology Education
- IT Business Value, Support and Solutions
- IT Engineering, Modeling & Evaluation
- Knowledge Management
- Mobile Commerce and Telecommunications
- Multimedia Networking
- Virtual Organizations and Communities
- Web Technologies and Applications

www.idea-group.com

Looking for a way to make information science and technology research easy? Idea Group Inc. Electronic Resources are designed to keep your institution up-to-date on the latest information science technology trends and research.

Information Technology Research at the Click of a Mouse!

InfoSci-Online
⇨ Instant access to thousands of information technology book chapters, journal articles, teaching cases, and conference proceedings
⇨ Multiple search functions
⇨ Full-text entries and complete citation information
⇨ Upgrade to **InfoSci-Online Premium** and add thousands of authoritative entries from Idea Group Reference's handbooks of research and encyclopedias!

IGI Full-Text Online Journal Collection
⇨ Instant access to thousands of scholarly journal articles
⇨ Full-text entries and complete citation information

IGI Teaching Case Collection
⇨ Instant access to hundreds of comprehensive teaching cases
⇨ Password-protected access to case instructor files

IGI E-Access
⇨ Online, full-text access to IGI individual journals, encyclopedias, or handbooks of research

Additional E-Resources
⇨ E-Books
⇨ Individual Electronic Journal Articles
⇨ Individual Electronic Teaching Cases

IGI Electronic Resources have flexible pricing to help meet the needs of any institution.

www.igi-online.com

Sign Up for a Free Trial of IGI Databases!